Poison and Poisoning

A Compendium of Cases, Catastrophes and Crimes

Celia Kellett

Published by Accent Press Ltd – 2009

ISBN 9781906373962

Printed and bound in the UK

Cover Design by Red Dot Design

What to do in a case of suspected poisoning

Seek urgent medical attention

If needed, dial 999 for an ambulance or
phone NHS Direct on 0845 46 47

Take the victim to Accident and Emergency or Primary
Care at your nearest hospital

DO NOT make the victim sick

If the poisonous substance was swallowed, wipe the lips and mouth
and remove any leftover substance

If the victim suffered a chemical burn, soothe the burn
with milk or water

If the victim is unconscious, place him or her in the recovery
position:on his side, but with one leg bent to support the body, so
preventing him from rolling on to his front, and his uppermost hand
placed under his chin, to support his head, which should be slightly
tilted back – this stops him swallowing his tongue, allows him to
breathe and allows any fluid to drain out of the mouth

If breathing stops, use resuscitation but take care to protect
yourself from poisoning

Get a sample of the poison to aid diagnosis

Contents

Preface

PEOPLE ARE ALWAYS SAYING how times have changed. But in the case of poison and poisoning, although few people can really be unaware of the danger of poisonous substances these days, still every year thousands are killed or made seriously ill by poison. Just like they were a century – or even a millennium – ago. Whatever happened to learning from our mistakes?

Socrates was put to death by a forced drink of hemlock; King Charles II was killed by his overzealous doctors; Napoleon was poisoned by his wallpaper ... These are just three of the many victims and their murderers that you will meet within the pages of this compendium of poisons and poisonings.

While most poisoning deaths are accidental, many victims are deliberately poisoned, and throughout recorded history, as this book will show, poison has been a favourite tool of murderers and assassins. Even in the 21st century, prisons worldwide still hold many murderers who used poison as their deadly weapon. Poisoning has the twin advantages for a killer of not requiring any direct physical violence and of there being a good chance that the crime will remain undetected. We know of course of many poisoners who have been detected and brought to justice, but how many have got away with it? We will never be sure.

While it is probably unlikely that you, the reader of this book, will be the victim of deliberate poisoning (unless you are very rich and extremely unpopular), there is no doubt that we are all, every day, at risk of becoming seriously ill, or dying, through contact with poison. Poisons are everywhere, as this book will demonstrate. Whether in the home or at work, in the countryside or in the sea, poisons pose a real threat to us.

To be fair, science has come a long way in learning the cause and treatment of a wide variety of poisons and poisonings, and many have been banned from use or assigned many precautions.

The purpose of this book is to clearly and accurately describe the many different kinds of poison that exist, what their effects are, how they can (sometimes) be treated and to show how, by design or by accident, they have led to the deaths of so many people.

Please note

This book is intended as a reference volume and informative guide for those wishing to know more about this fascinating subject. The author and publisher accept no liability for any claims arising from the use of this book and the information contained within it.

How this book is arranged

The contents are divided into sections and subdivided into chapters for ease of reference. There is a great deal of further information at the end of the book, and if you are looking for information about a specific poison (or poisoner) then you can find it by referring to the index on page 423. Many words are included in the glossary on page 387.

These are the sections to come:

Introduction

Before getting too deep into detail, this chapter will give you a broad overview of poison and poisoning, including types of exposure, side effects and many wicked stories of past poisoners.

The poison basics: sources, symptoms and treatments

We'll start with the sources of poison, covering mineral, vegetable, animal and man-made substances. These chapters will also explain how poisons work and the signs and symptoms of poisoning. A further source of poisoning – medical conditions caused by deficiencies, or excesses, of substances needed for the normal functioning of the body – will also be discussed, followed by a chapter on treatments.

The poison problems: at work and in the home

There are countless problems caused by specific poisonous substances, and we will discuss many of them here, from poisonous pigments and dangerous dyestuffs to occupational hazards, including fatal farming; from frightening foodstuffs to household horrors; from murderous make-up to the noxious nursery.

The ultimate poison legacy: catastrophes, accidents, murder and suicide

This chilling group of chapters covers the more gory details: catastrophes and accidents with poisons, intentional poisonings (including murder and warfare), self-inflicted poisonings, including abortion and, finally, suicide.

The poison penalties: uses, abuses and consequences

Delving even deeper, this final group of chapters begins with the medicinal uses of poisonous substances and the poisoning that results from misuse and abuse. We'll also look at the lessons learned: medicines that have been restricted or withdrawn from use due to their unacceptable side effects. Finally, we'll discuss the legalities that affect or have affected the sale, supply and use of poisonous substances.

Further Information

For those readers who would like more detail, this final section includes a series of appendices that list the full names of the living species mentioned in the text and their poisons, more information on natural dyestuffs, a more detailed cellular-level explanation of how poisons actually work, effects of deficiency and excess of vitamins and minerals in the body and a lethal lexicon. Finally, there is a list of abbreviations, a glossary of medical terms, references for each chapter, a bibliography and a list of websites for further reading, followed by the index.

Be prepared to see the world in a totally different (and perhaps more paranoid!) light by the time you've finished reading about the fascinating world of poisons.

Chapter 1

A General Introduction to Poisons and Poisoning

POISONS ARE EVERYWHERE. THOUSANDS of people die every year from accidental poisoning caused by household products, wild plants, venomous creatures, occupational hazards and many more.
In a one-year period from 2005 to 2006, the United Kingdom's National Poisons Information Service, which supplies specialist information and advice on the treatment of poisoning to doctors, received 90,000 telephone calls, of which almost 10,000 related to poisoning by household products alone.

All substances can be poisons, but the right dose distinguishes a poison from a remedy. So said the physician Paracelsus nearly 500 years ago. Even water can be a poison if you drink enough of it. But, by poison, we usually mean a substance that causes serious bodily harm or ultimately death from just a very small quantity.

Many of these poisonous substances have been abused throughout the years. For example, Dr Harold Shipman, a GP who killed at least 215 of his patients in the final decades of the twentieth century (more about him in Chapter 20), and other medical professionals took advantage of their special position and ease of access to poisonous medicines, using them deliberately to kill. Lesser mortals used whatever was available: rat poison such as the tasteless and odourless arsenic was frequently added to a variety of food and drinks without conjuring suspicion.

Many different foodstuffs have been used by murderers to hide poison, from scones and cakes to spaghetti Bolognese. Murderers are inventive, and where a stronger flavour is needed to mask the poison, such as with phosphorus, HP sauce and curry have also been used.

Poisoning may be deliberate like this, but it may also be unintentional. Of the thousands of cases of poisoning every year, the majority are accidental. The incidence of poisoning continues to rise annually, due to the increasing availability of potent medicines, the illicit use of so-called recreational drugs like ecstasy and an ever-growing number of toxic substances used in industry. Luckily, the number of antidotes is increasing as well: until recently there had been no antidote to cocaine overdoses, but in September 2008 scientists from Kentucky University in the USA announced that they had discovered one that they will develop. And more sophisticated medical treatments used today mean that, although incidences of poisoning are increasing, fewer of them result in fatalities.

Cleaning substances, painkillers, cosmetics and plants are the most common causes of accidental poisoning, and small children constitute the largest group of patients involved in these cases. Fortunately most of them survive because of prompt treatment. The majority of deaths from poisoning are of adults, and the largest number each year are caused by antidepressants, painkillers and street drugs.

Many examples of poisoning were first made known by their publication in medical journals. The *British Medical Journal* (BMJ) and *The Lancet* are probably the oldest and best known in Britain. In the United States, the *New England Journal of Medicine* and the *Journal of the American Medical Association* (JAMA) are considered the most prestigious. There are a great many medical journals published regularly throughout the world on every conceivable specialist topic, and, through these journals, cases and causes of poisoning are still frequently made known to the medical profession and the wider world.

Dr Thomas Neill Cream was a physician born in Glasgow who studied medicine at McGill University in Canada and graduated in 1876. He practised as a physician in Chicago, where he was also involved in murder, arson, abortion and blackmail. In 1881, he poisoned his mistress' husband with strychnine. Showing rather odd behaviour for a murderer, he wrote to the district attorney suggesting that the body be exhumed. As a result he aroused suspicion and was arrested, tried and sentenced to life imprisonment.

His sentence was shortened and, upon release in 1891, he came to London, where for several months he preyed on prostitutes, giving them pills laced with strychnine. He managed to kill four of them. Being cross-eyed, he was somewhat noticeable in appearance, and so became known as the Cross-Eyed Lambeth Poisoner. As he had done in Chicago, he drew attention to himself and was soon arrested.

A prostitute who survived his attentions remembered the cross-eyed doctor well. After she came forward and identified him, he was charged with murder. A chemist told the court that Dr Cream had purchased Nux Vomica (which contains strychnine) and some gelatin capsules from his shop. A search of Dr Cream's lodgings produced no less than seven bottles containing strychnine. The jury convicted him and he was hanged in November 1892.

What's your poison?

Not surprisingly, along with the many naturally occurring poisons – animal, vegetable and mineral – there are many more man-made ones. All poisons are chemicals of one sort or another, and work in a wide variety of ways.

With the wide variety of poisons and poisoning techniques comes a wide spectrum of undesired effects. Many chemicals, while not toxic themselves, can be converted into toxic breakdown products – metabolites – by enzymes in the body. This is what happens with alcohol, a poison that occurs many times in this book, in both good and evil guises. The degree of poisoning of a person will depend on how fast the toxic metabolite is produced, compared to the speed with which the body can break it down into something less toxic, or excrete it.

In 1914 the French government banned the liqueur absinthe, claiming that it caused hallucinations and epilepsy. But this was not true. The ban was really an attempt to reduce the number of alcoholic poisonings and deaths, for the common good.
Absinthe is a highly alcoholic drink, containing some 68 per cent alcohol. It is made by infusing double-distilled spirit with wormwood flowers and other herbs. The wormwood plant contains a bitter alkaloid called thujone. The essential oils of aniseed, coriander, fennel, peppermint, hyssop, angelica and lemon balm are used to mask its bitter taste in the drink. The finished product, absinthe, is a pale green liqueur, which has traditionally been served by pouring iced water over a sugar cube placed in a perforated spoon, above a glass containing the absinthe. The alcohol and water mix results in the traditional cloudy drink.
While the alkaloid thujone certainly is a narcotic poison, the quantity to be found in absinthe is unlikely to cause any problem. The high alcohol content in absinthe was the real culprit in this particular case.

Exposure – it will get you either way

Poisoning can result from either acute or chronic exposure to a noxious substance. Acute exposure refers to when the poison is delivered as a large single dose. Chronic exposure is usually due to a low level of exposure over a prolonged period of time – anything from days to decades.

About 40 years ago mentholated cigarettes were very popular. These were manufactured and packaged in the same way as any other cigarette, except that the tissue and foil wrapping encasing the cigarettes within the carton were impregnated with menthol. Only a short period of storage time was needed for the menthol to suffuse the cigarettes.

A woman who smoked 80 mentholated cigarettes daily for three months developed insomnia, an unsteady gait, thick speech, a tremor of the hands, mental confusion, depression, vomiting and cramps in her legs. Her heart rate was found to be very slow, at only 44 beats per minute. All these symptoms rapidly disappeared when she smoked ordinary cigarettes, without menthol. Later a test dose of 65mg three times a day for seven days produced the same slow heart rate and evidence of toxicity.[1] Now that's a lesson learned the hard way.

Chronic exposure can result in a slow accumulation of a poison within the body, as has happened to workers in a number of industries in the past. For example, the lung disease silicosis can have a latency period of about ten years before the symptoms begin to show themselves. With asbestosis, this period may be as long as 25 years.

In recent years there has been a massive increase in the number of cases of skin cancer, not only due to chronic exposure to sunlight but also to chronic sunbed tanning in beauty parlours. Young ladies who try the acute approach to tanning in coin-operated booths usually end up on the news, having required hospital treatment for burns.

Ultra-violet or even visible radiation from sunlight acting on the skin, in conjunction with a number of drugs, can trigger photo-toxic and photo-allergic reactions. Contact with some plants can cause allergic reactions too – the sap of spurges, *Euphorbias*, can cause contact dermatitis if it gets on the skin, and is much worse if it happens while the skin is exposed to sunlight, often resulting in an acute condition requiring hospital treatment. For gardeners, this can have serious consequences as initial slight acute reactions can become serious and chronic, as we shall see in Chapters 8 and 9.

Poisoners of the deliberately murdering kind tend to match their motive with the type of poisoning. For example, those who are in a hurry to collect the life insurance of their victim prefer a massive one-off acute dose while others use the slow insidious method of chronic exposure, in the hopes that they are less likely to be discovered. In recent times, both techniques have been used and abused.

Dena Thompson was nicknamed the Black Widow because she chose lonely single men as her victims. She preyed on them physically, sexually and financially, tricking them out of a total of more than £500,000. It was in 1982 that she met her first husband, Lee Wyatt, whom she married two years later.
In 1990 she met and bigamously married Julian Webb. Only a few months later, she told him that she had cancer, a lie created to extort money from him. In 1994 she murdered him by mixing some tricyclic antidepressants and painkillers into a home-made curry for a quick kill, the spices in the curry masking the bitter taste of the drugs. When he died, she persuaded the police that he had committed suicide due to depression. However, her mother-in-law was suspicious and refused to allow her son's body to be cremated, which would have destroyed any evidence. She insisted that he be buried, but no investigation was attempted for years.
Dena then rekindled an old friendship with a teacher and obtained money from him by using the cancer story once again.
In 1995 she was sent to prison for stealing money from the building society where she worked. At that trial, she tried to blame the theft on her original husband, Lee Wyatt. Upon her release from prison she had a relationship with a prison officer and then with an army officer. In 1997 Lee Wyatt finally divorced her, and she then set off to find another victim.
In 1999 Richard Thompson became her third husband and the cancer story was used yet again. She spent over £12,000 on his company credit card, emptied his savings account and then attempted to kill him. At a trial the following year, she was cleared of the attempted murder but found guilty of stealing the money from him as well as from two former boyfriends, and she was sent to prison yet again.
Julian Webb's body was exhumed in 2001 and re-examined, largely as a result of his mother's insistence over the years that he would never commit suicide. This belated forensic investigation proved that Webb had indeed been poisoned by curry. At a further trial, the Black Widow was sentenced to life imprisonment for murder, and she will serve at least 16 years in prison before her sentence is re-evaluated.

Different poisons tend to build up in specific parts of the body: for example, lead gets into the bones while insecticides prefer body fat.

But the central nervous system is by far the most frequently affected by poisoning, followed by the circulation, the liver and the kidneys. Less frequently affected are the lungs and skin, and bones and muscles least of all. Forensic examination of a body can uncover poison in any of these places, as in the case of the Black Widow, Julian Webb and the killer curry.

In December 2007, Heather Mook, who had previously been convicted in 1981 for poisoning her seven-year-old daughter Theresa with antidepressants, was found guilty of trying to murder her husband using antidepressants and rat poison in his spaghetti Bolognese. She had cooked the poisoned pasta in an attempt to get away with the theft of £43,000 worth of life savings intended to pay for a nursing home for her husband's mother.
Like the Black Widow, she too had tricked friends into handing over £20,000 by pretending she was being treated for cancer. At her trial, many other scams and deceptions came to light, together with a string of previous convictions, including a £5million luxury car and bogus hotel property scam. When the jury found her guilty, she was sent away for psychiatric reports before being jailed indefinitely, although she will be considered for parole after serving a minimum of five years in prison.

Going cold turkey

We all know that the morning after can be mildly unpleasant after a few glasses of wine, but a nightmare after a few bottles. If the body can break down toxins rapidly enough, then the ill effects of a poison will be minimal, as with the occasional glass of wine. But when the quantity of the noxious substance outweighs the speed of the body's ability to break it down, as with binge drinking, an awful hangover is only the best-case scenario.

Exposure to a number of poisons taken together may sometimes change the speed of absorption, breakdown or excretion. The effect may be a simple additive one, where the total effect is equal to the sum of the effects from each substance on its own. But often there is a synergistic effect, where the total combined consequence is far greater. Such synergy is seen, for example, with alcohol and cocaine, which both affect the brain on their own but produce much more serious effects when taken together. Sometimes even the presence of a non-poisonous substance can increase the toxicity of a poison, an effect called potentiation, such as in Chapter 3, where a perfectly edible mushroom should not be eaten by beer drinkers.

The opposite effect, antagonism, is far more useful, as this is how an antidote works. Antidotes are used in the treatment of poisoning to interfere with or counteract the action of toxins. Drugs derived from opium, such as morphine, and synthetic drugs made from it, like heroin, are called opioids. Opioids all act at specific receptors in the body, particularly in the brain, to depress the central nervous system. The effects of an opioid overdose can be reversed within seconds by an intravenous injection of an opioid antagonist drug, such as naloxone, which is used to compete with the opioid drug at these receptors, thus rapidly reversing their effects – a sort of instantaneous cold turkey.

Everything in moderation

A controversial theory called hormesis[2] suggests that low-dose stimulation by some poisons may lead to protection against their toxic effects. Dioxins are environmental poisons and carcinogens produced during the manufacture and incineration of plastics. But in minute doses in rats, these dioxins have been shown to reduce certain cancer rates. It has been known for centuries that very small doses of cyanide or strychnine act as tonics; indeed, many medicines in the past included them as 'must-have' ingredients.

Curiously, nuclear workers have been found to have lower rates of some cancers than the general population, perhaps due to their exposure to poisonous substances – in moderation – at work. In Cornwall, the radioactive gas radon, escaping from the underlying rock, can build up in houses that are not adequately ventilated. But this county's rate of childhood leukaemia is lower than the national average. This controversial theory even appears to hold with certain plants, which have been found to flourish when given low doses of certain herbicides. As a result of this theory, some doctors suggest that a moderate intake of alcohol is healthier than none at all.

The effects of some poisons are reversible, so the effect of a moderate dose will wear off with time, as the body breaks down – that is, metabolises – the poison into less toxic substances, which can then be excreted. Fortunately this is the case with alcohol. With some toxins the body responds by way of rapid evacuation, as in the case of food poisoning, where vomiting and diarrhoea quickly remove the cause of the problem. Other poisons, like the heavy metals such as lead and mercury, tend to accumulate because our body has no use for them, and so it stores them away until eventually the quantity is so great that the poison starts to interfere with vital functions.

Unintended side effects

Poisoning produces a number of common and expected symptoms: abdominal pain and vomiting, progressing in more serious cases to delirium and loss of consciousness, and even slowing respiration and death. A host of other symptoms can also occur during this progression, some of which may be highly characteristic to the poisonous substance.

George Hersey lived and worked in South Weymouth, Massachusetts. This young widower became engaged to a young lady called Mary Tirrell. Unfortunately she died soon after the announcement of their engagement. Her parents, seeing the young man's grief, offered George a home with them. He accepted and soon befriended their eldest daughter, Betsy. She was plain, unmarried and 25 years old. She always seemed somewhat low in spirits. In those days, she was likely to remain a spinster. Betsy began to look after George, mending his shirts and caring for his appearance. George did not work for a time, but he studied chemistry and, in particular, poisons, about which he could talk quite knowledgeably.

As the months went by, Betsy and George embarked on a secret love affair. But even so, Betsy's depression continued. One day George went to Boston and bought some strychnine from a pharmacy, explaining that he wanted it to put down a dog. In early May 1860, he took Betsy out for a carriage ride and on returning home went straight to bed, complaining of a headache. Betsy sat with the family that evening, reading aloud from the newspaper, before she too retired.

Half an hour later she was thrashing about in bed, twitching, convulsing and screaming. Her back was arched and she was in terrible pain, symptoms suggestive of strychnine poisoning. George was sent to fetch the doctor, but Betsy was dead before he arrived. A post-mortem showed that Betsy was three months pregnant. The doctor thought that she had been poisoned. The coroner was notified and tissues from various parts of the body were removed for further analysis. Betsy's parents, upon learning of the pregnancy, ordered George out of their house.

A search of the dead woman's room produced a spoon with traces of jam. Subsequent analysis showed this to be strychnine-laced jam. George Hersey was arrested and put on trial. He pleaded not guilty but a Boston doctor appeared at the trial and testified that George had requested that he perform 'an operation' on a lady friend who was

pregnant. The doctor had refused, of course, and had also refused to supply strychnine that was also requested.
The court decided that George had supplied Betsy with the poisoned laced jam, telling her that it would 'help' her out of her trouble, and she had taken it in the hopes of getting rid of her baby. The defence tried to suggest that Betsy had committed suicide because of her depression, but the jury found George guilty of murder, and he was sentenced to death. Before his execution, he admitted to causing Betsy's death but denied killing his first wife and his fiancée Mary Tirrell, Betsy's sister.

While some symptoms of poisoning are expected, others can creep up on us, sometimes even revealing seemingly harmless substances as poisonous. Medications prescribed by doctors can produce unwanted and even harmful effects, described as iatrogenic. This does not imply that there has been any lack of care by the doctor. Adverse, unwanted reactions may simply be a side effect of the medication, and may be unavoidable and mild. For example, the drowsiness caused by taking one of the older type of antihistamines is inevitable, as these older antihistamines cross the blood-brain barrier while the newer non-drowsy type do not, or do so to a far lesser extent. In recent years, this drowsiness side effect has been put to good use in over-the-counter remedies sold to treat insomnia.

We all know that you can have too much of a good thing, and sometimes iatrogenic illness can be caused by the interaction of several different types of well-intentioned medicines taken together. Every year, many elderly patients fall ill and are admitted to hospital due to interaction of their prescribed medicines – either with one another, or more likely with over-the-counter medicines or herbal remedies purchased by the patient, or their well-meaning relatives. This is not a new phenomenon: the physicians who treated King Charles II went overboard in their eagerness to cure their royal patient, and they killed him in the process.

It is widely believed that King Charles II was killed by the treatment he received from his physicians during his final illness. Charles II was born in 1630. His father, Charles I, was beheaded in 1649, and Charles II was forced to live in exile in Europe for nine years until the collapse of Cromwell's regime, after which he was restored to the throne. He reigned from 1660 to 1685.
His final illness began with a sore foot when he retired to bed on Sunday 1st February 1685. By the next morning he appeared to be

quite ill. The royal physician, Sir Edmund King, attended to the royal foot, but while a barber was preparing to shave the king, Charles II was struck with an apoplexy (now called a stroke). In an attempt to purge his leader of the illness, Sir Edmund immediately bled the king of 16 ounces of blood – the better part of a pint. He risked his own death with this act, as he should first have obtained the permission of the Privy Council.

By now, a dozen physicians were gathered about the king's bed, all of them anxious to draw off the 'toxic humours' assailing their monarch. He was bled, he was purged, his head was shaved, and cantharides – Spanish Fly – plasters were applied to his scalp to cause blistering; plasters of spurge were also applied to his feet to induce yet more blisters. When they didn't work, red-hot irons were applied to his skin. The large royal bedroom became very crowded, with 75 or more people now present: not only the physicians and their assistants, but also family members and many state officials.

The physicians worked on their monarch for five days, applying enemas of rock salt and syrup of buckthorn and an 'orange infusion of metals in white wine'. The king was treated with white hellebore root, Peruvian bark (now better known as quinine), white vitriol in paeony water, a distillation of cowslip flowers, sal ammoniac, julep of black cherry water – in addition to even more bizarre medicaments such as oriental bezoar stone from the stomach of a goat and boiled spirits from a human skull. Nothing worked, but despite all this 'treatment', the king apologised to those about him: 'I am sorry, gentlemen, for being such an unconscionable time a-dying.' It seems hardly surprising that he then added, 'I have suffered much more than you can imagine. '

By the following Friday (6th February 1685), the king was totally exhausted. His body was raw and aching with the burns and inflammation caused by his treatment, so his physicians gave him heart tonics, but it was all to no avail and the king lapsed into a coma. He died at noon on the following day. The historian Thomas Babington Macaulay recounted at the time that the poor king was 'tortured like an Indian at the stake'. In an extreme case of iatrogenic effects, the king was killed with kindness. Fortunately, today, doctors know far more about how medicines work and such polypharmacy is now much rarer – though not unknown.

So that's poisoning in a nutshell, but this chapter only skims the surface. Read on to discover where poison comes from, how to treat poisoning, how to avoid it – and some of the people who didn't.

The Poison Basics: Sources, Symptoms and Treatments

Chapter 2

Momentous Minerals and Deadly Dusts

OUR OWN PLANET CAN cost us our lives. For thousands of years, we have known that there are many poisons of mineral origin, from the very earth beneath our feet. These substances cause poisoning in humans because the body absorbs them in error, mistaking them for those essential elements that the body really needs, such as minerals to incorporate into enzymes. This mix-up happens because of the similarity of the poisonous minerals' atomic size to that of the essential elements.

Some deadly natural chemicals

Antimony is a metallic element widely used in modern industry in alloys and semiconductors. Its compounds are used in flame-proofing and rubber technology, paints and dyestuffs, as well as ceramics, enamels and glass. Even in the Middle Ages it was known that pigs thrived if a little antimony was added to their feed, given as the compound antimony potassium tartrate. However, the same feed supplement, if given to humans, caused people to vomit, which explains this compound's name, tartar emetic (emetic means to cause vomiting). Medicines containing antimony have been used for centuries, but few have been used as a means of poisoning, as in the case of Charles Bravo.

Charles Bravo was an ambitious 31-year-old barrister who was poisoned with tartar emetic in 1876. It took him three days to die. A number of suspects were involved in the case: his beautiful heiress wife, Florence, aged 29; her physician lover; and the stableman George Griffiths, who bore a grudge, having been dismissed by Mr Bravo. George also had a supply of the poison, which he kept for worming the horses.

Florence had been married before, and during her previous marriage she had taken to drinking heavily, so she had gone to Malvern to take the water cure. It was there that she began an affair with the physician, James Gully, who was twice her age. Dr Gully made her pregnant and then performed an abortion, an illegal act at the time for which Florence was ostracised by her social circle.

Her second marriage to Bravo had reinstated her position in society, but it made her lover very jealous. Florence was very rich, and for Bravo, this marriage offered the possibility of access to her fortune. He had political ambitions and hoped to enter parliament. However, the night before the poisoning, there was a furious row between husband and wife as Florence had invoked the Married Women's Property Act of 1870 to prevent Bravo from getting his hands on her wealth.

In her first marriage, Florence had been abused, and she now found herself abused again by Bravo. He was autocratic, and she claimed he made her perform 'un-natural acts'. She had by now suffered two miscarriages, and had come to dread his visits to her bedroom. On the night in question, Florence drank too much and retired to her bed, as did her housekeeper. Bravo then retired to his room and drank a glass of water from the carafe beside his bed. This water was poisoned: it contained tartar emetic in sufficient quantity that Bravo was quickly writhing in agony and vomiting severely.

Jane Cox, Florence's housekeeper and companion, a Jamaican woman with three sons, lied to the police and destroyed some evidence, fearing that she and her sons would be left destitute following their master's death. As a consequence, with so many suspects who all had reasons to kill Charles Bravo, it is no surprise to discover that the Coroner's Court was unable to apportion blame to any one person. However, following the inquest, Florence was disowned by her family and died, an alcoholic, at the age of only 32.

Arsenic and its compounds are extremely poisonous and were used in the past as weed killers, insecticides and wood preservatives. Despite its poisonous nature, arsenic is actually present in all our bodies, but only at a very low level, measured in millionths of a gram. It appears to be an essential trace element in some animals, and may also be so in humans. Experiments found that chickens that were fed on an arsenic-free diet had their growth stunted, and a similar result was found when testing rats.

Arsenic attaches to sulphur-containing enzymes and, by blocking their action, causes the toxic effects seen in poisoning. The body can excrete arsenic quite easily, so low doses obtained in the normal diet can be readily disposed of. Shellfish can contain quite high levels, but they can store it safely by converting it to a harmless form called arsenobetaine, which does not affect the shellfish. It does not harm us either, when we eat shellfish, as the arsenobetaine is readily absorbed from the gut and equally readily excreted through the urine.

The symptoms of acute arsenic poisoning are vomiting and diarrhoea, progressing to numbness and tingling of the feet, followed by muscular cramps, suppression of the urine, intense thirst, prostration and collapse. Chronic, long-term, low-level arsenical poisoning is characterised by rather different symptoms: there is a tendency to oedema (swelling up), due to the accumulation of fluid in the tissues, especially of the face and eyelids, with a feeling of facial stiffness, itching, tenderness of the mouth, loss of appetite, nausea, sickness and diarrhoea.

A hundred years ago, a common method of poisoning made use of arsenic-containing flypapers soaked in water, which provided a toxic solution. This solution could then be added very easily to food and drink, as it was both tasteless and odourless. Such a simple method was used by many murderers, and one instance, that of Frederick Seddon, eventually led to a change in the law.

Florence Elizabeth Maybrick was born in Alabama, USA, and was the daughter of a titled lady, the Baroness von Roques. She married James Maybrick, a Liverpool trader, in 1881. James was 23 years older than his wife and was a hypochondriac. He regularly took small doses of arsenic, which he considered to be an aphrodisiac, as did many others in those days. The Maybricks lived in a large house in Liverpool with their two children, five servants and a nanny.

In 1887 Florence discovered that her husband kept a mistress, so she took a lover, a family friend named Alfred Brierley. In March 1889 they spent a weekend in London together, but James found out about it. The Maybricks quarrelled and James gave his wife a beating. Two weeks later, Florence bought a dozen arsenic-based flypapers from the local chemist and, at the end of April, her husband became ill. She told the doctor that her husband was taking 'a white powder'. He rallied briefly but died on May 11th.

The nanny was suspicious and handed over some compromising letters to the dead man's brother. The house and particularly Florence's room were searched, and a packet labelled 'Arsenic poison for cats' was found. Traces of arsenic were found in James Maybrick's body, and so the coroner's jury returned a verdict of murder. Florence was tried in Liverpool in July 1889. The defence proposed that this was a natural death. Traces of strychnine, hyoscine and morphine, as well as arsenic, were found in his stomach, suggesting that James, the known hypochondriac, had poisoned himself.

However, the unfaithful wife was found guilty by the court which was prejudiced against her for her waywardness. The death sentence was passed but was later commuted to life imprisonment, of which Florence served some 15 years before being released. She died, aged 76, in Florida in 1941.

Arsenic has a prolonged action, so in many criminal cases of arsenical poisoning, where the poison was administered over a period of time, the symptoms shown may be those of acute poisoning caused by the final dose, or they may be the more chronic symptoms of long-term nerve, kidney or liver damage. The symptoms shown will vary from patient to patient, depending on the doses used and the length of time involved for the chronic poisoning. Continued use of small doses over long periods may also cause dryness, pigmentation and thickening of the skin, which may be accompanied by peripheral neuritis and alopecia. The breath may smell of garlic. Chronic inhalation may also result in perforation of the nasal septum, an effect also seen today in cocaine addicts.

Frederick Seddon worked for an insurance company. In 1911, he and his wife took in a lodger, Miss Eliza Barrow, a 49-year-old spinster. She owned substantial property and also had a cash box that contained hundreds of pounds in gold, which she insisted was kept under her bed because she had no faith in banks. As an officer of an insurance company, Seddon persuaded her to let him handle her affairs.

And so, within a short time, Barrow had signed over most of her property to Seddon in return for a small annuity and remission of her rent. It was not long afterwards that she fell violently ill, with vomiting and diarrhoea. The illness continued for several months until she eventually died. Despite there being ample money for a decent funeral, Seddon gave her only a pauper's grave, and even managed to earn 12 shillings in commission from the undertaker in the process.

Miss Barrow's cousin only learnt of her death from the local newspaper. After speaking to Seddon and getting only evasive answers from him, he then found that all her money had apparently disappeared. At this point the cousin went to the police with his suspicions. The body was exhumed and body tissues analysed: arsenic was found in sufficient quantity to suggest murder. Mr and Mrs Seddon were both arrested and put on trial. The evidence against them was somewhat flimsy as the only arsenic that could be found was in the flypapers that had hung above Miss Barrow's bed.

According to Seddon, they were used instead of carbolic, by which he meant carbolic acid, the old name for phenol. Carbolic acid was first used as an antiseptic and disinfectant by Joseph Lister in 1867 and was widely used for almost a hundred years. Frederick Seddon tried to use a lot of bluff and bluster in court, where he even used the Freemason's Oath in a desperate attempt to evade justice. His manner throughout the trial led the jury to convict him, while they acquitted his wife.

It was found that Seddon obtained the arsenic that he had used to poison Miss Barrow by extracting it from flypapers. White arsenic, as its name suggests, is white and opaque, like flour. It has little taste and no colour and so can be mixed into food by anyone with murderous intent. In the nineteenth and early twentieth centuries, white arsenic was combined with potash or soda to make flypapers. By soaking the flypapers in water, a strong solution containing arsenic could easily be obtained, and Seddon used this simple method. He was duly hanged at Pentonville Prison in 1912. The law relating to the sale or supply of flypapers containing arsenic was amended as a consequence of this and too many other similar cases of poisoning.

As little as 70mg of arsenic trioxide may be fatal as a single dose, and if adequate treatment is not given, death can occur within as little as an hour, although it more often takes one to two days.[1] Fortunately, arsenic and its compounds are far less common and much more strictly controlled today, detection is now easy and treatment with an antidote is readily available.

Cadmium is widely used in industry today, but many of its compounds have been found to be poisonous. This metal has similar properties to zinc, and so can be absorbed in error by the body, mistaking it for zinc. Zinc-containing enzymes perform many vital functions, as they are involved in many systems, such as regulating growth, development, longevity and fertility, as well as digestion, nucleic acid synthesis and the immune system. Interference with these key proteins can have dire consequences for the normal functioning of the body.

The presence of cadmium in the body triggers a defence mechanism in which production of a special enzyme is stimulated. This enzyme contains a lot of sulphur atoms to which the cadmium atoms can attach: seven cadmium atoms to each enzyme molecule. These conglomerates are then transported to the kidneys for excretion. Unfortunately, the cadmium atoms are so strongly bound that they tend to accumulate there, causing damage to the kidneys and, eventually, kidney failure.

The natural rate of excretion of cadmium is so slow that the harmful mineral can stay in the body for about 30 years. Consequently, even the smallest possible intake can cause problems. The replacement of zinc by cadmium in the testes, for example, damages viable sperm production.

Chromium or its salts, when taken by mouth, may lead to serious gut and kidney problems. This results in greatly reduced urine production, with large amounts of sugar in the urine. The blood supply to the limbs is also greatly reduced. The acute symptoms are intense thirst, dizziness, abdominal pain, vomiting and diarrhoea, leading to liver and kidney damage, which may prove fatal. Less than half a gram of potassium dichromate can cause serious poisoning, and the fatal dose is less than 8g.

Skin contact with chromium salts causes a widespread dermatitis, usually on the hands and other exposed parts of the body, frequently accompanied by deep perforating ulcers known as 'chrome holes'.

Chrome dust, if inhaled, causes rhinitis – inflammation of the mucous membrane of the nasal passages – resulting in perforating ulcers of the nasal septum. Inhalation may also cause severe lung damage and bronchitis as well as gastritis and other inflammatory conditions, particularly of the eyes. There is also an increased risk of lung cancer, and there may even be damage to the central nervous system.[2]

Copper is an essential element, being part of more than ten enzymes in the body. A body deficient in copper is unlikely, as we receive all our needs from our diet and drinking water (more so in soft water areas with copper pipes). A trace of copper is essential for a number of our enzyme systems, the most important of which is the 'cytochrome c oxidase' enzyme system, which is required by all cells to produce energy. During trials with copper-plated door handles and other fittings at Birmingham Children's Hospital, it was recently shown that the bacteria causing MRSA – Methicillin (or Multiple) Resistant Staphylococcus Aureus – infections are killed in less than a minute when in contact with copper.

Copper is part of other enzymes, including those that are involved in repairing connective tissue, those that produce hormones and the skin pigment melanin and those that remove certain harmful protein breakdown products called amines. However, poisoning can occur, as happens in the rare inherited condition Wilson's disease, in which copper accumulates because its excretion is impaired. Providing the excess copper is removed from the body by regular treatment, both the physical and mental development of sufferers may be unaffected. When it is not properly removed, that's when copper becomes a real poison.

Iodine is essential to life in animals, but not so in plants or algae, although they absorb it from soil and seawater respectively. Iodine is essential to humans for the production of the thyroid hormone, which regulates the metabolic rate of the body. Kelp is included as an ingredient in herbal slimming preparations because of its iodine content, in the hopes that extra iodine might speed up the metabolism a little and thus help with weight loss.

However, you can have too much of a good thing. Too much iodine can be toxic and may interfere with female fertility. In addition, when heated, iodine forms a vapour that irritates the eyes and lungs. In industry, there are now limits placed on iodine concentration in the air.

Iron, although poisonous, is essential to human life. People with a reduced iron level in their blood (a medical condition called anaemia) feel tired, because a lack of haemoglobin (haem is an iron-containing compound) means there is insufficient oxygen being carried around the body. Haem is responsible for both the red colour of blood and its oxygen-carrying capacity.

Not all blood is red though. While our red blood contains iron as part of the haemoglobin molecule, some animals like octopuses, oysters, snails and spiders have blue blood, because they have copper, rather than iron, contained at the centre of their oxygen-carrying molecule, haemacyanin.

Iron is also a part of various enzyme systems, including those involved in the synthesis of DNA. The brain needs iron to function properly. Excess iron is stored in the liver, where it is held as the iron storage proteins ferritin and haemosiderin. The bone marrow, where red blood cells are made, is another part of the body that is naturally rich in iron.

Although rare, some people are allergic to iron. In the 1970s, a 66-year-old toolmaker suffered from an allergic contact dermatitis due to exposure to iron.[3]

Oral doses of iron, such as ferrous sulphate tablets, are best absorbed on an empty stomach, but they are usually taken after food as they can be an irritant to the stomach of some patients. In overdose they can be corrosive to the gastro-intestinal tract, where necrosis and perforation may occur. The symptoms may be delayed for some hours and can include epigastric pain, diarrhoea and vomiting blood. If the diarrhoea and bleeding are very severe, circulatory failure can result, leading to death.

Poisoning from swallowing iron tablets can be particularly dangerous to small children, as it can result in a severe electrolyte imbalance that develops soon after the tablets have been absorbed. Thus, speed is essential in treating such children, as a dose of as little as five ferrous sulphate tablets can be considered toxic in small children.[4] This topic will be discussed in more detail in a later chapter about paediatric poisoning.

Lead, used by humans for thousands of years, is easily extracted from the sulphide ore galena by heating. Lead is a very useful metal; because it is so soft, it can easily be worked, bent and shaped even at room temperature. It has been used for pipework since Roman times, despite its poisonous properties, which were known even then. As a heavy metal, lead is also widely used in industry, paints, batteries and as tetraethyl lead, an antiknock agent which used to be added to petrol.

Lead's poisonous qualities have been known since ancient times. Lead poisoning, or plumbism, may be acute or chronic and can be caused by either organic or inorganic lead. Lead poisoning causes gastro-intestinal and cardiovascular effects, as well as liver and kidney damage, but it does the most damage to the central nervous system, causing mental disturbances and even convulsions.

Symptoms of acute lead poisoning include intense thirst, a metallic taste in the mouth, a burning abdominal pain called lead colic, vomiting and diarrhoea. Longer exposure leads to anorexia, anaemia, peripheral nerve damage and brain damage, together with convulsions and coma. There can be permanent kidney damage and impairment of mental function.

Chronic poisoning is due to the accumulation of small quantities over a prolonged period of time following inhalation, ingestion or absorption through the skin, leading to the development of a characteristic black or blue lead line on the gums. The fatal dose is estimated to be 500mg.[5]

This infamous poisonous metal – lead and its compounds – will be discussed in many chapters of this book.

Lithium's mode of action is not yet fully known. However, it is confirmed that it competes with sodium at various sites in the body and acts on the brain, where it was found that those suffering from bipolar disorder, formerly called manic depression, have an excess production of a chemical messenger. Lithium appears to interfere with the production of this messenger, reducing it to normal levels, and so making life more bearable for those suffering from bipolar disorder.

The therapeutic dose of lithium is not much less than the toxic level, so patients must have regular blood tests to check for toxicity. Symptoms of toxicity are tremor, weakness, nausea, thirst and excessive urination. Lithium interferes with thyroid function and long-term use can cause kidney function changes; too high a dose of lithium can ultimately lead to renal (kidney) failure. An overdose can be fatal.

Manganese is yet another essential element, but we need only a very small quantity for a number of important enzymes, such as those involved in glucose metabolism, and for the functioning of thiamine (vitamin B1). Most manganese is found in our bones, but it also concentrates in the pituitary and mammary glands, the liver and the pancreas. Inhalation by miners in under-ventilated mines causes brain damage and symptoms similar to Parkinsonism.

Mercury is a heavy metal that we acquire from our food, although our bodies have no use for it at all. Mercury can be absorbed through the lungs, the skin and the digestive tract. Most mercury in the body is found stored in the kidneys, liver, spleen and brain. It is the only metal that is liquid at room temperature, which explains why, in the past, it was called quicksilver. Unfortunately, it gives off mercury vapour into the air, which is an added problem with its use. Much used in medicine in the past, mercury and its compounds are now known to be very toxic.[6]

Arsenic and mercury seem to have been the most popular poisons of choice in the past, as was the case with Lady Frances Howard, both an adulteress and a murderer. Her victim is better remembered for his death by poisoning than for his career as a theatre critic during the reign of James I.
In 1607 a marriage was arranged between Frances, who was then

only 13 years old, and the 14-year-old Earl of Essex. This betrothal was a political move to advance the interests of the Howard family. Once the teens were married, the bridegroom was sent off to the continent to complete his education, and his bride was left behind.

By the time she was 16, Frances had developed into a beautiful young woman. She now became infatuated with a courtier, Robert Carr, who was a favourite of King James I. When she heard that her husband was returning to claim his bride, she fled to her great-uncle's house, knowing that she would be safe there. She then heard that her husband had taken ill and was confined to bed, so she quickly returned to court to pursue her beloved Robert Carr.

Realising that a divorce on the grounds of non-consummation of her marriage to the Earl of Essex was a possibility, providing Frances could keep her husband from making love to her, her lover enlisted the help of Anne Turner. This lady, as well as running a number of brothels, was a court dressmaker. She took Frances to see a fashionable quack, Simon Forman, who gave her various devices and potions to use so that her husband's sexual ardour might be reduced. Whatever was used worked so effectively on his virility that it was said the Earl of Essex became impotent as a result.

In the winter of 1611, Frances and Robert at last became lovers, using Mrs Turner's brothels for their assignations. Frances wanted a divorce and Robert asked the king for his help. The king agreed that perhaps an annulment would be possible. The clergy and lawyers, who considered the matter on the king's behalf, wanted proof of the lady's virginity before allowing a divorce. As Frances was no longer a virgin, she used the young daughter of a courtier to take her place for the examination under cover of many veils on the bed. The deception was successful and she was granted a divorce on the grounds of her husband's impotence.

The ultimate victim of this case, Sir Thomas Overbury, was a great friend of Carr. He knew all about what had been going on and thought that once Carr was married to Frances, and thus also to the powerful Howard family, there would no longer be a place for him. He also realised that if he could tell the king about the affair, Frances would be disgraced and the power of the Howard family would be broken.

Frances's great-uncle, the Earl of Northampton, realised the threat that Sir Thomas Overbury posed with his knowledge of events. He managed to convince Carr that his friend Thomas should be imprisoned in the Tower of London to keep him safely out of the way until the divorce was granted, after which he could be released.

Sir Thomas was in a very powerful, but also a very dangerous, position.
He then publicly called Frances a whore and, by way of retaliation, she decided that he would never get out of the Tower alive. Together with Anne Turner, she visited a number of alchemists, with a shopping list of poisons such as arsenic, cantharides and sublimate of mercury. These were put into tarts and jellies, which Anne Turner then took to the Tower for Sir Thomas to eat.
Meanwhile Carr decided that if his friend's health was in danger, then his release might be secured on compassionate grounds. Accordingly, without telling Frances, he began to send his friend enemas and 'vomits' to make him violently ill and thus provide compassionate grounds for his release. So it was that while Frances sent him poisoned food, her lover sent him purges that expelled the poisons and consequently kept him alive. When Frances finally found out what Carr was doing, she sent Thomas an enema, but this one was made with sublimate of mercury. It was enough to kill him, and he died in agony on 15th September 1613.
The divorce came through and the lovers Frances and Robert then married. The mysterious death of Sir Thomas Overbury seemed to be forgotten until about a year later, when the apothecary's assistant, who had delivered the fatal enema, became seriously ill himself. Believing he was about to die, he confessed to what had happened. Word of his confession reached the king who called in his chief justice. Frances and Robert were duly tried and both were found guilty and sentenced to death.
They were later pardoned by the king but confined to the Tower of London. After a short time they were released and moved to live at Rotherfield Greys in Oxfordshire, in a house called Greys Court, which is now in the care of the National Trust. By this time, Frances and Robert were barely on speaking terms. Each lived in a different wing of the house until the end of their lives, meeting only at meal times.

Micro-organisms, including algae, use mercury to produce organomercury compounds, which then move up the food chain where they are consumed by larger creatures such as fish and crustaceans. Methyl-mercury is an example of such an organomercury compound, and it can cross the blood-brain barrier in humans to affect the central nervous system, including the brain. It can also cross the placenta to affect an unborn child. Consequently, the use of organomercury compounds, as pesticides and seed dressings, is now severely limited. We will return to this in Chapter 9.

The effect of mercury poisoning on children used to be called acrodynia by doctors but was more commonly known as 'pink disease'. This severe illness suffered by children of teething age was marked by pink, cold, clammy cheeks, hands and feet, heavy sweating, raised blood pressure, rapid pulse, photophobia, loss of appetite and insomnia.

It was not until the 1950s that 'pink disease' was recognised as a form of mercury poisoning due to the mercury in teething powders, as well as the absorption of mercury from dusting powders and ointments used to treat nappy rash. Many fatalities had occurred over the years, particularly as the affected infants were also prone to secondary infection. It was suggested at the time, by some doctors, that the fatalities were due to an allergic reaction to mercury-containing products; however, 'pink disease' disappeared once the offending products were banned and removed from continued use.[7]

The more soluble forms of compounds containing mercury are, as would be expected, the more toxic ones. Mercury has a specific affinity for sulphur and so attaches itself to the sulphur atoms of certain amino acids and, if they happen to be part of an enzyme, this enzyme will stop working. Acute poisoning causes severe headaches, nausea, vomiting, stomach pains, diarrhoea and a metallic taste in the mouth. After only a few days there will then follow excessive salivation, swelling of the salivary glands and then, later, loosening of the teeth.

Mercury has been, and still is, widely used in industry in electrical equipment such as mercury vapour lamps, switches, pumps, thermometers and batteries as well as pesticides and dental fillings. Chronic poisoning, as in industrial exposure, causes a different set of symptoms: fatigue, weakness, loss of memory, insomnia, irritability, depression and a paranoid belief that people are persecuting the victim. Tremor of the hands and increased salivation are also seen.

Occupational exposure is now carefully regulated, although it was not so in the past and many cases of work-related chronic poisoning occurred as a result. Because of the volatility of mercury, it can be breathed in (and out), so the level allowed in the atmosphere of the workplace today is carefully regulated, and must not exceed 0.1 milligram per cubic metre.[8] The body can rid itself of these small amounts of mercury not only through the urine and faeces, but also via the lungs and sweat. It also locks some away in hair and nails.

Though mercury poisoning used to be quite common, due to improved health and safety legislation, it is now rare in the Western world.

In October 2008, a Russian lawyer who defends cases against the Kremlin was living in Paris with her family. Karinna Mosakalenko alleged that enemies in the Kremlin were trying to poison her. She claimed that her family suffered from headaches and nausea. Liquid mercury was found in her car, and she believed it had been put there to frighten or poison her. But the French police found that the car's previous owner had been an antique dealer, who when tracked down said that the mercury came from a barometer that broke while being transported in the car. So mercury poisoning, yes, but a Kremlin conspiracy, no!

Molybdenum is another essential element, although again only a trace is required in the body, and larger quantities are toxic. For example, too much molybdenum can damage an unborn child. Molybdenum is contained in a number of enzymes, including aldehyde oxidase, which is involved in the metabolism of alcohol. So for many people, this is an absolutely essential enzyme, particularly at the weekends.

But drinkers beware! A high intake of molybdenum can result in an attack of gout. This is due to another enzyme, also containing molybdenum, called xanthine oxidase. Increased activity of this second enzyme causes a rise in the uric acid level in the blood, and of the uric acid crystals being deposited in the tissues, particularly the joints, and this increase causes the excruciating pain of gout. A high molybdenum intake may also impair the body's utilisation of copper.

Nitrogen oxide, better known as **nitric oxide,** is toxic and regarded as a major pollutant found in the exhaust fumes of traffic-congested cities. So it may come as a surprise that this poison is actually produced in our bodies and acts as a chemical messenger for relaxing our muscles. This natural occurrence was only relatively recently discovered – in 1987 – by Salvador Moncada. We now know that nitric oxide is involved in the cardiovascular system, the immune system and both the central and peripheral nervous systems.[9] The scientific paper describing this discovery became the most cited of all time, because of its link to a certain drug and how it works: Viagra!

Sildenafil, better known as Viagra (the drug used for erectile dysfunction), works by releasing nitric oxide, which activates an erection by relaxing the muscle controlling the flow of blood into the penis, thus allowing this organ to become engorged, enlarged and erect. Sildenafil is also used to treat altitude sickness and pulmonary arterial hypertension with great success, although for these purposes it has a different brand name, Revotio, and a different strength.

Though nitric oxide has been found to have these positive uses, it has always been considered too toxic to use as a gas, and in the past amyl nitrite vapour was used instead, as a treatment to relax muscles and lower blood pressure during angina attacks. This became the standard treatment for angina for many years.

During World War I a better treatment was discovered when doctors noticed that the munitions workers who were packing shells with the explosive nitroglycerin had very low blood pressure. This was a result, they discovered, of absorption of the nitroglycerin through their skin.

And so, under its alternative name, glyceryl trinitrate, it came into use for angina, and this treatment is still used today, as tablets, skin patches, ointment and a spray. However, it must be used sparingly, because too much can result in a pounding headache. Alfred Nobel suffered from the most terrible headaches in his explosives factory, and now we know why: the TNT he discovered and made is chemically similar to nitroglycerin.

Phosphorus may be essential to life as phosphate in our bodies, but the element phosphorus is very toxic. Phosphorus poisoning has three distinct phases when acute. In the first phase, there is intense thirst and pain, with nausea, vomiting and diarrhoea, due to gastric irritation. The breath smells of garlic and tastes of it too, and any vomitus and excreta glow in the dark. Sometimes shock, delirium, convulsions, coma and even death may occur in this first stage. But usually the symptoms now subside as the poisoning enters the second phase, which may last for several days, when the patient appears free of symptoms, and his or her recovery may be expected.

But this is false hope. For then the third phase, due to systemic toxicity, begins: liver and kidney damage and circulatory problems, which include haemorrhage and cardiovascular collapse. Central nervous system involvement leads to confusion, convulsions and coma. Death may occur during either the first or the third phase, the fatal dose being about 1mg per kilogram of bodyweight.

Chronic poisoning symptoms are very slow in onset and result in reduced resistance to infection and tissue repair defects, with poor wound healing. Phosphorus can cause severe burns to the skin on contact, and poisoning can also occur following absorption through the skin.[10]

A most bizarre accidental death caused by phosphorus was reported in *The Lancet* in 1890. A fake medium had smeared his assistant, a young girl, with some phosphorus rat poison so that she would glow in the dark during a séance and appear to be a luminescent spirit during the entertainment. She died, of course, having absorbed through her skin a more-than-lethal dose of phosphorus.

Selenium is used in enzymes in every cell of our body, and yet it was only about 30 years ago that it was recognised as an essential trace element with important antioxidant properties. We now know that selenium is vital for the proper functioning of our immune system and is involved in our cardiovascular health, embryology, sperm production and the ageing process. Despite its importance, we only need selenium in microgram quantities in our daily diet. It can be very poisonous in larger amounts – as little as 5mg could be fatal.[11]

Selenium sulphide was used as a shampoo in the treatment of dandruff and seborrhoeic dermatitis of the scalp for many years, and dermatologists also prescribed it for a range of other skin complaints. In America in 1961, the *New England Journal of Medicine* reported a case of poisoning in a 46-year-old woman with open eruptions on her scalp. She had used a shampoo containing selenium sulphide two or three times a week for the previous eight months. During this time she developed weakness, anorexia, tremors, sweating, a metallic taste in her mouth and breath smelling of garlic. However, once she stopped using the shampoo, her symptoms all disappeared within ten days.[12] Some patients who used this type of shampoo also found that it caused their hair to fall out in varying degrees, though this effect could be halted by changing the shampoo.

Silver has no biological role, and indeed solutions of silver salts are irritants to the skin and mucous membranes and can sometimes be fatal if swallowed. Fortunately, the hydrochloric acid in the stomach can react with silver solutions to produce silver chloride, which is not soluble and so passes safely through the gut, unabsorbed, to be excreted in the faeces. Any silver that does get absorbed tends to lodge in the liver and skin.

In the past, silver poisoning, or argyria, resulted from prolonged treatment with silver salts such as silver nitrate. Poisoning could also occur as a result of ingestion or industrial exposure to these salts. With argyria, a slate grey pigmentation of the skin develops slowly, accentuated in those areas exposed to light.

However, silver usage now is more positive than negative. Both colloidal silver and silver protein have been used in skin lotions and washes for the treatment of acne, as they are antibacterial. This antibacterial action is also utilised in first aid plasters and in wound dressings containing silver and charcoal for malodorous pressure sores and ulcerated areas. Athletes and walkers can now buy socks made with silver fibres woven into them to prevent growth of the bacteria that produces the sulphur-containing substances that make feet smell. Silver acetate was used in anti-smoking preparations, as this addition produces a bitter taste in the mouth when a cigarette is smoked. Silver is also used in homoeopathic medicine both as the metal and silver nitrate.

Sulphur is a non-toxic element, but many of its compounds are toxic, including sulphur dioxide, SO_2, which is a colourless, poisonous gas with a characteristically pungent smell that irritates the nose and lungs and can even trigger asthma attacks in those with that predisposition. It is a by-product of fossil fuel combustion and a major air pollutant, but it has many industrial uses including bleaching and food preservation.

Sulphur is needed by all living things as it is part of the amino acids methionine and cysteine. Cysteine is the amino acid that links peptide chains together by forming sulphur bonds between them. These bonds are found particularly in our hair and nails, in keratin. When ladies have their hair permed, the chemicals used first break these sulphur bonds and then later reform them differently, producing ‘permanent’ waves and curls in the hair.

Sulphur dioxide dissolves in the water vapour in the atmosphere to form sulphurous acid and is then oxidised to sulphuric acid, producing acid rain, which causes environmental damage. The term ‘acid rain’ was first used as long ago as 1872 by Robert Angus Smith, the first British pollution inspector, when he discovered that the rain falling on Manchester contained sulphuric acid. Today it is used more generally to refer to any deposition of pollutants.

Sulphuric acid, or Oil of Vitriol, as it used to be called, is a very strong acid that, in concentrated form, causes terrible burns if it comes into contact with the skin. Blue, green and white vitriol were the old names for the sulphates of copper, iron and zinc, respectively.

Hydrogen sulphide is a colourless but very poisonous gas that is present in some mineral waters and in volcanic emissions. It can also be produced by putrefying organic matter and has the characteristic smell of rotten eggs. Even very low levels are easily detected by our noses, which instantly causes us to move away from the area, as nature intended: a built-in safety mechanism. In August 2009, a French vet reported the death of his horse as he rode along a beach in Brittany. He lost consciousness himself and only survived because he was rescued by local residents. These events were due to hydrogen sulphide gas emitted from rotting seaweed, a phenomenon that is spreading across the region's beaches as nitrates pollute the water supply from intensive agriculture.

Radium, Uranium and Plutonium are elements (among others) that are radioactive and, although they can be very useful to both science and industry, they are very dangerous if allowed to escape into the atmosphere, as they can cause cancer. Poisoning caused by radioactivity appears several times in this book, including the chapters about occupational poisoning and accidents.

Polonium is one of the deadliest substances known, being about a trillion times more toxic than hydrogen cyanide, weight for weight. This volatile substance is a powerful and highly dangerous alpha emitter. Marie Curie, who discovered radium, died in 1934 of leukaemia, caused by so many years of exposure during her work. Like mother like daughter, her daughter Irène Joliot-Curie, who also won a Nobel Prize, also died of leukaemia, in 1956. But her death was attributed to exposure to polonium some 15 years earlier, when a capsule of this substance exploded on her laboratory bench.

Alexander Litvinenko died in agony in a London Hospital on 23rd November 2006. He had been poisoned by drinking a cup of tea into which a small quantity of radioactive polonium had been added. He had been poisoned during a meeting with two former KGB contacts. These Russian agents, Andrei Lugovoy and Dmitri Kovtun, had been sent by President Leonid Putin to execute him. Their meeting at the Millennium Hotel had taken place on 1st November 2006 and their victim was dead 22 days later. Scotland Yard found that they could trace the route of both poisoners and their victim around London by following the trail of radioactivity.

Thallium is yet another element similar to lead, which, together with its compounds, is poisonous. Thallium is so toxic that it is readily absorbed through the skin, and handling its compounds without protective gloves can easily lead to loss of the fingernails. Thallium mimics potassium in the body, as it is of similar size and ionic charge – but the two have dramatically different results. The symptoms of thallium poisoning are lethargy, numbness, tingling hands and feet, tremor, slurred speech, debility and hair loss. The poisoning can lead to cardiovascular collapse, delirium, convulsions, paralysis and coma, resulting in death in as little as one or two days.

Graham Young was a compulsive poisoner who used thallium as his 'modus operandi' in his later years. When he was only 15 years old, Graham was committed to a hospital for the criminally insane for confessing to poisoning his stepmother when he was 13, attempting to poison his sister with belladonna, and later poisoning his father and a school friend with antimony. This confession was in 1962, and some nine years later, after his committal and several years' confinement in the Broadmoor facility, the doctors decided Graham had made a full recovery, so he was released back into society.

He was now 23 years old and, following his release, he got a job with John Hadland Ltd at Bovington, Hertfordshire. The day after he got the job, he went to a chemist in Wigmore Street in London and purchased 25 grams of tartar emetic, which contains antimony, and he signed the poisons register in the name M E Evans. He told the chemist that the solution was for chemical analysis purposes. Some time later he also purchased a quantity of thallium from the same chemist.

In April 1971, Young began working with a company that made photographic equipment. Within a few months, a mystery illness began to affect the workers; several of them had stomach pains, vomiting and diarrhoea. In November, Frederick Biggs, a department manager, died in hospital. The chief storeman then died from what was certified to be broncho-pneumonia, and two other workers were admitted to hospital. The management launched a medical enquiry and, at a meeting of the workforce, a doctor reassured them that none of the chemicals they used in their work were to blame. The young storeman Graham Young stood up at the meeting and asked if thallium could be to blame. Needless to say, this strange suggestion aroused suspicion.

The police were called in and the chief storeman's body was exhumed. Analysis of his tissues and those of Biggs both showed that they had died of thallium poisoning. The police then checked into the background of Young and, following a search of his possessions, they found the thallium. He was then arrested for the murder of his two workmates. He pleaded not guilty at his trial but his diary full of incriminating entries was found and used in evidence against him. Former workmates testified that he was an enthusiastic tea-maker, but that it always tasted very bitter when he made it. The chief storeman always drank the tea, and he had died. Others who left their tea untouched had survived. It transpired that both the tea and the coffee made by Young were liberally laced with either antimony or thallium. Among the symptoms his workmates suffered was loss of hair, not only from the scalp but from beards too.

In July 1972, Young was convicted of murder, attempted murder and administering poisons and was sent to Parkhurst Prison for life. This was not for as long as might have been expected, as he died of a heart attack in prison at the age of only 42in July 1990, although there were rumours that he had committed suicide.

In November 2005 a Japanese schoolgirl was arrested for gradually poisoning her mother, using thallium, just as Young had done. Following in his footsteps in more modern boots, this young lady even kept a diary as he had, although she used her computer to keep a blog of her progress, over several months, for all the world to read via the internet.

Tin varies in toxicity depending on its form. Inorganic tin compounds are regarded as non-toxic; indeed, stannous fluoride was used in the same way as sodium fluoride, as a supplement and in toothpaste to strengthen teeth. The metal itself has been used to coat the iron used in making tin cans, a long-lasting form of food preservation.

However, organic tin compounds such as trimethyl-tin and triethyl-tin are very toxic to the liver and kidneys and can cause severe neurological damage due to oedema, swelling of the white matter of the brain. Also, these compounds can cause severe burns if put in contact with skin. Organotin compounds have been used in wood preservatives, as anti-mould agents applied to stonework and as anti-fouling paints used on the hulls of boats, although the use of these has now been restricted due to environmental damage to the food chain. Many more problems in the food chain will appear again later.

Zinc is a trace element involved in many enzyme systems in the body, and deficiency of this mineral tends to be rare. However, it can happen and a deficiency of zinc in the long term can result in stunted growth, retarded sexual development and taste disturbance.[13] Too much zinc can and does cause poisoning, as we shall see in later chapters.

Minerals are by no means the only natural occurrence of poisons: many plants can pack a deadly punch too.

Chapter 3

Poisonous Plants and Frightful Fungi

WHEN OUR MOTHERS TOLD us not to put just anything in our mouths, they had a point. Many plants, even those in the garden, that might tempt a young child's curiosity and need to touch and taste, contain poisonous substances. Even the most innocent looking of these plants can in fact be a double-edged sword: for example, rhubarb stalks are eaten, but their leaves are discarded, as these leaves contain highly poisonous oxalic acid.

A plant poison may be concentrated in a specific part of the plant, such as the fruit or the root, or it may be distributed throughout the plant tissue.[1,2] For thousands of years, the effects of these plants have been explored. Like most poisons, in moderation some can have positive effects on various body functions, while large doses can be deadly. For example, alkaloids (nitrogen-containing substances produced by plants) have been used as medicines for thousands of years in drugs such as atropine, morphine and quinine, with positive results. (See Appendix I.) Even today these substances are used, administered in carefully calculated doses. It is wise to remember, though, that there are no safe drugs, only safe ways of using them. We all know what happens from a morphine overdose. And we're about to find out what happens when these drugs end up in the wrong hands.

Poisonous plants that paralyse

Curare is a crude extract made from a woody vine called Pareira, mixed with extracts containing **strychnine**, which is derived from various plants of the *Strychnos* species. Around the world, plant products have been used as arrow poisons, for hunting, execution or warfare, and curare is one such example.For instance, South American natives have used it for centuries to poison the tips of their arrows. Curare blocks the nerve signals to muscles, causing relaxation and paralysis of the voluntary – skeletal – muscles, which can kill. It docs not affect the involuntary – smooth – muscle, which produces the slow, long-term contractions of which we are unaware, such as in the gut, the bladder or blood vessels.

Miss Christina Edmunds, a spinster living alone in Brighton, became infatuated with her doctor, Dr Beard, in 1870. The doctor began to receive a stream of romantic letters from her, which he kept hidden from his wife. In September that year, Miss Edmunds obtained a quantity of strychnine on the pretext that she was being plagued by stray cats.

One evening when the doctor was out, she took some poisoned chocolates round to his house and offered them to Mrs Beard. The doctor's wife selected one but quickly spat it out because it tasted unpleasant. The next day she suffered diarrhoea and felt rather unwell but said nothing to her husband. Miss Edmunds, however, did tell the doctor about it. He felt obliged to tell her she was no longer welcome in his house and warned his wife to be on her guard. Shortly afterwards the doctor went away for three months.

During this time, Miss Edmunds tried to show that the doctor's wife had been poisoned by accident. She felt that this would be a way of getting back in his good books. Unfortunately her method of achieving this was to buy even more chocolates, which she also laced with strychnine. She then handed them out to children in the town. As a result, a number of children became ill and one child died. Poisoned fruit and cakes were then sent out, and the people who ate them were lucky and survived. Investigations began and eventually the sender of the poisoned chocolates, fruit and cakes was discovered to be the infatuated spinster.

Edmunds was arrested in August 1871. At her trial in the following year, at the Old Bailey, her mother told of a long history of insanity in the family. The jury, however, were unimpressed by this and found her guilty of murder. She was sentenced to death but the doctors who examined her all agreed that she was quite mad. So the Home Secretary commuted her sentence and sent her to Broadmoor, a hospital for the criminally insane, and there she died, some 35years later, in 1907.

Strychnine is an alkaloid produced by the seeds (nuts) of *Strychnos nux vomica*. The symptoms of strychnine poisoning are mainly those due to overstimulation of the central nervous system. This poison competes with glycine, an inhibitory neurotransmitter, which normally switches off nervous impulses. This competition with strychnine results in a continual stimulatory effect, which can eventually result in death.

The first signs of poisoning are tremors and slight twitching; then stiffness of the face and limbs develops within 15 to 30minutes of ingestion. These preliminary symptoms are followed by the sudden onset of painful convulsions. Initially the movements are intermittent but spinal tetany (twitching and spasms) quickly appears, involving all the muscles in a symmetrical fashion. The body becomes rigid, and the contraction of the facial muscles produces a characteristic grinning expression, known as 'risus sardonicus'.

Eva Rablen was the attractive second wife of Carroll Rablen. They lived in Tuttletown in California. Mr Rablen was deaf, following an injury during World War I. His new wife loved dancing and despite his deafness, he took her to dances, sitting it out while she danced with other men. In April 1929 the couple attended the regular Friday night dance. Carroll waited out in the car while Eva danced the night away.

At about midnight, she took him a cup of coffee and some refreshments, handed them to him in the car and then returned to the dance. Within seconds of drinking some of the coffee, Carroll was writhing on the floor of the car in agony. People heard his cries and came running. He managed to tell them how bitter the coffee had tasted, but was dead before the doctor arrived. His wife seemed heartbroken; the police considered suicide and on analysis of the stomach contents by a local chemist apparently found no trace of poison.

However, Rablen's father told the police that he suspected his daughter-in-law Eva because of a number of insurance policies she had taken out on Rablen's life. A police search of the area around the schoolhouse, where the dance was held, found a bottle bearing the label 'strychnine'. This bottle had come from a pharmacy in a nearby town. The pharmacist traced the sale of the strychnine in his poison register to a Mrs Joe Williams. This lady had said, at the time of purchase, that she wished to use it to kill gophers.

In due course, Eva was identified as Mrs Joe Williams, and she was arrested. An eminent chemist and criminologist, Dr Edward Heinrich, was called in and found traces of strychnine in the stomach of the dead man and on the coffee cup he had drunk from. Strychnine was even identified from a coffee stain on the dress of another woman who was at the dance. Eva had spilt coffee on this woman's dress as she carried it out to her husband. At her trial Eva decided to change her plea to guilty and was sentenced to life imprisonment.

All forms of sensation are heightened by the action of strychnine, and since consciousness is not impaired, the victim may become extremely distressed. Even the most minor of sensory stimuli can trigger painful spasms and convulsions. The body arches backwards due to contraction of the muscles of the diaphragm, together with a spasm of the chest and abdominal muscles. This spasm can even result in respiratory arrest. The convulsions usually last for a minute or two and are then followed by a period of depression, with a second convulsive seizure usually following within the next ten to 15 minutes.

Death results from medullary paralysis in the brain, usually following the second seizure, although in some cases it has not occurred until during the fifth seizure. The fatal dose, by mouth, is usually about 100mg but can be as little as 16mg. As strychnine is excreted very slowly by the body its action can be cumulative.

Jean Pierre Vaquier was a Frenchman and a poisoner. He was a wireless operator and met Mrs Mabel Jones when she was staying in Biarritz, alone, on holiday. The Frenchman didn't speak a word of English, and Mabel didn't speak French, but they became lovers – her first affair in some 19 years of marriage. Alfred Jones, her husband, was the landlord of the Blue Anchor Inn at Byfleet, Surrey.
After a few weeks, Jones sent his wife a telegram asking her to return home, which she did a few days later. Shortly after her return, she received a telegram from her lover, telling her that he was now in London. She went to his hotel and they resumed the romance for a day and a night while her husband was away on business.
She then returned home to the Blue Anchor and, shortly afterwards, Jean Vaquier appeared, requesting a room. Jones, who knew nothing of the romance, got on very well with him, so much so that he even lent him some money.
Two weeks after his arrival in London, the Frenchman went to a chemist's shop near Holborn and produced a list of chemicals, which he said he required for experiments concerning the wireless. The list included chloroform, mercury perchloride and strychnine. The chemist was somewhat reluctant to supply what amounted to enough poison to kill at least 12 people. However, the sale was made and the customer duly signed the poison register.
Every night, the landlord Alfred Jones got drunk. The French lover had become rather useful during his stay, carrying the inebriated Alfred up to his bed each night. To treat the resulting hangover, Jones normally started each day by taking a teaspoonful of salts mixed in water. One morning, Jones came down to breakfast and as

usual prepared his hangover cure. He drank the solution, but this time exclaimed at its bitterness. His wife dipped her finger in the salts and tasted them; they were very bitter indeed and looked different too.
She gave her husband table salt and water to make him sick but his body went into spasm and he died in agony later that day. The next day Mrs Jones accused her lover of killing her husband. Vaquier admitted that he had done it – for her. A post-mortem proved that poor Alfred had died of strychnine poisoning. Although Vaquier had carefully washed out the salts bottle, he did not notice that some crystals had fallen from the bottle onto the table; on analysis they were found to be strychnine.
Later when Vaquier's picture appeared in the newspaper, it was recognised by the chemist who had sold him the strychnine, even though he had used a false name when he signed the poison register. Vaquier appeared in court in July 1924, claiming that he bought the poison at the request of Mrs Jones' solicitor, but he was found guilty of the murder of Jones and hanged at Wandsworth the following month.

In bygone days strychnine was included as an ingredient in a number of 'tonics', albeit in a very small quantity. Today, synthetic versions of curare are a major component of modern anaesthesia, paralysing the patient and allowing the anaesthetist to use less of other drugs, like muscle relaxants, while keeping the patient unconscious but still very much alive.

Thomas Griffiths Wainewright, English art critic, painter, forger and alleged poisoner, was born in Chiswick, London, in 1801. At 18, he joined the army but quickly tired of that and so became a painter and writer. He managed to earn a little by writing art criticisms under a number of pseudonyms. In 1821, he married Eliza Frances Ward and then committed forgery on a document, with which he defrauded the Bank of England of more than £2,000, a great sum of money in those days.
In 1829, his grandfather died of a fit and Wainewright inherited all his wealth. As he was in severe financial difficulties at the time, this inheritance was immediately seized upon by his creditors to pay off his debts. Years later, it was thought that Wainewright probably poisoned his grandfather with strychnine.
His mother-in-law came to live with Thomas and his wife Eliza, bringing two other daughters along too. Within a year, the mother-in-

law was dead, no doubt poisoned by Wainewright. He almost certainly poisoned his sister-in-law with strychnine in 1830, as she had been fraudulently insured for £16,000. Two actions were brought by Wainewright to enforce the payment of this money but both failed because the insurance company had become suspicious about the claim. Wainewright probably poisoned his uncle and several other people too.

He then fled to France, where he poisoned the father of a girl he had met, but not before insuring the man's life for £3,000. In 1837, he returned to England but was soon recognised and arrested. He was put on trial and sentenced to transportation for life as punishment for the forgery of the bank draft in 1821. Transported to Van Diemen's Land, which today is called Tasmania, he painted portraits and ate opium for some years, before dying in Hobart Hospital in 1847. On his deathbed, he finally admitted to poisoning his mother-in-law but left the rest of the deaths unaccounted for.

Poisonous plants that affect the heart

Digoxin is a type of sugar derivative, called a cardiac glycoside, made from the effects of sunlight within the leaves of the **Foxglove** plant. These leaves have been used to treat heart failure (long ago called dropsy) for hundreds of years since they work as a cardiac stimulant: digoxin acts on the heart to increase the force of contraction of the heart muscle.

Drugs called beta-blockers, widely used in medicine for the last 50 years, have the opposite effect. However, digoxin is still used to treat heart failure today, but because the margin between the therapeutic dose and the toxic dose is small, toxic side effects are common.

Early symptoms of digoxin toxicity are nausea, vomiting and anorexia, accompanied by diarrhoea and abdominal pain. The heartbeat also becomes slow and erratic. Then headache, fatigue, weakness, disorientation and confusion occur. There can also be visual disturbance, in which everything appears to have a yellow-green halo around it. However, careful monitoring of the blood level can prevent this.

A number of other plants are used for similar purposes. The **Hellebore** produces cardiac glycosides in its rhizome. In Russia, **Lily of the Valley** is used as a source for a similar cardiac stimulant. **Ouabain**, with an action like that of foxglove leaves, has been used as an arrow poison. This is the poisonous principle from various species of **Oleander**, which grows in Africa. One species even became known as the 'Hottentot's Poison Bush'. Mass executions in parts of Asia used to involve the **Upas Tree**, as the sap of this tree also contains some of these highly poisonous heart poisons. In the eighteenth century the natives used this sap to poison the wells that provided drinking water for some unwelcome Dutch colonists.

Poisonous alkaloids: black sheep of the family

Remember that alkaloids are nitrogen-containing substances produced by plants. They are many and varied, and they have potent effects on the body.

Belladonna, another name for **Deadly Nightshade,** contains the alkaloids **hyoscine** and **hyoscyamine**, as well as **atropine**. Deadly Nightshade, as well as henbane (discussed below), is a member of the *Solanaceae* family of plants, which also includes the potato and tomato, which of course we eat.

Belladonna poisoning is essentially due to the atropine it contains. The symptoms of poisoning are excessive dryness of the mouth, burning and constriction of the throat, dilatation of the pupils, nausea and sometimes vomiting. There is excitement of the central nervous system, with hallucinations, leading to delirium, giddiness and a staggering gait, passing into drowsiness and stupor. The pulse becomes rapid, the face flushed and the breathing jerky and laboured.

In fatal cases, death is preceded by a rapid intermittent pulse, with coldness of the extremities and coma. The symptoms set in rapidly and may last from ten to 15 hours or even several days in fatal cases. The usual fatal dose is only about 100mg of atropine in adults and about one-tenth of that for children. A single nightshade berry can be fatal, but recovery has occurred with prompt treatment after taking as much as 1g. Indeed, one woman in the 1940s survived after taking 50 times the maximum therapeutic dose.[3]

The temporary muscular paralysis that atropine can cause is dose-related and reversible, and treatment for atropine poisoning is fast and effective with an antidote. Atropine is still used today in various medicines, although newer, safer, synthetic analogues have now superseded them for some uses.

One other juicy titbit: the juice of deadly nightshade berries (belladonna), when put into the eyes, causes dilatation of the pupils. This was all the rage in Renaissance times because it supposedly made the ladies look more alluring. We will return to belladonna eye drops in Chapter 12, Murderous Make-Up.

Physostigmine, contained in the **Calabar Bean**, is the antidote to atropine, and has itself been used as a poison, for which, of course, atropine is its cure.

Its own poisonous use was the reason for the Calabar Bean's alternative name, the Ordeal Bean. Found near the Calabar Coast in West Africa, these beans were used in trials by ordeal, in executions and even for the divination of witches. Eight beans were ground and added to water as a drink. If lucky, the recipient vomited and survived.

Other related species such as **Woody Nightshade**, **Henbane**, **Mandrake** and **Datura** are similarly poisonous plants. All of them contain a number of toxic alkaloids, including hyoscine. **Hyoscine** is still used today in travel sickness tablets and transdermal patches, as well as in other uses such as an antispasmodic in the treatment of irritable bowel syndrome, and sometimes even as premedication prior to surgery.

Dr Hawley Harvey Crippen was a patent medicine salesman and a doctor of sorts. Born in Michigan, USA, Dr Crippen was a mild-mannered man who did not get on very well with women. He found his first wife to be a worrier, and he got on just as badly with his second wife, Cora Turner, a music hall artist whose stage name was Belle Elmore. Dr Crippen and Cora moved to London in 1910, but his American qualification in medicine did not allow him to practise in England, so he came to be employed selling patent medicines.

Cora was a strong-willed woman who took in paying guests at their home, 39 Hilldrop Crescent. She made her husband do the housework and generally kept him under her thumb. After a time, Crippen became infatuated with his secretary, Ethel le Neve. He later said that with her he was happy, for the first time in his life. In 1910, he poisoned Cora with hyoscine, giving her deadly symptoms similar to heart failure from this drug overdose, and he then cut up her body and buried it in the cellar.

Crippen explained Cora's absence to friends by telling them that his wife was on holiday in America. However, the police began an investigation after neighbours saw Ethel wearing some of Mrs Crippen's jewellery and became suspicious. The doctor and Ethel fled to Canada on board the SS Montrose. He pretended to be a Mr Robinson and Ethel disguised herself as his son. The captain became suspicious of the pair and telegraphed Scotland Yard, its first use for police purposes, and they were both arrested.

Dr Crippen was tried for murder, found guilty and hanged at Pentonville Prison in November 1910. Investigations showed that he had purchased five grains of hyoscine (this is about 300mg and would be enough to make 1,000 travel sickness tablets). Ethel was also tried but was acquitted. She later remarried and died in 1967.

Henbane is an Anglo-Saxon plant name that means 'killer of hens', which should indicate, to any passing Anglo-Saxon, that this plant is definitely not edible. However, in a magazine article about foraging for wild herbs in the summer of 2008, the celebrity chef Antony Worrall Thompson mistakenly recommended henbane to eat in salads instead of the wild herb, fat hen. The main poison found in henbane is hyoscine.

Mandrake is yet another member of the *Solanaceae* family that produces hyoscine. This plant was associated with magical or curative properties in the Middle Ages. According to the 'doctrine of signatures', this plant, with its Y-shaped root said to resemble the human form, was associated with enhanced fertility and reproduction.

Thornapple, another plant of this family, is also rich in hyoscine and other alkaloids. In toxic doses, it can render the victim unconscious and result in an almost painless death. Thornapple also came to be called Jimson Weed because in the seventeenth century some of the early settlers near Jamestown, Virginia, mistook it for spinach and were poisoned after they ate some of it, only narrowly avoiding death.

Aconite, commonly called **Monkshood** or **Wolfsbane**, is a beautiful plant when in flower, but deadly. The active component, an alkaloid called aconitine, is present in all parts of the plant, although only the root is used medicinally. Aconite can cause almost instantaneous death with a large dose. A smaller dose will initially cause tingling of the tongue, mouth, stomach and skin. This tingling was an important diagnostic feature in the past, when aconite was still in regular use as a medicine.

The tingling would be followed by numbness and anaesthesia. Other symptoms of aconite poisoning are nausea, vomiting and diarrhoea. There follows excessive salivation, with an irregular, weak and slow pulse, which may later become rapid, and difficulty in breathing, with a cold and clammy but livid-looking skin. The victim develops muscular weakness, incoordination and vertigo.

Death occurs as a result of paralysis of either the heart or respiratory system. The symptoms appear almost immediately and are rarely delayed beyond an hour. In a fatal poisoning, death usually occurs within two to six hours. A dose of as little as 20-40ml of the tincture has proved fatal in the past.

Hemlock is an umbelliferous plant, belonging to the same family of plants as parsley, parsnip and carrots. The leaves and unripe fruits of hemlock contain very poisonous alkaloids, including coniceine and coniine.

In Ancient Greece, poisoning was a common method of execution. The philosopher Socrates (469-399 BC) was accused of impiety and of corrupting youth by his enemies. At the ripe old age of 70, he was summarily tried and convicted. He refused the option of merely paying a fine, and declined a later opportunity to escape from prison. He was sentenced to die by the traditional Greek method of execution – by drinking hemlock. His death was due to asphyxia, because the alkaloid coniine contained in hemlock causes respiratory arrest.

Theramenes (441-403 BC) was an Athenian statesman and general who committed suicide, also using the traditional drink of hemlock. He may have been forced to drink it, having incurred the displeasure of the most notorious member of the government of Thirty Tyrants, Critias.

Poisonous poppies and other flowers

Opium Poppies are the source of a number of well-known drugs. Raw opium is the dried latex collected from the unripe capsules of the opium poppy. This latex, or sap, contains a mixture of **morphine**, **codeine** and a number of other alkaloids. Laudanum, the popular cure-all in the nineteenth century, was simply opium tincture by another name.

Like all drugs, an overdose can be poisonous. Initial symptoms of poisoning with opioids include some confusion and nausea, and possibly skin rashes and urticaria (hives). Excitation and euphoria then follow, then depression, and then possibly restlessness, vomiting and delirium. A characteristic feature of this type of poisoning is the gradual development of coma. As the respiration rate slows, cyanosis develops, and later the blood pressure falls as a result of circulatory failure and deepening coma.

A morphine overdose causes symmetrical pinpoint pupils, which do not react to light, although as death approaches they dilate again. Death is usually due to respiratory failure. The toxic dose of morphine varies considerably with the individual and addicts can tolerate much larger doses than the average person. The fatal dose is usually between 150mg and 300mg.

The toxic dose of opium is about ten times larger than that of morphine, and since its absorption is much slower, the symptoms do not appear so rapidly. Opiate Squill Linctus, perhaps better known as Gee's Linctus, a cough mixture containing opium and squill, has been subject to abuse, because of its morphine content. There have even been reports of cardiotoxicity – a digoxin-like slowing of the heartbeat – caused by the action of the squill content on the abuser. Squill has been included in the preparation for its emetic qualities, specifically to prevent an opiate overdose,[4,5] but clearly causes its own problems too.

Other plants, or their fruits or berries, are also poisonous, including the seeds of **Lupins** and the **Laburnum tree**, which are both highly poisonous, particularly to children. Many spring bulbs and corms, such as **Daffodils, Tulips** and **Crocuses** are also poisonous if eaten, although it is interesting to note that daffodils contain a substance called galantamine, which is now used in the treatment of Alzheimer's disease.

Poisonous cyanide: deadly in any form

Cyanide is another poisonous plant product, found in the oil expressed from the nuts of **Almonds, Cherries, Apricots** and **Peaches**, and even from immature bamboo shoots and young fern leaves. The liquid form of cyanide, formerly called prussic acid, is an intensely poisonous, volatile acid that can cause death within a minute if drunk. The fumes given off from prussic acid are hydrogen cyanide gas, which are just as dangerous.

Cassava is the staple diet for many who live in the tropics. The root, which contains cyanide, is used to make tapioca. It is vital that the cassava is properly prepared to ensure it is safe to eat. Cassava must be cooked in a pot without any lid to allow hydrogen cyanide gas to escape. Some 250mg of hydrogen cyanide can be released from every 100g of fresh root, and this is more than a lethal dose. A lid on the pot would cause the cyanide to remain in the food, and this would poison the whole family when they ate it.

Cyanide interferes with the oxygen uptake of cells, and without oxygen, of course, we die. Poisoning can occur in several ways: by inhalation of the vapour, by ingestion or even by absorption through the skin. It can result from exposure to pesticides, accidental industrial exposure, inhalation of fumes from burning plastics in domestic house fires, or even from eating plants or fruits containing cyanide. Although cyanides have the characteristic smell of bitter almonds, this smell may not always be obvious and indeed some people cannot detect it.

Unconsciousness occurs within a few seconds of taking large doses of prussic acid or other cyanides, and death occurs within five minutes. With much smaller doses, the symptoms of poisoning, which begin within a few minutes, are giddiness, staggering, headache, dilated pupils, palpitations, and difficulty in breathing, then unconsciousness and convulsions, leading to death within 15 minutes to an hour later. If the patient survives for the first hour, they will usually recover. The fatal dose of prussic acid is said to be about 50mg and that of its salts, the cyanides, is about 250mg.[6]

Richard Brinkley lived in Fulham and tried to kill his friend, Reginald Parker, using prussic acid. Reginald, who was an accountant's clerk, knew that Richard had falsified a signature on a will, and also knew that Richard was desperately worried that he might tell someone about it. The will that he had falsified belonged to an elderly woman who had died in mysterious circumstances, within a few days of signing over all her worldly goods to Mr Brinkley. Richard Brinkley decided to deal with his concern by offering the clerk a bottle of stout that he had poisoned with prussic acid.

Unfortunately, it was drunk instead by Parker's landlord and landlady, Richard and Daisy Beck, and their daughter, who was also called Daisy. Prussic acid, taken on an empty stomach, can kill very quickly indeed. Within minutes, the mother, father and daughter were all writhing about on the floor. Another daughter hastily summoned the doctor but her parents were dead by the time he arrived. Her sister, Daisy, however, was lucky and survived the poisoning. Brinkley was arrested and claimed in his defence that he had only intended to kill Parker, which he had singularly failed to do. Further investigation showed that Brinkley had something of a reputation with poisons. A few years previously his wife had been found dead at home. At the time, the police had found a medicine case containing a number of poisons in Brinkley's study. The case

contained strychnine, arsenic, prussic acid, chloroform and ergot of rye. Mrs Brinkley was found to have died of arsenic poisoning, but with no proof to link her death to her husband, the coroner's jury gave a verdict of suicide.
Shortly afterwards, now widowed, Brinkley became friendly with a young lady named Laura Glenn, who also subsequently died of arsenic poisoning, in Brinkley's rented room in Chelsea. She left a suicide note, but she was in fact one of a number of women who had once been friendly with Richard Brinkley and had then died of poisoning – not suicide. For the murder of Mr and Mrs Becks, Brinkley was sentenced to hang in Wandsworth Prison in 1907. Right to the end, he refused to admit he had killed anyone at all.

Plants poisonous on contact: look but don't touch

Some plants can cause problems just by touching them. **Poison Oak** and **Poison Ivy** both cause burning and blistering of the skin on contact. The **Giant Hogweed**, which is classified as an official weed in the United Kingdom, also contains poisons that cause blistering and burning after the skin touches it and then is exposed to sunlight. In the herb garden, **Rue** can also cause a severe skin irritation in bright sunlight, as can **spurges (*Euphorbias*)**, which irritate both the eyes and skin and are intensely poisonous if eaten.

Even **Leylandii hedging** can be a skin irritant, and all parts of the **Yew** are poisonous, including its berries. Yew hedge clippings from stately homes throughout the UK are collected and processed by the pharmaceutical industry to produce a drug used in the treatment of breast and ovarian cancer called paclitaxel.

Stinging nettles have hollow hairs on their leaves, which are made of silica. Contact with the leaves causes the delicate hairs to break, and histamine leaks out onto and into the skin, causing the typical nettle rash, which is very itchy. However, help is close at hand, as the application of a crushed dock leaf to the affected area is frequently far more effective than using an antihistamine cream.

Toxic beans

The **Castor Bean**, from which we get castor oil, can also be a danger, as it contains the potent neurotoxin **ricin**. The beans must be cooked properly for the heat to destroy the ricin. This highly toxic agent initially causes flu-like symptoms, with abdominal cramps and diarrhoea, which develop rapidly and are followed in short order by respiratory and circulatory collapse. There is no antidote and death normally ensues within three or four days. It is so virulent that half a milligram, enough to just cover a pinhead, can be fatal.

One of the most famous victims of poison in modern times was Georgi Markov. One September afternoon in 1978, Markov, a Bulgarian dissident who worked for the BBC World Service in London, was waiting for a bus on Waterloo Bridge when he suddenly felt a sharp stabbing pain in the back of his right thigh. He turned to see a man carrying a furled umbrella, which he assumed had caused the pain. The man mumbled an apology in a thick accent and hurried off to catch a cab. When he got home, Markov looked for and found a small puncture wound on the back of his thigh.

By the next morning he felt very unwell, had a high temperature and was vomiting. He was taken to hospital, and the wound, which was now rather inflamed, was X-rayed. Nothing showed up on the scan but by now his blood pressure and temperature were both dropping. Additionally, his white blood cell count had jumped to three times the normal level, which is an indication of an infection. His doctors suspected blood poisoning and began treating him with antibiotics, but he then became delirious and began having violent convulsions.

He was dead just four days after the wound was inflicted. A post-mortem was held and a lump of tissue containing the puncture wound was removed and sent to the Porton Down chemical warfare research laboratories for examination. The experts at Porton Down found, buried beneath the skin, a small spherical pellet about the size of a pinhead. The pellet had two tiny cavities drilled into it, but there was no trace of any poison that might have caused Markov's illness.

The pellet was then sent to the Metropolitan Police forensic laboratory, where it was examined using a scanning electron microscope. This showed that the pellet was made of an alloy of platinum and iridium, which was very hard and immune to corrosion. Investigators found that it was also almost invisible when X-rayed. However, the holes in the pellet were considered large enough to have held minute traces of poison, but none now remained.

Investigators quickly deduced that the pellet had been fired by some form of gas-powered device hidden in the furled umbrella, but identifying the poison itself was much more difficult and became a process of elimination. The small amount used and its effects on Markov led the experts to consider the most likely poison to be ricin, a neurotoxin found in the castor bean, about 500 times more poisonous than cyanide. They tested their theory by injecting a pig

with a quantity of ricin similar to what could have been contained in the holes in the pellet. The poor animal died within 24 hours and a post-mortem found similar organ damage to that of Markov.
At the time of the attack, Eastern Europe was governed by communist regimes and although the Bulgarians strenuously denied any responsibility for the murder, in fact another Bulgarian émigré had been similarly attacked in Paris the previous year but had fortunately recovered. When a surgeon examined him, an identical pellet was found. This man had been lucky, as the pellet was fired into his back, well away from major blood vessels.[7] It was not until 1991, following a change of regime in Bulgaria, that the new government finally admitted that assassination attempts were made upon a number of dissident former citizens who had moved to live in the West, including Georgi Markov.

Another bean, the Tonka Bean, is used for benefit rather than harm. The Tonka Bean produces a substance called coumarin, or tonka bean camphor, which is used as a flavouring in some foodstuffs. Coumarin derivatives are used as anticoagulants in medicine. This type of anticoagulant is used long term by patients who have had heart or bypass surgery, to prevent any future formation of blood clots, or thromboses. Taken daily as tablets, this type of anticoagulant is said to act indirectly, as it only stops clots from forming – it cannot dissolve any clots that have already formed. To dissolve existing blood clots, which can cause thrombosis or a stroke, a direct-acting anticoagulant such as heparin must be used. Heparin must be injected, because, as it is a protein, if it were swallowed it would be digested like any other protein we eat.

Frightful fungi

Mushrooms and toadstools are the fruiting bodies of fungi that grow in the soil or on wood – particularly wood that is dead and rotting. While some fungi are edible, many are inedible or poisonous. There is no biological difference between mushrooms and toadstools – they are all umbrella-shaped fungi – but generally speaking the edible ones are called mushrooms and the inedible or poisonous fungi of this type are referred to as toadstools. Morels are cup-shaped edible fungi, and the similar-looking but poisonous fungi are called false morels. Many people have been accidentally poisoned because they mistook one for the other. As a rule of thumb: **if you are in any doubt, don't eat it.**

Some mushrooms, although edible, are best not eaten by beer or wine drinkers as they contain **coprine**, an unusual amino acid which inhibits one of the enzymes involved in the metabolism of alcohol. Anyone eating these mushrooms and drinking alcohol to wash them down would get a terrible headache and feel very nauseous, a sort of instant hangover. Some edible fungi also contain hallucinogenic substances. The so-called 'magic mushrooms' contain **psilocybin** and **psilocin**, which have effects similar to those of LSD.

In August 2008, the novelist Nicholas Evans, author of *The Horse Whisperer*, and his family fell seriously ill after eating some poisonous mushrooms while on holiday in Scotland. They unfortunately ate a fungus known as the Deadly Webcap, probably in mistake for the similar-looking, but edible and delicious, chanterelle. It was two days later before they became ill.

The toxins in this mushroom attack the kidneys in particular, and the family all ended up in hospital, receiving dialysis and other forms of treatment to support their kidney functions. One month later, neither Evans nor his brother-in-law had any kidney function at all while his wife and sister had only a little. The doctors do not know yet when, if ever, they will get it back.

While some toadstools signal their poisonous nature visually, such as the **Fly Agaric** – the red toadstool with white spots often depicted in children's fairy tales – others are the visual equivalent of a silent killer. The **Death Cap** is far less noticeable than the fly agaric, being a pale olive colour, yet it is this species that is considered to be the cause of 90 per cent of all the fatalities of fungi poisoning, due to the **amatoxins** and **phallotoxins** that this species contains.[8]

A single death cap (or amanita) mushroom, which weighs only about 20g, can be fatal. After eating this particular mushroom, the symptoms of poisoning are somewhat delayed, not occurring until from four to 24 hours after ingestion. Then the initial symptoms include abdominal pain, nausea, vomiting and diarrhoea. This last symptom can sometimes be so severe that in the past it was confused with cholera. Dehydration and vascular collapse may follow.

This initial phase of amanita poisoning can then be followed by a period of apparent improvement, when recovery may be foreseen, but two or three days later the more serious phase begins, due to the damaging effects that the amatoxins have had on the liver. Jaundice develops, and there may also be lapses in heart, kidney and central nervous system functions.

Symptoms also include urinary problems, circulatory collapse, convulsions and coma. There is no specific antidote or treatment available. The mortality rate can be as high as 90 per cent if left untreated, but this can be greatly reduced to 15 to 30 per cent with symptomatic and supportive treatment in an intensive care unit.[9]

Another poison victim, Claudius the Roman emperor (10 BC-AD 54), was married several times. His fourth wife, Agrippina the Younger, persuaded him to adopt her son by an earlier marriage, Nero, as his successor, even though Claudius already had a son of his own. Agrippina is believed to have poisoned Claudius with a dish of mushrooms in order to secure the succession of Nero to the imperial throne.

Ergot is a fungus that can infect rye, and sometimes other cereals too. This fungus produces ergot alkaloids, which contaminate the rye flour and so too any bread baked from it. Ergotism was suggested as a possible cause for the so-called 'bewitchings' that occurred in Salem, Massachusetts, in December 1691 and even for outbreaks of plague in medieval times. More recently, the medical journal *The Lancet* carried a report of an outbreak of ergotism in 1979 that occurred in Ethiopia and was attributed to eating infected wild oats.[10]

Two forms of epidemic toxicity of ergotism have been described in the past: the gangrenous and the convulsive. It is rare to find both types occurring together. In the more common gangrenous form, the early symptoms of chronic ergotism include a headache, vomiting, depression, muscular twitching and a staggering gait. Later symptoms are gangrene of the limbs and agonising burning pains, thus the old name 'St. Anthony's Fire' to describe the poisoning. These symptoms occur because the ergot alkaloids are vasoconstrictive and reduce or even totally cut off the blood supply.

In the much rarer convulsive type, symptoms include writhing and shaking with muscle spasms and sometimes extend to hallucinations and delusions, caused by other ergot alkaloids.

A number of ergot alkaloids have been used medicinally in the past. For example, ergotamine was used in the treatment of migraines and ergometrine during childbirth, both given in carefully controlled doses. These drugs have now been superseded by newer, far less toxic, but more effective, agents.

Other fungal toxins can be equally nasty. **Aflatoxins**, produced by the Aspergillus fungus ***Aspergillus flavus***, can cause gross liver damage and even cancer. A high incidence of cancer was found to be linked to prolonged use of groundnut (peanut) meal and maize, both of which have become contaminated with this fungus and its aflatoxins. Unfortunately both of these crops are the staple diet of many people in Africa[11], where poor harvests cause widespread malnutrition and infant mortality in rural areas. It is likely that many other mycotoxins (a toxin produced by a fungus) infecting foodstuffs also contribute to human disease, as a result of low-level, long-term exposure. Research in this area continues.

A different species, ***Aspergillus fumigatus,*** causes devastating infections of the respiratory system. Infection by Aspergillus can cause a severe acute pneumonia, which can then spread to invade the heart, kidneys, bone, brain, liver and skin, with devastating consequences. Aspergillosis is usually associated with impaired immunity and is due to aflatoxins and other mycotoxins.

If you've found fungi fascinating, you'll be happy to know that fungal toxins will appear again later in this book, in Chapters 9 and 10.

Chapter 4

Venomous Vipers, Irritating Insects and Deadly Bacteria

MANY PEOPLE HAVE FEARS of snakes or spiders or other creatures. Are these fears unfounded or rational? Depending on your fear of poison and your risk-taking sense, you might understand these anxieties. Many creatures, including various species of snakes, insects, spiders, fish and amphibians, produce poisonous substances, either as a means of disabling their prey and/or as a defence mechanism (see Appendix II). Many of them produce venom, which can result in little more than localised pain and a small area of swelling in humans; however, some venoms may have a general but more profound effect on the whole body, which, in certain circumstances, even prove to be lethal.

Algae and bacteria may be microscopic in size, but they can also produce potent poisons with dire effects. These tiny killers have caused many fatalities in the past.

Various snake venoms

There are about 3000 species of snake, but most of them are non-venomous. They kill their prey by constriction, as the boa constrictor does, or by engulfing the prey and then swallowing it whole. However, there are three families of front-fanged snakes that are venomous.

The first of these, the **elapids**, includes cobras, mambas, kraits, coral snakes and some Australasian land snakes. These snakes' venom tends to cause neurotoxicity by affecting the nervous system. This group includes the **Black Mamba**, a very aggressive snake whose bite can kill a person within about four hours. More serious still is the **Green Mamba**, whose bite can be fatal within half an hour because its venom is so toxic. The most venomous snake in the world is thought to be the **Inland Taipan** (also known as the fierce snake) from central Australia. This snake produces the most toxic venom on the planet, and each bite contains enough venom to kill up to 100 people.

The **vipers** belong to a second family of snakes, and their bites tend to be vasculotoxic; that is, they affect the circulation. The **Carpet Viper** is widespread, found from West Africa to India. It is less toxic than the taipan but is more dangerous as it kills more people than any other species. Its venom rapidly affects the body's blood-clotting mechanism, which can lead to septicaemia and organ failure.

Cleopatra, Queen of Egypt, is said to have committed suicide by allowing an asp viper to bite her breast, rather than be shamed following her army's defeat at the Battle of Actium in 31 BC by Octavian the Emperor. Cleopatra's asp was probably a horned viper.

The **sea snakes** belong to the third family of venomous snakes, and their bites are of the myotoxic type, with the venom affecting the muscles.

Depending on the type of snake venom, the effects of the venom may involve one or more of the body's systems – sensory, motor, cardiac, renal or respiratory. Snake venom can cause a massive drop in blood pressure due to relaxation of blood vessel walls (vasodilatation), and this may affect the heart. There may also be local swelling around the area of the bite or puncture site, as well as bleeding from the gums. Additionally, there is a risk of the life-threatening allergic reaction anaphylaxis. In the United Kingdom, this extreme reaction by the body to an allergen is more commonly experienced from the sting of the wasps and bees described in the next section.

India's snakes are said to be the most deadly in the world, killing over 50,000 people every year. Fortunately in Britain we have only one venomous snake, the **Adder,** whose bite is not usually very serious.

Insects and spiders

Insects in the United Kingdom are relatively harmless; however, the stings of **wasps and bees** do contain venom, and some people are allergic to it. Stings introduce poisonous substances into the body, causing local pain and swelling. Sometimes there are also systemic effects, which may be life threatening. Insect stings, particularly in the mouth or on the tongue, can cause local swelling, which may be severe enough to threaten the upper airway. This should always be treated as a medical emergency. Because of this danger, bees and wasps are probably the most dangerous venomous creatures in the world. Thousands of people around the world are killed by these insects each year.

While many species of spider are venomous, only a few of them are dangerous to humans. Spiders indigenous to the UK are non-poisonous, as their bite is too weak to break our skin. This is not the case elsewhere in the world, and some foreign insects and spiders, which can bite or sting us, are arriving as 'stowaways' in goods imported into the UK. Some of these foreign spiders are taking up residence since, because of climate change, they are able to acclimatise to living here.

The bites of the **American Black Widow Spider**, its relative the **Red Back Spider**, the **Australian Funnel-Web Spider** and the South American **Banana Spider** can all have a neurotoxic effect causing severe pain, headache, vomiting, raised blood pressure and heart rate, muscle spasms and sometimes coma. This type of bite can be treated by injection of antisera, in addition to other symptomatic treatment.

Other spiders such as the **Brown Recluse Spider** can also give a nasty bite. This bite has a necrotoxic effect, with local pain and swelling, which can be extensive and results in necrosis (death of the tissues) with a black eschar, or scab, forming. When the scab eventually comes off several weeks later, it leaves behind an ulcer that heals very slowly. In rare cases, the symptoms of such a bite can be life threatening. Antisera are also available for this necrotoxic type of spider bite, but they have been found to be rather less effective than those used for the neurotoxic type.[1,2]

Marine creatures and Amphibia

There are only a few marine creatures in UK waters that sting: severe pain is the main symptom caused by **Weever Fish** as is the case with **Jellyfish** and **Portuguese Man-O'-War**. The most venomous creature in the world is considered to be the **Australian Box Jellyfish** or **Sea Wasp**. Each one contains enough poison to kill 60 people in as little as four minutes.

The popular first aid remedy of urinating on jellyfish stings has a basis in fact. One medical treatment involves dilute acetic acid (or more commonly, vinegar), so urine, which is sterile, slightly acidic and at body temperature, is a good, instantly available substitute.

Marine snails such as ***Conus geographus*** produce a mixture of **conotoxins**. These are polypeptides that act in various ways so that their combined effect produces total muscle paralysis. The sea snails inject these conotoxins into their prey by means of a hollow harpoon-like tooth. There have been occasional fatalities recorded, sometimes of swimmers, but more often of beach-combing shell collectors who received an injection from within their bucket of collected shells. So do take care! Fortunately for us, these dangerous creatures tend to live far away in the sub-tropical seas near Indonesia.

The **Stonefish**, found in both the Indian and Pacific Oceans, is the world's most venomous fish. It hides in shallow coastal waters and coral reefs and, if stepped on, erects the spines on its back. These act like hypodermic needles, injecting poison into the wound. This is agonisingly painful and can lead to death within six hours. Although such deaths are rare, stonefish stings can lead to amputation of the affected limb.

The **Puffer fish**, also called the globefish, and some of its relatives, including the **Porcupine fish**, all produce a deadly neurotoxin called **tetrodotoxin**. It is said to be 1,000 times more deadly than cyanide. Every year, a number of people in the USA are poisoned as a result of eating puffer fish. This is also the case in Japan, where it is regarded as a great culinary delicacy. The chef must be trained in the preparation and cooking of the puffer fish and must hold a licence before the item can be included on the menu because of its extremely poisonous nature.

The poisonous nature of the puffer fish was known to both the ancient Egyptians and the Chinese over 2000 years ago. It is now known that tetrodotoxin and related poisons are quite widely distributed in nature, having also been found in a number of species of **newts, frogs, lugworms** and **crabs**.[3] This wide distribution suggests that this neurotoxin is produced by a small organism at the bottom of the food chain and then simply accumulated and stored by the larger creatures, who are immune to its effects.

Some species of **frogs**, **newts** and **toads**, even in the United Kingdom, produce a variety of toxins, secreting them through their skin, but usually only as a defence mechanism. The **Californian Newt** is just one example that, like the puffer fish, produces tetrodotoxin. Some species of South American frog only secrete their poison when placed under stress, as a defence mechanism, which the local natives have used for centuries as very effective arrow poison.

The various species of the **phyllobates** genus of frogs living in Panama, Costa Rica and Columbia produce the toxin **batrachotoxin**, which is about five times as potent as tetrodotoxin. Other species of frogs have been found to produce many different types of related toxins. These toxins all seem to affect the nervous system in a variety of ways.[4,5]

Algal toxins

In the Caribbean and around the Hawaiian islands certain **seaweeds** are prevalent and produce a poison called **palytoxin**, which was traditionally used by natives as an extract to smear onto their spears for hunting and warfare.

Palytoxin was first isolated in 1968, and when its chemical structure was finally worked out, it was found to be very complex. Although it is neither a carbohydrate nor a protein, it is one of the most toxic substances of marine origin yet found. This very potent toxin acts by paralysing all tissues: heart muscle, skeletal muscle, smooth or involuntary muscle and nerves, in addition to causing the rupture of red blood cells. These effects are all caused by disrupting the normal controls of the cell membranes, throughout the body, with disastrous consequences. For more details see Appendix IV.

Some algae, including the **dinoflagellate plankton**, produce **brevetoxins** or other related toxins. Every year, more than 20,000 people are poisoned, usually while on holiday, by eating coral reef fish that have fed on the algae that produces these toxins. Paralytic shellfish poisoning is also due to these same algae. The shellfish themselves appear to be immune to the brevetoxins. The toxin affects both nerve and muscle cells and results in some of the symptoms of tetany.[6] Fortunately, fatalities are rare, as consumption usually only causes dizziness and tingling of fingers and toes.[7]

Some rather well-known bacteria

Many bacterial names are now familiar to the general public after numerous press stories about a variety of infections such as **MRSA, Clostridium, E. coli and Salmonella**. Some of these bacteria produce toxins, which are of two types: endotoxins and exotoxins.

Arthur Warren Waite was a dentist in New York who succeeded in poisoning his mother-in-law by lacing her food with a mixture of the influenza virus and diphtheria germs. This unlucky lady died in January 1916. Dr Waite then decided to deal with his father-in-law, John Peck, by the same method, but with the addition of a nasal spray contaminated with tuberculosis germs. It took a couple of months, but eventually the father-in-law died. Following the old man's death, forensic tests for arsenic were used and showed that Dr Waite had become somewhat impatient and so had added a little arsenic to speed things along. He was duly convicted of John Peck's murder.

Endotoxins are attached to the inside of the bacterial cell wall and are only released when the bacteria dies and its tissues break down, releasing the cell contents into the surrounding tissues, with devastating effects. **Typhoid fever** is an example of an illness caused by an endotoxin, which is produced from the bacteria *Salmonella typhi*.

Other types of bacteria produce the far more common exotoxins. **Exotoxins** are released by the bacteria into the surrounding medium or tissue during the growth phase of bacterial infection. Many organisms that cause food poisoning produce enterotoxins, which are exotoxins acting specifically on the small intestine of the gut.

They usually cause massive fluid loss leading to copious diarrhoea. *Staphylococcus aureus, Clostridium perfringens, Bacillus cereus, Vibrio cholerae, Salmonella enteriditis* and *Escherischia coli (E. coli)* all cause food poisoning by producing these exotoxins. *Clostridium perfringens* can also cause gas gangrene by producing at least a dozen protein toxins, which wreak their havoc in a variety of ways.

C*lostridium botulinum* is yet another species that produces an exotoxin. This bacterium is spore-forming, and can be readily found in soil and mud. The botulinum toxin selectively affects the central nervous system and is the cause of **Botulism**. This extremely serious form of food poisoning requires intensive life support and treatment in hospital.

As well as the symptoms of vomiting and diarrhoea, the pupils become dilated, causing double vision or blurred vision. Difficulty in speech and swallowing, with shakiness and unsteady gait follow while bulbar palsy develops, resulting in a progressive flaccid muscle weakness and paralysis, which descends from the head and can result in respiratory paralysis and death from heart or lung failure within two or three days.

The botulinum bacterium thrives in improperly preserved foods, typically in canned raw meats. Honey has also been reported as a common source of infection, as it was found to contain Clostridium spores, which may be a particular danger to infants. The toxin, which is a protein, is destroyed by heating, so adequate sterilisation during the canning process will denature and thus destroy this toxin and others like it. This may be a powerful nerve poison that can be lethal, but, when used with great care and precision, it removes tics, spasms and, as a beauty treatment, wrinkles too.

Anthrax is an acute infection of farm animals, particularly sheep. In humans, it is caused by the germination of *Bacillus anthracis,* anthrax spores in the lungs or on the skin, both of which can be fatal. Pulmonary anthrax, the infection affecting the lungs, initially causes flu-like symptoms, followed quickly by severe respiratory problems, septic shock and death. Cutaneous anthrax affects the skin, causing the eruption of painless papules, rather like insect bites, which then develop into fluid-filled blisters that become dry and ulcerated, and then become covered by black scabs, or eschars. Fluid then accumulates throughout the body producing swelling from fluid retention – oedema – and death then follows.

Anthrax, which in the past was called Woolsorter's disease, was well known to those in the weaving industries, as transmission to humans by contact with animal hair, hides or excrement was common in any trade involving wool or leather. Whether caused by working on cloth or on carpets, this is an example of an industrial disease. Even leather workers sometimes caught it. Such occupational hazards are no longer the problem they used to be, as today they are easily prevented or treated.

As well as these naturally occurring toxins, poisons have been born from synthetic substances, including many modern drugs. Some human creations meant to aid people have in fact created more problems. And even our own bodies can betray us, with excesses or deficiencies becoming poisonous. Read on for tales of man-made mischief and malfunctioning metabolism.

Chapter 5

Man-made Mischief and Malfunctioning Metabolism

Man-made mischief: signs and symptoms

YOU KNOW THE SAYING that man is his own worst enemy? In the world of poisoning, this is proven: many of the potentially lethal chemicals, including drugs that we use today, no longer come from the natural world but are man-made.

In August 2008, as soon as the Beijing Olympics had celebrated the closing ceremonies, a food scandal involving powdered milk laced with melamine surfaced. Melamine is used to make plastic cups and saucers and is also used in glues. It had been added to milk powder to give the illusion of an increased protein content. Melamine contains nitrogen in its chemical formula, and the protein content of foodstuffs is measured by checking the quantity of nitrogen present. This had first come to light about six months earlier but had probably been going on for years.

The Chinese authorities managed to keep it hushed up until babies who had been fed with it started dying. Thousands of babies and young children had fallen sick with kidney stones and other complications. The most seriously ill were admitted to hospital with kidney failure. Six babies died and almost 300,000 more were sick as a result. This scandal had worldwide repercussions as so many other food products also use milk powder as an ingredient. Many of these products sourced from China were found to contain melamine and have now been recalled. Melamine was even found in sweets and chocolate.

The signs and symptoms of poisoning indicate what is happening to the body as a result of its contact with a noxious substance and seem rather like clues in a puzzle to discover the exact cause.

In the case of poisoning, while some victims may be able to say what has poisoned them, some may not know the cause or, if attempting suicide, may not tell the truth about what was taken. Consequently accurate observation and interpretation of the signs and symptoms by those treating the victim are paramount.

The signs are those indications, of disease or disorder, which are observed by the doctor during his or her examination of the patient.

The patient usually describes the symptoms they are suffering from, with a presenting symptom being the one that finally made the patient decide to see the doctor. With babies and children, their parents must describe the symptoms. These signs and symptoms together help the doctor to make a diagnosis.

Some signs and symptoms are useful indicators of specific drugs that may be the cause of poisoning. For example, respiratory depression, which causes abnormally slow and shallow breathing, can lead to an increased level of carbon dioxide in the blood, and suggests that the cause could be opioid or benzodiazepine toxicity. A fast or irregular pulse, which means that the heart rate is also fast and irregular, may be due to salbutamol, atropine and hyoscine, tricyclic antidepressants, quinine or tranquilliser poisoning. A reduced body temperature may be due to drugs such as barbiturates while an increased temperature may be due to different types of drugs such as amphetamines or cocaine. Kidney failure may result from salicylate, paracetamol or ethylene glycol (antifreeze) poisoning, and can also affect the acid/base balance in the body. This latter effect can be due to poisoning by alcohol, antifreeze, paracetamol or carbon monoxide.

Medicines gone wrong

Medicines themselves, meant to be treatments in their own right, can sometimes become a problem rather than a solution.

Sometimes even a small change in the manufacturing process of a medicine can have catastrophic consequences. This happened with one medicine, called tryptophan, about 20 years ago. In late 1989, a stream of reports began to emerge, initially in the USA, but then also from Europe and Japan, of patients developing a particular set of symptoms. Painful muscles that made movement difficult and abnormal white cell counts were side effects normally associated with some other drugs, but never before with tryptophan.

It was called the Eosinophilia Myalgia Syndrome by the doctors. The symptoms developed over a period of several weeks, and some patients developed multi-system organ problems as well as inflammatory involvement affecting the joints, skin, connective tissue, lungs, heart and liver. Some of these affected patients had been taking tryptophan for many years previously, as a treatment for depression, either alone or in conjunction with other types of antidepressants, without any problem.

Withdrawal of the tryptophan led to improvement in most patients, but not all of them. In some patients the disease progressed, and there

were even some fatalities. A worldwide withdrawal of all products containing tryptophan (or in some cases a set of very severe restrictions on its use) was imposed in 1990.
Investigations revealed that the whole of the world's supply of tryptophan was manufactured by a single manufacturer in Japan, and was then exported in bulk to other countries, where it was used to make pharmaceutical products.
It appeared that changes to the manufacturing process had been made prior to the reports. The strain of bacteria involved in the fermentation process had been changed and the amount of charcoal used in the purification stage of manufacture had been reduced. These production changes had led to a number of new contaminants being present in the tryptophan now produced.
Despite much testing and investigation, no single contaminant has yet been found to be responsible; however, refinements to the manufacturing process have been made and production improved. In 1994, tryptophan was reintroduced in the UK, for restricted use, under carefully controlled conditions.[1] In February 2005, the restrictions imposed were finally lifted as no further cases of this syndrome had been reported for many years.

All new drugs need to undergo trials. Initially, these are studies on tissue samples *in vitro* (outside the body) before then testing on animals. Only when a drug has been shown to work in these *in vitro* and animal tests, and only when its safety has been assessed, will the drug be considered for an initial clinical trial in human volunteers – normally a small group of fit, young, healthy people. Then, only if this initial human testing is successful will the new drug be tried out in a second phase of clinical trial, where the drug is tested on a few patients to see if it treats the condition for which it is designed, without significant side effects. A larger third-phase trial would then follow.

New drugs are always tested against placebos in clinical trials. A placebo is a medicine that is clinically ineffective, but may help to relieve a condition because the patient has faith in its powers. Placebos are made to have the same appearance as the drug on trial, and all supplies are coded so that neither the doctor nor the patient knows who is receiving the active drug and who the placebo. The code is only broken at the end of the trial to avoid any bias in the results. However, sometimes the results are so clear that a trial is stopped early, so that more patients can take advantage of a marked advance in the treatment of their condition.

These trials are very important, and only after a successful outcome on a larger group of patients can the drug company apply for a marketing authorisation, and only once that is granted are doctors allowed to prescribe it. Every patient, at every stage of the clinical trial, is carefully monitored for safety and efficacy. This topic appears again in Chapter 18.

It is not uncommon for a new drug to be launched on to an unsuspecting world, having been only tried out on a thousand patients or so – and sometimes a lot fewer – during the clinical trials.

Sometimes things go wrong, as they did on the 13th March 2006, when eight men entered a clinical trials unit at Northwick Park Hospital in North London. They were each being paid £2,000 to take part in a phase one clinical trial of a substance code-named TGN1412, which it was hoped could be used to treat Rheumatoid Arthritis, Multiple Sclerosis and Leukaemia.

Two patients received a placebo and came to no harm, but the other six suffered what the doctors described as a cytokine storm. Their bodies' immune response systems went into overdrive, with a massive systemic inflammatory response, quickly causing extreme pain and massive swelling, with high temperature and heart rate, but falling blood pressure. They had respiratory distress, kidney problems and even blood coagulating within their blood vessels. Initial treatment in the clinical research unit was followed by transfer of all six to the intensive care unit of the hospital.

Injections of high doses of steroids were used, together with other treatments, and all six men survived. The men were affected to varying degrees, with the one most severely affected losing his toes and parts of several fingers. Investigation by the MHRA considered the most likely cause of the adverse reaction was an unpredicted biological action of the drug in humans. The conduct of the trial on that day exacerbated the situation. If the injections had been given at timed intervals, rather than one immediately after the other, then once any of the men became ill, the trial could have been stopped, so saving the rest from being affected.

Aspirin and Ibuprofen are both commonly used painkillers that are very useful in reducing fever, pain and inflammation, but are also commonly misused. The symptoms of overdosage seen with salicylates, like aspirin and other non-steroidal anti-inflammatory drugs (NSAIDs) such as ibuprofen, include gastro-intestinal and haematological effects. There may also be kidney damage, as well as effects on the central nervous system. Nausea, vomiting, dehydration, hyperventilation and sometimes gastric pain are seen. Headache, drowsiness, dizziness, blurred vision, tinnitus, vertigo and sweating can occur. Rarely there may be lethargy or even coma, together with seizures. There may be a drop in blood pressure and heart block, with pulmonary oedema and hyperthermia. Other problems such as hyperventilation can occur and cause disturbances of the body's acid/base balance. The effects are dose-related and potentially fatal. Fortunately the average lethal dose of aspirin is well in excess of 100 tablets.[2]

There are rather different problems with paracetamol poisoning. **Paracetamol** is, like digoxin, another example of a drug where there is a relatively small margin between the therapeutic and the toxic dose.

Please note that paracetamol is called acetaminophen in the USA and in Ireland. Both names refer to the same substance.

Paracetamol poisoning is very serious, and a fatal dose can be as little as 20 tablets. However, prompt treatment with methionine or acetylcysteine can prevent liver failure and subsequent death. Paracetamol overdose initially shows no symptoms or perhaps just a little vomiting. It is only later that the liver damage caused by the overdose shows up as jaundice and brain damage – encephalopathy. Sometimes there may be kidney failure as well.

Occasionally, patients can die because of lack of treatment, even when they are in hospital. In 2003, a 26-year-old man took an overdose of painkillers, intending to commit suicide. He had taken 70 tablets but then changed his mind and went to hospital for treatment, telling a nurse in the Accident and Emergency Department what he had done. However, he died some eight hours later, still waiting to see a doctor. This unfortunate case came to light at the inquest on the poor man, which had happened at the University Hospital in Cardiff.

Early signs of overdose include nausea, vomiting, sweating and lethargy, and usually settle within 24 hours. Abdominal pain may be the first sign of liver damage but may not appear until at least a day later, and sometimes may be delayed even four to six days after the overdose was taken. Following liver failure, a whole catalogue of complications set in over the ensuing days and weeks, until death finally occurs.

As a rule of thumb, if treatment can be started within 12 hours of a paracetamol overdose, then the liver can be saved, but after that time it is likely to be too late, and only a liver transplant will save the patient's life.

That said, paracetamol is a very effective painkiller if used properly. The maximum dose is two tablets at a time and the frequency is every four to six hours up to four times a day. This means that no more than eight tablets should be taken in any 24-hour period.[3]

Carbon monoxide poisoning is another very lethal type of poisoning. It can occur accidentally when gas-fired central heating boilers, water heaters and fires are not regularly serviced. The colourless and odourless gas displaces oxygen from its binding sites on the haemoglobin molecules in red blood cells, forming carboxyhaemoglobin instead of the normal oxyhaemoglobin.

The symptoms of carbon monoxide poisoning include headache and vomiting with an increased pulse and respiration rate. Once more than half of the haemoglobin is bound with carbon monoxide, the patient has convulsions, lapses into a coma and then dies of cardiac arrest. Despite the resulting hypoxia (low oxygen level), the skin and flesh do not look cyanosed (blue) but are a bright cherry pink because of the carboxyhaemoglobin. The treatment is simple, if diagnosed in time – give the patient oxygen to breathe.

The introduction of catalytic converters to the exhaust system of cars has greatly reduced the number of attempted suicides made using exhaust fumes, which used to result in death from deliberate carbon monoxide poisoning in the past.[4]

Organic solvents, such as alcohol, acetone and benzene, are widely used throughout industry, where acute or chronic exposure, resulting from inadequate ventilation of the workplace, may lead to toxicity. Adverse effects may result from inhalation of the vapour, ingestion or even absorption through the skin. Organic solvents are irritant to both the mucous membranes and to the skin, and commonly affect the central nervous system.[5] They may also affect the heart, causing arrhythmias, ectopic (extra) heart beats, tachycardias, fibrillation and heart block.

Continued chronic exposure can then lead on to liver and kidney damage, as well as to both peripheral and central effects on the nervous system. Carbon tetrachloride, for example, which is used in many industrial processes, causes vomiting, abdominal pain, diarrhoea, seizures, coma, kidney failure and an enlarged liver with jaundice and liver failure. Intoxication leads to symptoms similar to those of alcohol intoxication, initially causing stimulation of the central nervous system, and then later depression, leading on to delirium, convulsions, coma and death.

Many different solvents have been the subject of abuse, as have been commonly seen in glue-sniffing. Many other products also containing solvents have also been subject to abuse, including lighter fuel, liquid paper solvents, dry cleaning fluids and nail polish removers.

Organophosphates, such as pesticides, like malathion, affect the nervous system and cause restlessness, anxiety, dizziness, confusion, hypersalivation, watering eyes and urination. Abdominal cramps, with nausea, vomiting and diarrhoea can also occur. There may also be sweating, constricted pupils, muscle twitching leading on from muscle weakness to flaccid paralysis, convulsions, coma, respiratory distress and cardiac depression causing a decline in blood pressure and a slow heart beat. All organophosphates are absorbed through the bronchi and intact skin as well as through the gut; toxicity varies between different compounds and the onset of symptoms after skin exposure may be delayed. Repeated exposure can lead to accumulation and ongoing symptoms.[6]

In the summer of 2008, tainted rice was found in Japan in massive quantities. Thousands of bottles of shoshu and saké (rice wine) were recalled along with many other products made from rice contaminated with pesticides or the mould aflatoxin (mentioned in the last chapter). This contaminated rice should have been used as animal feed, glue or other low-grade uses, but for five years it had been diverted, relabelled and resold many times, to make huge profits for an Osaka-based milling company, which had imported it from elsewhere in Asia.

Paraquat the weedkiller, in concentrated solution, can irritate, inflame and blister the skin. Prolonged skin contact with it results in delayed healing of cuts and wounds. Paraquat may also cause cracking of the skin and shedding of the nails.

Those paraquat preparations available to domestic gardeners are now considered to be safe, as they are only available in a reduced strength. However, products available for agricultural or horticultural use are far more concentrated and, if swallowed, can cause death, although this may be delayed for two or three weeks. Drinking paraquat causes oesophageal ulceration and gastro-intestinal effects, resulting in widespread organ damage, particularly to the kidneys and lungs – and there is no antidote.[7] The main danger is the lung damage it causes. In an attempt to deter ingestion, some preparations now contain an emetic or a laxative, or a malodorous agent, to try to dissuade or deter people from drinking it.

On 7th May 2007, the fashion icon Isabella Blow died. She ended her life by drinking paraquat, the deadly weedkiller. She died in hospital, several days after drinking it, having recently had ovarian cancer diagnosed. She was 48 years old. It is interesting to note that in the past both her grandfather and her father had killed themselves, and Isabella used the same method as her father.

Recreational drugs, such as ecstasy and amphetamines (stimulants), can cause serious poisoning, sometimes from as little as a single tablet. The acute effects of ecstasy are generally similar to those of amphetamines. The toxicity caused can be severe, affecting many organs or body systems.[8] The symptoms associated with fatalities are due to overstimulation of the central nervous system, and can lead to insomnia, night terrors, nervousness and euphoria.

Other symptoms experienced include cardiac arrhythmias, hyperthermia, convulsions, cardiovascular collapse, muscle pain, water intoxication and acute renal failure. Repeated use can cause personality changes and problems with long-term memory, as well as liver damage. Lesser symptoms include nausea, blurred vision, confusion and ataxia. Psychiatric effects, including psychosis, depression and brain damage from the accumulation of fluid – cerebral oedema – have all been reported.

In 1995 Leah Betts died during her 18th birthday party. She had taken a single ecstasy tablet and her death was caused by water intoxication. She drank glass after glass of water, so much that, in conjunction with the tablet, it caused her brain to swell and she collapsed in a coma. Within two hours of taking that single tablet she was in her local hospital, in Chelmsford, Essex, on life support, until her death five days later, when they switched off the life support machine.

Many deaths from recreational drugs have occurred, some even from taking a single ecstasy tablet, as shown in the case of Leah Betts where water intoxication was the ultimate killer. Many deaths from stroke, heart attack and kidney failure have also resulted from the use of these recreational drugs. There is no antidote, so the only treatment that can be offered must be both supportive and symptomatic.

In November 2008 a woman died of water intoxication, but no ecstasy tablet was involved this time. Jacqueline Henson, a forty-year-old mother of five, died after drinking too much water only three weeks after beginning a water-based diet. She was trying to lose weight using the LighterLife method, part of which recommended drinking four litres of water daily, in small amounts over the course of each day. But Henson drank four litres in less than two hours, which caused her brain to swell. She collapsed and was pronounced dead the following day, leaving her husband and children devastated.

.

One young man developed the symptoms of Parkinson's disease at the age of only 19. Before the onset of his symptoms, he had taken ecstasy about twice a month for six months. There was a family history of Parkinson's disease, and it was considered that he had a genetic predisposition to this condition. By taking the ecstasy tablets, he made himself even more susceptible at an unusually early age.

Recreational drugs, or even some conventionally prescribed drugs, can cause convulsions when taken in overdose, whereas benzodiazepines, alcohol, opioids, tricyclic antidepressants and barbiturates all cause the patient to fall into a coma. Examination of the pupils can be an important clue to a doctor examining a patient: constricted or pinpoint pupils, caused by the stimulation to nerves of the parasympathetic nervous system, would suggest opioid or organophosphate poisoning; the opposite effect, dilated pupils, would suggest a cause by drugs such as amphetamines, cocaine, quinine or tricyclic antidepressants.
Those pinpoint pupils will appear again in Chapter 15, where a couple of murdering doctors were both caught out by this important clue.

Malfunctioning metabolism

The human body is a very complex machine, needing the right sort of fuel to function properly. It needs vitamins, minerals and essential trace elements; water and dietary fibre; carbohydrates and fats to provide energy; together with proteins for growth and maintenance.[9]

The correct proportions of each of these nutrients form a balanced diet.

Too much or too little of one or more of these ingredients can lead to illness. Any problems in the body's use of nutrients can also result in toxicity. Such problems may be inherited or acquired during life as a result of infections or because of the malfunction of one or more body systems. Even one such problem can wreak havoc with the body's normal metabolism. Too little of something vital, such as insulin to the diabetic, can be every bit as harmful as too much. All such conditions, whether caused by too much or too little, can be regarded as yet another sort of poisoning. Further details and explanation relating to vitamins, minerals and trace elements are given in Appendix V.

The way the body works can differ widely from person to person, depending on genetic variations, sex, weight, age and other factors such as diet, any medicines being taken and even gut bacteria. All these factors have important implications for toxicity and effectiveness.

Take your vitamins and minerals – but not too many

These substances, needed in very small amounts for healthy growth and development, cannot be made by the body, and so must be obtained from diet. A healthy, varied diet will supply all the vitamins and minerals you need to stay healthy. vitamins generally act as co-enzymes, and without them certain enzymes would not function properly.

However, you can have too much of a good thing and hypervitaminosis is the name given to excessive consumption of vitamins, which has become fashionable with some groups of people, particularly in the USA. This is not usually serious with **water-soluble vitamins** such as vitamin C, where any excess is simply excreted in the urine, but occasionally even here things can go awry, as shown by a report of traces of blood in the urine of a patient who took large amounts of vitamin C in a soft drink.[10] However, if **fat-soluble vitamins** are taken to excess, it is usually very serious: the body will store them in the body fat, resulting in a toxic build-up, as with the polar bear's liver, mentioned below.

Polar bears, because of their exclusively fishy diet, have evolved so that they can tolerate very high levels of vitamin A, which they store away in their liver. In 1596, the Dutch explorer William Barents and his men nearly died after eating some polar bear liver. Apparently this delicacy, previously known only to the Inuit, was so potent that a mere mouthful was sufficient to cause abdominal pain, headache, nausea, dizziness and general ill health that lasted for several days. A larger quantity would cause the skin to peel off and might even be fatal. It was not until the 1940s that this toxic ingredient found in liver was identified as vitamin A. It is interesting to note that drinking a glass of wine with your meal of polar bear liver will further intensify the toxic effects of the vitamin A overdose.

A more modern tale shows that vitamins can't work miracles. vitamin E has been hailed as a wonder vitamin because it is an antioxidant. It was claimed that a daily supplement of vitamin E would reduce the risk of coronary thrombosis in those with heart disease. However, in the USA, after seven years of follow-up, a long-term trial has shown no evidence for this claim. Indeed, patients with vascular disease or diabetes who took vitamin E supplements appeared to be at even greater risk of heart failure, not less.[11]

Certain minerals are essential to the human diet: calcium, phosphorus, potassium, sodium, iron, chlorine, sulphur and magnesium, as well as trace elements including manganese, zinc, copper, iodine, cobalt, selenium, molybdenum, chromium and silicon. Minerals frequently act as cofactors in enzyme systems or as part of other complex molecules.[12] Frequently, it is the interplay of both vitamins and minerals that can prevent deficiency diseases, such as rickets and osteomalacia.

Rickets is a disease of childhood, once very common, in which the bones do not harden due to lack of vitamin D. Without this vitamin, insufficient calcium is deposited in the bones, so they are soft and liable to bend as a result. This is particularly noticeable in the long bones of the legs, resulting in bowed or 'bandy' legs, and also in misshapen ribcages.

Osteomalacia is the adult equivalent to rickets, with softening of the bones leading to progressive decalcification of the bones and bone pain. In the United Kingdom, rickets and osteomalacia are far more common in Asian immigrant families than in native British people. This is particularly true of the women and girls who continue to swathe themselves in their traditional all-enveloping mode of dress, which allows very little sunlight on to their skin. This would be fine if they were in Asia, but in the UK, even in a heatwave, our sunlight is nowhere near as intense, resulting in problems for these women. Other conditions involving calcium, which is the major component of our skeleton, include Paget's disease and osteoporosis. Paget's disease, otherwise known as the medical condition *osteitis deformans*, is a chronic disease of the bones occurring in the elderly. This disease was named after the British surgeon Sir James Paget (1814-1899). It usually affects the skull, backbone, pelvis and long bones. The affected bones become thickened and their structure becomes disorganised. There are often no specific symptoms, but pain, deformity and fractures can occur. If the skull is affected, blindness and deafness can also occur due to bone thickening. Nowadays treatment is readily available using medicines such as the bisphosphonates or calcitonin. **Osteoporosis** – the loss of bony tissue which causes bones to become brittle and liable to fracture – is also common in the elderly, and particularly in women after menopause. This is also treated with the bisphosphonates, together with calcium and vitamin D tablets.[13]

Anaemia is a well-known iron deficiency disease, but not many people know that all the iron in each of our bodies would only amount to enough to make a small nail. The absorption and loss of iron by the body are finely controlled.[14] Too little iron results in anaemia, of which there are a number of different types. Treatment of anaemia depends on the cause and in many cases is symptomatic. Too much iron can also cause problems; details of both anaemia and excess iron are given in Appendix V.

Some doctors have suggested that ME, **myalgic encephalomyelitis**, otherwise known as chronic fatigue syndrome, is a form of magnesium deficiency as it appears to clear up when sufferers are injected with magnesium in saline solution. When oral supplements were tried as a treatment they were found to be poorly absorbed and thus far less effective than injections.[15] It may be that this is due to the absence or a vastly reduced level of some substance, as yet unknown, which is needed to effect absorption of magnesium from the gut, as is the case with the intrinsic factor that is needed for the absorption of vitamin B_{12}. This is also explained in Appendix V.

Water is a vital part of the human diet; without a continuous supply, death will occur within days. Clean drinking water is taken for granted in the Western world today, but in the developing world it is estimated that a billion people still face daily problems in obtaining clean, uncontaminated drinking water. Despite walking many miles every day to reach their nearest water source, regardless of its purity, many women (for this is women's work) spend most of their days simply fetching water for their family. The World Health Organisation (WHO) has, for decades, spent millions annually (in dollars or pounds) providing wells and pumps to communities in the developing world.

In Britain and other parts of the developed world, chlorine has been used to disinfect water for more than a century and is the main means of disinfecting drinking water, preventing outbreaks of water-borne diseases such as typhoid, cholera and meningitis. It is also used to disinfect swimming pools.

But even today, infected water is still a problem, and many millions of people in Bangladesh and Thailand are suffering from arsenic poisoning because the tube wells which were sunk to give them clean, safe drinking water, as part of a World Health Organisation initiative, were sunk into arsenic-bearing rock strata. We will return to this unsettling topic again in Chapter 10.

Dietary fibre, more commonly known as roughage, is the indigestible component of our food, but nonetheless an essential part of our diet. Dietary fibre is helpful in the prevention of a number of digestive problems such as diverticulitis and constipation. This plant material falls into four groups: the lignins, the pectic compounds, the hemicelluloses and cellulose itself, which is the largest component of dietary fibre. The lignins help strengthen the cellulose, which is the main structural material in plant cell walls; the hemicelluloses act as food reserves in seeds; and the pectic compounds form gels and are the basis of fruit jellies. Highly refined foods such as sugar contain no dietary fibre at all, while wholemeal cereals such as oats, wheat and barley have a high fibre content, as do many fruits and vegetables.

Carbohydrates are important as an energy source, being broken down in the body to the simple sugar glucose, which can then take part in energy-producing metabolic processes. Excess carbohydrate is stored in the liver and muscles as glycogen, which is composed of branched chains of glucose molecules. These can be readily broken down to glucose whenever our body needs some extra energy.

Milk and other dairy products are essential to our health, although humans are the only species known to continue to include these products in our diet after weaning.

Lactose is the sugar found in milk, and consists of glucose joined to galactose – one molecule of each is joined together to make one molecule of lactose. An intolerance to lactose occurs when there is a deficiency of the enzyme lactase, which is needed to break down the lactose present in milk into glucose and galactose. More than 90 per cent of Asians and Africans are lactase deficient, to varying degrees, and only those of northern European descent seem to retain the ability to produce lactase into adulthood. Other racial groups possess a partial ability to a greater or lesser extent. Abdominal pain, diarrhoea, distension and flatulence in the intestines result from this deficiency. These symptoms can also occur in those who ingest excessive amounts of lactose. An alternative to cow's milk, such as soya milk, should be drunk by sufferers of lactase deficiency, as lactose-containing foods are contra-indicated and so should be avoided.

Infants born with the rare inherited metabolic disease galactosaemia lack the enzyme required to convert the galactose in milk sugar to glucose in their liver. If left untreated the galactose builds up to a toxic level in the blood. The baby then fails to thrive and becomes mentally retarded. Fortunately the elimination of galactose from the diet of such babies can result in normal growth and development, if the condition is diagnosed early enough.

Fats contain fatty acids, in the form of triglycerides, and these are the main form of energy storage in the body. Fats are needed in our diet as a source of the three essential fatty acids: linoleic, linolenic and arachidonic acids. They are all unsaturated fatty acids, which are necessary for growth, but which cannot be made by the body. However, as long as we get a supply of linoleic acid in our diet, we can make the other two essential fatty acids from it. Corn oil and soya bean oil are both rich sources of linoleic acid.

Proteins are essential constituents of the body, forming the muscles, tissues and organs as well as enzymes and hormones. They are complex structures made up of one or more chains of amino acids, linked together by peptide bonds. Proteins are manufactured in the body from their constituent amino acids, which are obtained from the diet.

Amino acids are the basic building blocks of all proteins, but of the 22we need to make our proteins, only nine of them are classed as essential. These essential amino acids are just that: they are amino acids that we must obtain from our diet because we are unable to make them ourselves. They are tryptophan, lysine, phenylalanine, histidine, threonine, methionine, valine, leucine and isoleucine. The rest can be synthesised by the body. We obtain the nine essential amino acids from protein-rich foods such as meat, liver, eggs and dairy products.

Hormonal disorders

Hormones are another form of protein. There are a number of hormonal disorders which, whether congenital or acquired, can result in poisoning. The best known of these is diabetes. This disorder can be either inherited or acquired, and long-term complications can lead to circulatory problems, including thickening of the arteries, which can lead to eye problems and blindness. Circulatory problems may also affect the lower limbs, which could ultimately need amputation. Regular visits to the optician and the chiropodist are a must for all diabetics.

Insulin is the hormone that regulates our blood sugar level. Diabetics with Type 1 diabetes need injections of extra insulin, as their pancreas cannot produce sufficient amounts for them. Without insulin, diabetics would lapse into a coma and die. Those who develop diabetes later in life, now called Type 2 diabetes, can usually be managed with tablets, as there are a number of drugs, called hypoglycaemics, that can act to reduce the blood sugar levels, although some Type 2 patients will eventually progress onto insulin injections after a period of some years.

As insulin is a protein, it must be injected – swallowing it would lead to it being digested like any other protein in our diet. However, a nasal spray formulation of insulin has recently been launched, but this will not totally replace injections, which will still be needed even by those using the spray.

The blood sugar level regulated by the insulin hormone is normally maintained within the range 3.5 to 5.5 mmol/l (millimoles per litre). Measuring the blood sugar level several times a day is a vital part of treatment for all diabetics. Low blood sugar – hypoglycaemia – where the level falls well below 3.5mmol/l, can be caused by too much insulin, or by taking oral hypoglycaemics, alcohol or salicylates. The opposite effect – hyperglycaemia – where the level is far in excess of 5.5mmol/l, can be caused by too much insulin, organophosphate pesticides or some drugs, such as theophylline.

Kenneth Barlow was a nurse who worked at Bradford Royal Infirmary. One evening in May 1957, his wife Elizabeth felt unwell and went to lie down in the bedroom. Later she decided to have a bath, and some time later Kenneth said he found her drowned there. He said that he then pulled the plug to drain the water and went for help. A neighbour called the doctor, who examined the dead woman and noticed that her pupils were widely dilated. However, the doctor could see no signs of violence, although before drowning Elizabeth had obviously vomited while in the bath.

The doctor called the police, who in turn called in the forensic experts. It was noticed that Barlow's pyjamas were completely dry, despite him saying he had tried to get his wife out of the bath and had tried artificial respiration in an attempt to revive her. A search of the house revealed two syringes found in the kitchen, which Barlow said he had used to inject penicillin for himself when he had a carbuncle. He pointed out that, as a nurse, he used syringes at work every day, but he denied giving his wife any injections.

The post-mortem examination revealed that Mrs Barlow had been two months pregnant, but the doctors could find neither cause of death nor any sign of injection marks on her body. The syringes were tested and did indeed show traces of penicillin as the husband had told them. The doctors were puzzled, and made an even more thorough search of the body for needle marks. They eventually found two on the right buttock and another two more recent ones in a fold of skin under the left buttock – but what had been injected?

Barlow had said that her symptoms were vomiting, sweating and weakness. The doctors also needed to consider the cause of the dilated pupils. They decided that such symptoms could be caused by hypoglycaemia, that is, low blood sugar. Mrs Barlow was not a diabetic but blood taken from her heart during the post-mortem had a higher than normal level of blood sugar.

This finding was later found to be a natural reaction of the liver, only happening in circumstances where a violent death is imminent. This reaction was an automatic survival mechanism that came into play, resulting in a surge of blood sugar being released by the liver, but it only got as far as the heart before death occurred.

The police then discovered that on every working day Nurse Barlow gave many patients their insulin injections at the hospital where he worked. The police spoke to another nurse who told them that Barlow had once suggested that injecting insulin could be a way to commit the perfect murder.

Further investigation by the police revealed that Barlow's first wife had died at the age of 33 and that no cause of death had ever been found. Experiments with mice injected with tissue extracts taken from the injection sites found on the body of Elizabeth caused the mice to go into a diabetic coma, as they would do when they were injected with insulin. In July 1957, Kenneth Barlow was arrested and charged with murder by the use of insulin; he was found guilty and sentenced to life imprisonment.

Another hormonal disorder, Addison's disease, named after the British physician Thomas Addison (1793-1860), is a syndrome caused by inadequate secretion of corticosteroid hormones by the adrenal glands. The symptoms include weakness, loss of energy, low blood pressure and dark pigmentation of the skin. This, like diabetes, used to be fatal but today can be treated by replacement hormone therapy.

Hypothyroidism, due to too little thyroid hormone, is caused by a deficiency of iodine, makes a person feel listless and cold and tends to cause weight gain. Any deficiency is easily treated with daily thyroxine tablets.[16] At the opposite extreme, too much thyroid hormone can cause **hyperthyroidism**, resulting in restlessness and hyperactivity. Hyperthyroidism, also called Graves disease, after the Irish physician Dr R. J. Graves (1850-1932), is also referred to as thyrotoxicosis. Further details of symptoms and treatment are given in Appendix V.

Inborn errors of metabolism

Unfortunately, there are sometimes metabolic problems in the manufacture, maintenance or disposal systems in the body that can be inherited or acquired throughout a person's life. Haemophilia and Wilson's disease are just two such problems, but more than 1,500 inborn errors of metabolism are known to exist.

Haemophilia is a genetic disorder in which the blood clots extremely slowly because of a deficiency of one or other of these clotting or coagulation factors. Prothrombin is essential not only for blood clotting but also for regulating the synthesis of several other clotting factors. These vital factors are normally present in the blood plasma.

When an injury leads to bleeding, these coagulation factors undergo a series, or cascade, of chemical reactions, which result in the liquid blood being converted to a solid. Lack of any one of these factors results in the inability of the blood to clot. Two such factors are Factor VIII, the antihaemophilic factor and Factor IX, also known as the Christmas factor.

Haemophilia is a sex-linked hereditary disease in which women carry the disease down the generations, but sufferers are almost exclusively male – that is, they are the sons of the carriers. Treatment may be by the simple transfusion of blood plasma, which contains the missing factor, or alternatively by the use of concentrated preparations of Factor VIII or Factor IX, obtained by freezing fresh plasma.

Exact copies of these same factors have been made using recombinant DNA technology, and although they are many times more expensive than the natural products, they are much safer to use, as they do not carry the possible risks of infection with hepatitis, HIV or CJD that may be present in the natural product.

Wilson's disease, an inborn metabolic disorder, is caused by the body's inability to utilise copper properly. Free copper is deposited in the liver, causing jaundice and cirrhosis. It also accumulates in the brain, resulting in mental retardation and other symptoms similar to those of Parkinson's disease. The corneas in the eyes become stained with a characteristic brown ring.

In 1993 there was a report of a patient who had developed acute liver failure with cirrhosis. This failure was attributed to the patient taking excessive copper supplements for a long period of time, which resulted in chronic copper intoxication.[17]

In **Menke's disease**, an X-linked genetic disorder, the body is unable to make the copper-transporting protein required. This results in a chronic lack of copper, which leads to retarded growth, cerebral degeneration and death in infancy. A therapy for Menke's disease, introduced in the 1970s, is available but is not very effective, with most sufferers dying by the age of six.[18]

In many of these inherited conditions, there is a disturbance in the structure, synthesis, function or transport of proteins. Phenylketonuria, the porphyrias and hypogammaglobulinaemia are inborn errors of amino-acid metabolism, and there are many others, which, although not as common, are nonetheless serious and can be fatal. Usually there is no treatment available, other than the use of a specially modified diet in addition to vitamin supplements.

PKU (phenylketonuria) is one of the best-known metabolic conditions.[19] It is an inherited defect of protein metabolism, usually due to a defect in the enzyme phenylalanine hydroxylase. This results in raised blood concentrations of phenylalanine in the newborn baby, which if untreated will produce a characteristic pattern of signs and symptoms, including skin rash, hypertonia and seizures which can lead to severe mental retardation.

All newborn babies are tested for PKU with the Guthrie test. Named after its originator, American paediatrician Dr Robert Guthrie (1916-1995), the Guthrie test is a heel-prick blood spotting test in which a drop of the newborn's blood is placed on a special card, which is then tested for the presence of phenylalanine. This test enables the condition to be detected as soon as possible after birth so that affected infants can be given a diet in which the offending amino acid, phenylalanine, is present at the lowest possible level. The gene responsible for PKU is recessive, which means that a baby will only be affected where both parents carry the offending gene.

Maple syrup urine disease is another inborn defect of amino-acid metabolism. This condition results in excess of the amino acids valine, leucine, isoleucine and alloisoleucine in the urine of sufferers, which gives it the odour of maple syrup. The only treatment is dietary, and if the condition is left untreated it can lead to mental retardation and death.

Fish odour syndrome, as the name suggests, is characterised by a foul body odour due to a problem in the liver from a genetic impairment of the oxidation of **trimethylamine** derived from the diet. Trimethylamine is formed in the gut by bacterial degradation of foods rich in choline and carnitine, and is readily absorbed from the gut into the blood supply, which carries it to the liver.

If it cannot be oxidised in the liver, the trimethylamine is excreted on the breath or in the urine, sweat and vaginal secretions. It smells of rotting fish and has a high olfactory potency – which is a polite way of saying that it is very noticeable. Treatment of someone with this problem involves modifying the diet to avoid choline-rich foods, such as eggs, liver, peas, soya beans and sea fish. Sometimes drugs such as metronidazole, neomycin and lactulose are used to reduce the intestinal flora and decrease production of trimethylamine.

Porphyrias are one of a group of inherited or acquired disorders, and are due to deficiencies in the enzymes involved in the biosynthesis of **haem**, the iron-bearing red pigment in blood. Deficiencies of specific enzymes result in the accumulation of various porphyrins and their precursors.[20] The deficient enzyme may be missing from the liver (hepatic porphyria) or from the bone marrow (erythropoietic porphyria) or both.

The main features of porphyrias are highly coloured urine or urine that changes colour on standing, and skin that is very sensitive to sunlight, due to a build-up of metabolites in the flesh, which may cause chronic inflammation or blistering. There may also be neuritis, due to inflammation of the nerves, and mental disturbances, as well as abdominal pain. King George III (1738-1820) suffered from bouts of insanity believed to be due to porphyria.

Haem is responsible not only for the red colour of our blood, but also for its oxygen-carrying capacity. Haem is used in the treatment of porphyrias, as sufferers cannot synthesise it properly themselves. Tin, like iron, is also able to combine with porphyrin to produce substances that can inhibit the enzyme involved in the breakdown of haem, haem oxygenase, to bile pigments. These substances have been tried in the treatment of porphyria as a means of delaying the normal breakdown of haem.

Hypogammaglobulinaemia is due to a deficiency of the protein **gammaglobulin** in the blood. It can occur in a variety of inherited disorders, or as an acquired defect, as can happen in certain lymphomas. Gammaglobulin consists mainly of antibodies, also called immunoglobulins, so it is hardly surprising that the condition hypogammaglobulinaemia results in an increased susceptibility to infections in those who inherit or acquire it. Treatment is by replacement therapy of special formulations of immunoglobulins which are prepared from donations of human plasma.

Coeliac disease is an inflammatory disorder of the small intestine that results from a sensitivity to gluten and causes an immunological reaction. Gluten is a mixture of two proteins, gliadin and glutenin, present in wheat flour and to a lesser extent in barley, oats and rye. The first, gliadin, belongs to one of the main groups of plant proteins known as the prolamines while the second, glutenin, belongs to another group, the glutelins. Those who suffer from coeliac disease are sensitive to the gliadin fraction of gluten.

Treatment is simple: the affected person must avoid wheat products, and hence gluten, in the diet.[21] Coeliacs need to ensure that the flour, bread, pasta and biscuits they eat are gluten free. There are a number of manufacturers who specialise in making this type of foodstuff, which are available to buy freely or on prescription. This type of diet has also been found to be of use in patients with dermatitis herpetiformis, who also suffer from gluten hypersensitivity.

Treatments aren't always this simple and can sometimes be downright traumatic, as the next chapter will reveal.

Chapter 6

Traumatic Treatment

What to do in a case of suspected poisoning

Seek urgent medical attention

If needed, dial 999 for an ambulance
or phone NHS Direct on 0845 46 47

Take the victim to Accident and Emergency or Primary Care
at your nearest hospital

DO NOT make the victim sick

If the poisonous substance was swallowed,
wipe the lips and mouth and remove any leftover substance

If the victim suffered a chemical burn,
soothe the burn with milk or water

If the victim is unconscious, place him or her in the recovery position:on his side, but with one leg bent to support the body, so preventing him from rolling on to his front, and his uppermost hand placed under his chin, to support his head, which should be slightly tilted back – this stops him swallowing his tongue, allows him to breathe and allows any fluid to drain out of the mouth

If breathing stops, use resuscitation but take care to protect
yourself from poisoning

Get a sample of the poison to aid diagnosis

A CENTURY AGO, WHEN medicines were nearly all of natural origin, it was possible for doctors, travellers and missionaries to purchase a variety of cases of drugs, including a poison antidote case. Such antidote outfits contained cork-stoppered glass vials of substances carefully labelled and arranged in trays. The case also included a toxicological chart and comprehensive list of known poisonous substances and their antidotes, together with any necessary medical equipment, such as a stomach syphon tube, a catheter or syringe. But today, things have changed....

Anyone showing signs of poisoning should seek urgent medical help. This generally means going to the nearest hospital with an accident and emergency department, where you will be admitted immediately and treated as a matter of urgency.

Medical staff in the UK have access to the National Poisons Information Service, which has local poison information centres available for consultation round the clock and a clinical toxicology database available on the Internet. Doctors can also consult a computer-aided tablet and capsule identification system in cases of overdose.

Apart from accidental or deliberate ingestion, inhalation or absorption, poisoning can also result from the toxic side effects of prescribed medicines or from toxic substances accumulating in the body due to a medical or metabolic disorder. But whatever the cause, poisoning can be treated in a number of different ways. The method used will depend on a number of factors, such as the nature of the poison, the amount taken and the time since it was taken. Broadly speaking, all poisons can be divided into those for which there is a specific treatment and/or antidote and those for which there is not.

Unfortunately the vast majority of poisons fall into this latter category and so symptomatic treatment – mainly supporting the vital functions of the patient – is the only way to deal with the situation. In the treatment of poisoning, the maintenance of respiration and circulation takes precedence over everything else.

Wrongs and rights of treatment

It is a common misconception that the first thing to do in a case of poisoning is to make the patient vomit, as a means of emptying the stomach of any unabsorbed poison. Although this may be indicated in some cases, in others it can be very dangerous. With corrosive poisons, such as strong acids or alkalis, which can be ingredients of products such as drain cleaners, making the patient vomit may do far more harm than good. It could lead to an increased risk of gastric perforation, and could cause further damage to the oesophagus, throat and mouth.

If the patient is comatose, vomiting could result in the gastric contents being aspirated into the lungs. This could cause further problems such as aspiration pneumonia. If a stimulant poison was taken, this may precipitate convulsions. However, once the patient has arrived at the hospital, gastric lavage (stomach wash-out) can be performed quickly and safely by the medical staff, if it is considered necessary, as a means of removing any remaining unabsorbed poison.

A number of substances are used in the treatment of poisoning to counter the toxic effects caused. These treatments may work as adsorbents, antagonists or chelating agents. Adsorbents are used to mop up the toxin in the gastro-intestinal tract to prevent further absorption by the body. Antagonists act as an antidote, while chelating agents react with the toxin to form a less harmful, or even an inactive, complex which can then be excreted safely from the body.

Some drugs, such as atropine, are used in specific types of poisoning to block the action of essential nerve receptors which are affected by a poison. Some act by competing with the toxin, and so reduce the rate at which a toxin may be converted to an even more toxic substance in the body. For example, in the treatment of methanol poisoning, ethanol is used to compete with the methanol, reducing the rate of breakdown of methanol to the even more poisonous formaldehyde.

Some drugs, such as calcium folinate, are used to bypass the effects of a drug, as in the case of methotrexate overdosage. Drug antagonists work in a variety of ways. Some work by protecting enzyme systems in the body, while others react directly with the poison to produce a non-toxic compound, which can then be excreted safely. Others work by reversing the poison-induced effect; methylene blue is an example of this type of antagonist.

Mopping up with adsorbents

There is a large and fatal difference between adsorbents and absorbents: Adsorbents bind atoms or molecules of liquids or gases in a thin layer upon their surface, as happens with detergent molecules on the surface of water, while absorbents soak up fluid or other substances, such as water and salts by tissues of the body. Adsorbents effectively mop up poison in the gastro-intestinal tract to prevent it from being absorbed by the body.

The most popular adsorbent used in the treatment of poisoning is activated charcoal, a highly adsorbent, odourless and tasteless black powder, which can be given by mouth, mixed with water for the patient to drink, or given via a naso-gastric tube, as soon as possible after the poison was taken. The dose given is 50g every four hours. Activated charcoal is given to adsorb a wide range of poisons in the gut and vastly reduces the amount of poison that is absorbed systemically into the body.

Among the many drugs it can be used to adsorb are the salicylates, paracetamol, the barbiturates and tricyclic antidepressants. Additionally, it can be used for many plant and inorganic chemical poisons.[1] Activated charcoal is also used in the haemoperfusion method of treating poisoning, which was used, for example, in severe acute cases of poisoning by drugs such as salicylates and barbiturates.

Activated charcoal can also be used orally in the treatment of some medical conditions, such as porphyria and for the pruritis, or itching which occurs in patients with renal failure.[2] Vegetable charcoal is used in dressings for foul-smelling wounds and ulcers, where it adsorbs the chemicals produced by some bacteria, which cause a foul smell. Other adsorbents sometimes used include Fuller's Earth, Kaolin and Bentonite.

Adsorbents can be very useful in appropriate cases, but they are of no use in treating poisoning by strong acids, alkalis or other corrosive materials. Neither are they of much use in cases of poisoning by organic solvents, organophosphates, cyanide or iron salts.

Counteracting with antidotes ...

The antidotes acetylcysteine and methionine are both used in the treatment of paracetamol poisoning. Acetylcysteine is the antidote of choice, and it should be administered as soon as possible after the overdose was taken.

As few as 20 paracetamol tablets taken within a 24-hour period can cause severe liver damage and sometimes kidney damage too, because of the production of a minor but highly reactive metabolite. After a normal dose of one or two tablets, this metabolite can be completely detoxified and excreted by the body before the next dose is taken. But, following an overdose, this metabolite accumulates, because the body cannot detoxify all of it fast enough, and it then causes cell damage in both the liver and kidneys.

The antidotes acetylcysteine and methionine both work because they are able to replenish depleted stores of substances involved in the overloaded detoxification system. Acetylcysteine may be involved in the repair of tissue damage caused by the overdose as well.[3] Both antidotes are very effective if given to the patient soon enough. And most likely, if not, the patient will need a liver transplant to survive.

... and antagonists

Opioid poisoning is due to an overdosage of morphine, codeine or other similar synthetic drugs like diamorphine (heroin). These drugs all act at opioid receptors, which are to be found both in the brain and elsewhere in the body. Overdosage can result in coma, respiratory depression and pinpoint pupils. Fortunately, a specific antidote called naloxone can be injected.

Naloxone, this specific opioid antagonist, acts competitively at opioid receptors in the central nervous system and also on those to be found on the smooth muscle of the blood vessels and intestines.[4] Repeated doses of naloxone may be needed as this antidote has a shorter duration of action than most opioids. Alternatively, it can be given as an intravenous infusion, running continuously for as long as required.

Many dyestuffs are poisons, as we shall see in Chapter 7, but some of them are used as antidotes to poisons. Ferric ferrocyanide is a dyestuff with the impressive name of Prussian Blue. It was so named because it was first produced in Berlin, in the nineteenth century, and at that time Berlin was part of Prussia. As its chemical name suggests, ferric ferrocyanide contains both iron and cyanide. Prussian Blue is used in the treatment of thallium poisoning.

In thallium poisoning, the body tries to excrete thallium from the intestines, but the body tends to mistake it for potassium and so absorbs it again. This cycle can only be broken by the use of the antidote Prussian Blue, to which the thallium binds very strongly. The antidote forms an insoluble and non-absorbable complex with the thallium in the gastro-intestinal tract, which can then be excreted in the faeces.

It was not until the 1970s that the German pharmacologist, Horst Heylauf, discovered this treatment for thallium poisoning, which had previously been thought to be incurable and previously had no treatment at all.[5]

In addition to thallium, Prussian Blue has also been used to remove radioactive caesium from the body and was used for this purpose

following the Chernobyl disaster. The Chernobyl disaster in 1986 released a vast quantity of radioactive caesium-137, which contaminated the whole of western Europe; the land and the animals, as well as the people.
People were successfully treated with Prussian Blue, which acted chemically to trap the caesium and so rendered it inactive. It was given in a slow-release form, remaining in the stomach for several weeks before being passed, without being absorbed by the body. Some countries – Austria, Germany and Norway – gave the same treatment to their animals, while many other people in these countries received a much simpler treatment of an increased amount of potassium in their diet. This high level of potassium competitively reduced the amount of caesium absorbed from the intestines.

Reactivating enzymes

This method is used in the treatment of poisoning caused by organophosphates, such as pesticides and related compounds. These poisons cause muscle paralysis by preventing the activity of the enzyme cholinesterase, which breaks down the neurotransmitter acetylcholine at the neuromuscular junction. Treatment with pralidoxime reactivates the enzyme, and so restores this ability to break down this neurotransmitter. This in turn relieves the muscle paralysis caused by the organophosphates.

Pralidoxime must be given with atropine to prevent the reactivation resulting in a dangerous accumulation of the neurotransmitter acetylcholine, as this has effects at the respiratory centre in the brain.[6] Pralidoxime is of no use in the treatment of phosphorus poisoning, or that caused by inorganic phosphates, or for those organophosphates which do not inhibit this enzyme.

Methylene blue is another dyestuff used to treat poisoning. This dyestuff is used in the treatment of methaemoglobinaemia, which can be due to certain drugs or to an inherited problem with haemoglobin. Its mode of action is by activating a normally dormant reductase enzyme system. This reduces the methylene blue to leucomethylene blue, which in turn is then able to reduce methaemoglobin back to haemoglobin.

Unfortunately it is not possible to use methylene blue as a treatment for all cases of methaemoglobinaemia. For example, those patients with glucose-6-phosphate dehydrogenase deficiency have a diminished capacity to reduce methylene blue, and so this dyestuff could be harmful to them, because a build-up of methylene blue could result in the development of haemolytic anaemia in their case.[7]

Neutralising the poison

Some treatments neutralise the poison, as an acid would neutralise an alkali. For example, heparin is an anticoagulant which must be given by injection in the treatment and prevention of deep vein thrombosis (DVT). Protamine sulphate is a basic protein that can combine with heparin to form a stable but inactive complex. For this reason protamine is used to neutralise the anticoagulant action of heparin, in the treatment of haemorrhage, which could result from severe overdosage of heparin.[8] Protamine specifically works by neutralising the anti-thrombin activity of heparin.

Protective drugs

In some cases, a victim can be given another substance as a means of protection from the effects of a poison.

Much radioactive Iodine-131 was also released in the Chernobyl disaster, and this too was absorbed by those in contact with the radioactive fall-out when the nuclear power station blew up. The thyroid gland needs iodine to function, and we normally obtain it from our diet, usually as the iodide salt. The recognised treatment in a nuclear accident is to take potassium iodide tablets, which provide the body with an extra supply of iodine to protect the thyroid gland from the release of radioactivity. This treatment came too late for those at Chernobyl and a massive increase in the incidence of thyroid cancer has occurred as a result.[9]

Some drugs are used to treat side effects that result from the treatment of cancer. Amifostine is one such protective agent, which is converted in the body to the active form. This active metabolite of amifostine then protects noncancerous cells from the harmful effects of radiotherapy and from some of the other drugs used to treat cancers, like the platinum-containing cisplatin. Mesna is also used to prevent toxicity in the treatment of certain cancers.

Folinic acid is the active form of folic acid, one of the B group vitamins. It is mainly used in the treatment of cancer as an antidote to drugs such as methotrexate. In treating cancer, methotrexate is used to interfere with cell growth, by blocking the formation of DNA and RNA within the cells. Folinic acid neutralises the toxic effects of drugs like methotrexate, whether acting on the bone marrow or by preventing damage to normal tissue.[10]

Chelating agents

Chelating agents are widely used in the treatment of poisoning, particularly metal poisoning. Chelators form complexes by binding the metal to them, and can then be excreted safely from the body. Many enzymes in the body have a chelating agent at their active centre, without which they could not work.

Desferrioxamine is a chelator used to treat iron poisoning as well as disorders where iron overload is a problem, such as bronze diabetes (haemochromatosis) and thalassaemia. Desferrioxamine has a high affinity for ferric iron in the body's tissues and fluids.[11] When it is given, by injection, it forms a stable water-soluble complex with iron called ferrioxamine, which can then be readily excreted in both the urine and bile.

It removes both free and bound iron from the body's iron-storage proteins, haemosiderin and ferritin, but not from haemoglobin, transferrin or the cytochromes. It can also be used for some other metals, such as aluminium. This chelator has proved useful for dialysis patients with kidney failure, where the treatment of aluminium overload (toxicity, fortunately now rare, but see Dialysis Dementia in Chapter 18) can be done during the routine dialysis sessions.[12]

Dimercaprol is a different chelating agent used to treat other types of metal poisoning – those caused by antimony, arsenic, bismuth, gold, mercury and thallium. It is also used to treat Wilson's disease and as an adjunct in the treatment of lead poisoning. Dimercaprol is also known as British Anti-Lewisite (BAL), because it was originally developed as an antidote to Lewisite, a poison gas that contains arsenic.

Lewisite was in use towards the end of World War I. It was first produced by a Belgian-American chemist called Julius Nieuwland who, realising how poisonous it was, refused to continue to work with it. The poison bears the name of an army chemist, Lewis, who subsequently devised methods of preparing it. The arsenic in Lewisite binds on to proteins, including enzymes, within the body. Fortunately the antidote dimercaprol, when given by injection, can remove the arsenic.

Dimercaprol contains sulfhydryl groups that compete with naturally occurring sulfhydryl groups on proteins, such as enzymes, in the body. It works by combining these groups with the metal, such as arsenic, causing the poisoning. This chelation then prevents or even reverses any inhibition of the sulfhydryl-containing enzymes by the metal poison. A dimercaprol-metal complex is formed, which can then be readily excreted from the body in the urine.

BAL is still used today to treat patients who have been poisoned by arsenic and other heavy metals. The initial treatment of this type of acute poisoning is by gastric lavage, followed by administration of activated charcoal to adsorb any arsenic or heavy metal poison which is still in the gut. This is then followed by chelation therapy using injections of dimercaprol, or other chelators, to remove any of the arsenic or heavy metal poison that has already been absorbed. These treatments are also used for chronic cases of poisoning, which may occur due to long-term low-level exposure.

Unithiol and **Succimer** are also chelators, similar to dimercaprol, that have been used in the treatment of lead, arsenic and mercury poisoning.[13]

Penicillamine is a chelator that can aid the elimination of a number of heavy metals from the body, including copper, lead and mercury. It acts by forming stable water-soluble metal complexes that can then be readily excreted by the kidney. It is also used in the treatment of Wilson's disease to prevent copper accumulation, in metabolic disorders such as cystinuria to reduce the urinary concentration of cystine, in the treatment of severe active rheumatoid arthritis and in the treatment of chronic active hepatitis.[14]

Sodium calcium edetate, dicobalt edetate and disodium and trisodium edetate are all chelators but each of them has a specific use.

Sodium calcium edetate is used in the treatment of lead poisoning. It forms a stable, water-soluble lead complex that is then readily excreted in the urine.

Dicobalt edetate is a chelator used in the treatment of severe cyanide poisoning. This particular agent is used because cobalt forms a relatively non-toxic ion complex with cyanide to allow its excretion. As dicobalt edetate is relatively toxic itself, it is only used once cyanide poisoning has definitely been confirmed. It is never used as a precautionary measure.

Disodium edetate and **trisodium edetate** are both chelators with a high affinity for calcium. They form stable, water-soluble complexes that are readily excreted by the kidneys. Di- and trisodium edetate cannot be used for heavy-metal poisoning because of their effect on the body's calcium levels.[15]

Ion-exchanging resins are small granules of insoluble plastic polymer that have active sites where specific, electrically charged atoms (ions) can be exchanged. For example, in water softeners sodium ions (which do not cause hardness) can be exchanged for magnesium and calcium ions (ions in the water that cause hardness). Ion-exchanging resins are used not only to soften and to purify water,

but some can also be used to treat specific types of poisoning that may result from medical conditions.
Sodium cellulose phosphate has an affinity for calcium ions like the edetates, but works in a different way. It is an ion-exchanging resin that exchanges sodium ions for calcium. When taken by mouth, it binds to calcium in the stomach and intestine to form a non-absorbable complex, which can then be excreted in the faeces, but is no longer in use as more effective treatments are available.
Sodium cellulose phosphate was used in the treatment of high calcium levels in the blood, which may result from conditions such as marble-bone disease and vitamin D intoxication. It was also useful in the treatment of hypercalcuria and the recurrent formation of kidney stones.[16]
Calcium polystyrene sulphonate and **Sodium polystyrene sulphonate** are both ion-exchange resins, which, when used in treatment, exchange calcium (or sodium) ions for potassium ions in the gastro-intestinal tract. This is used by mouth (in water) or rectally (in solution), when the potassium level in the blood is too high.

Some more specific treatments

Sodium nitrite and **sodium thiosulphate** are used in combination to treat cyanide poisoning. Sodium thiosulphate can even be used alone in less severe cases of cyanide poisoning, while the chelating agent dicobalt edetate is used in the treatment of more serious, confirmed cyanide poisoning, as has already been mentioned.[17]
Digoxin, a cardiac glycoside, is used in the treatment of heart failure. It is very potent, with the tiny dose needed for the treatment each day being measured in micrograms (millionths of a gram). Toxicity from digoxin can easily occur, especially in the elderly, who may be taking a number of different medicines. This is a particular problem if the elderly are taking water tablets, also known as diuretics, which may result in a low blood level of potassium, due to the loss of potassium through the extra urine produced. This can predispose the patient to digoxin toxicity, which can be managed by giving a potassium-sparing diuretic, or by potassium supplementation.

Digoxin-specific antibody fragments given by injection are used to treat this toxicity. The formula is derived from antibodies produced by sheep that have been immunised to digoxin because digoxin has a greater affinity for these antibodies than for binding sites in the tissues of the body. The resulting digoxin-antibody complex is then rapidly excreted in the urine.[18]

In Sri Lanka, in the last 25 years, it has become a popular practice among young women and children to eat the seeds of the Yellow Oleander as a method of self-harm in response to stressful events. This plant contains at least eight different cardiac glycosides, and all parts of the plant are poisonous, especially the seeds. Currently there are several thousand cases of this type of poisoning every year. Treatment with antibody fragments, as used in digoxin overdose, was found to be effective in cases of oleander poisoning and other cases of cardiac glycoside poisoning. However, cost precludes its use in many of these cases in Sri Lanka, although the use of activated charcoal may improve the outcome in all overdoses here.[19]

Haem gives blood its red colour as well as its oxygen-carrying capacity. Haem is useful in the treatment of porphyrias, as we have seen in Chapter 5. Porphyrias can be due to hereditary or acquired disorders of haem biosynthesis. Great care must be taken not only by affected individuals, but also by their relatives or carers, as the use of certain medicines could induce a life-threatening acute porphyric crisis.[20]

Glucagon is a hormone produced in the pancreas, by the alpha cells in the islets of Langerhans. It is used to treat diabetics whose insulin levels are too high and who have fallen into an insulin-induced coma, which if not treated could be fatal. Insulin, produced by the beta cells in the islets of Langerhans is the hormone which regulates the amount of sugar (as glucose) in the blood. Production of insulin is stimulated by a high blood sugar level, and lack of it gives rise to diabetes mellitus, which can be treated by insulin injections.

Too much insulin causes the blood sugar level to fall too low (hypoglycaemia) and sucking a glucose tablet or two can quickly solve the problem if noticed in time. But if the patient falls into a coma (and so is 'nil by mouth') an injection of glucagon is used instead. Glucagon causes an increase in the blood sugar level by activating the breakdown of glycogen in the liver, to produce glucose. Thus glucagon has the opposite effect to that of insulin, by raising the blood sugar level to counteract hypoglycaemia.[21]

Flumazenil is a benzodiazepine antagonist that is used as an antidote to diazepam, lorazepam, temazepam, and the rest of this group of drugs. Flumazenil acts competitively at the benzodiazepine receptors in the central nervous system. It must be administered by injection and is used to treat overdosage and also to reverse benzodiazepine-induced sedation in patients undergoing sedation with anaesthesia in operating theatres and intensive care units.[22]

Alcohol intoxication (alcohol poisoning) is common in adults and sadly also increasingly in children. In recent years there has been a massive rise in patients, both adults and children, requiring hospital treatment for alcohol poisoning. Such patients are treated supportively, with the maintenance of a clear airway being paramount. Measures to reduce the risk of vomiting and aspiration of the gastric contents must be taken. There is no antidote available, so sleeping it off under medical supervision is best for severe cases, with glucose being given as needed, if blood glucose levels are seen to fall.

The symptoms of hangover following the drinking of too much alcohol are due to the chemical acetaldehyde produced in the body as the alcohol is broken down. A recent study in Japan has shown that the worst hangovers are suffered by those people who lack the enzyme that breaks down the acetaldehyde to acetic acid, and who consequently have much higher levels of acetaldehyde in the body.

Far more serious is the poisoning produced by drinking meths (methyl alcohol) or antifreeze (ethylene glycol). There have been many cases of poisoning caused by people drinking these other alcohols, deliberately or accidentally, with dire consequences. Drinking methyl alcohol, also known as methanol and methylated spirits, and an ingredient of surgical spirit, initially produces intoxication resembling that caused by drinking ethyl alcohol, or ethanol (drinking alcohol, on sale as beers, wines and spirits). However, problems result from methyl alcohol because of the production of toxic metabolites, which develop over the 12 to 24 hours following the ingestion.

When ethyl alcohol is metabolised, this results in the formation of acetaldehyde which is then converted to acetic acid. This can be readily utilised or excreted by the body. Methyl alcohol is oxidised by the same enzyme, alcohol dehydrogenase, but in this case it results in the formation of formaldehyde and formic acid instead, mainly in the liver but also in the kidneys. These are the toxins that cause tissue damage.

Formaldehyde is used, in solution, as a preservative for tissue specimens in laboratories. It has a pungent characteristic smell. When ants bite, the pain and itching is due to the formic acid they inject into us. Neither formaldehyde nor formic acid is recommended for application to humans, other than carefully applied in certain skin diseases. So it is hardly surprising that these metabolites of methyl alcohol can both cause extensive tissue damage.

Drinking antifreeze similarly produces toxic metabolites,

including aldehydes and oxalate, and a fatal dose is about 100ml.

In the treatment of poisoning by meths or antifreeze, ordinary ethyl alcohol is given by mouth or even intravenously to act as a metabolic competitor to the toxic alcohol.[23]

Infection control

Treatment following bacterial infection requires antibiotic therapy, and some prophylactic vaccines are also available. Poisoning resulting from viral agents can only be treated by supportive care to maintain the circulation, respiration and hydration.

Anthrax infection, whether pulmonary or dermatological, can be effectively treated with antibiotics, but treatment must be started before the appearance of the symptoms to be fully effective. A vaccine is available to immunise anyone likely to come into contact with anthrax spores during their work. For **botulism,** antibiotics are ineffective, but there is an antitoxin available, which is effective against five of the seven forms of toxin produced by this organism.

Cholera is caused by an enterotoxin that causes vomiting, watery diarrhoea and severe dehydration, leading, if untreated, to death. There is an oral vaccine available for travellers to those parts of the world where cholera is endemic or epidemic.

Clostridium infection can cause food poisoning symptoms, including abdominal cramps and diarrhoea, as well as gas gangrene, which manifests itself as intensely painful lesions in the soft tissue, which contain a dark red fluid and emit a foul-smelling gas, leading to toxaemia and death. Replacement of fluids and electrolytes is vital, with antibiotics needed to treat the causative agent of the infection.

Marine bacteria produce a variety of neurotoxins, such as saxitoxin, which is the main cause of paralytic shellfish poisoning, resulting from eating clams and mussels from tropical waters. Shortly after eating these infected shellfish, the lips, tongue and extremities begin to tingle and then go numb. This is followed by a general muscle weakness and then paralysis, with respiratory arrest and death within two to 12 hours, unless supportive care of circulation and respiration is available, such as in intensive care facilities. If the patient survives for 24hours, he or she will usually make a full recovery.

Insect bites and stings

Bee stings are acidic and so best treated by first removing the sting and its attached venom sac, and then bathing the affected area with a solution of sodium bicarbonate, which, being alkaline, will neutralise the venom. Wasp stings are alkaline and so best treated by using vinegar, which is a weak solution of acetic acid. Ants (or formica, by their Latin name) produce formic acid when they bite, so treatment is with a solution of sodium bicarbonate to neutralise the formic acid. This is the same irritant that used to be found, in more concentrated form, in some kettle descalers, which we shall meet later.

Marine creatures

Weever fish, jellyfish and Portuguese Man-o'-War stings can be very painful. Any bather who is stung should be removed from the water before being treated. Any tentacles or hairs should be scraped or washed off with seawater. Those stung by weever fish should apply heat, or soak the affected part in hot water, as this destroys the poison.

The stings of other marine creatures found in UK waters should be treated with a mixture of sodium bicarbonate (or baking soda) and water; vinegar, which is acid, should not be used. The application of alcoholic suntan lotions should not be used on jellyfish stings, as this can exacerbate the situation by causing any stinging hairs still attached to the skin to discharge even more. Sea urchin stings can be treated with acetone.

In other parts of the world, jellyfish can produce more severe stings. The box jellyfish. or sea wasp, has a sting that can be rapidly fatal.[24] Fragments of tentacle adhering to the skin after being stung need to be inactivated by application of acetic acid solution (vinegar), and the antiserum should be administered as soon as possible.

For snake and spider bites, venom antisera are available in countries where bites from venomous species are a hazard to life.[25,26] Snake bites are uncommon in the United Kingdom as there is only one indigenous species, the adder, that is poisonous. However, many exotic species are kept, some illegally, by members of the public, and some of these are also venomous.

A hundred years ago in tropical South America and Mexico, a snake bite antidote called 'Guaco' was carried by all Indians travelling to that area. Its efficacy was unknown. In the first half of the twentieth century a snake bite remedy called 'Tincture of Life' was in use in Africa. This was a carminative stimulant mixture, which would stimulate the expulsion of any wind or flatulence from the stomach or intestines. Quite how this would counteract the venom from a snake bite is unknown. The mixture was made from four parts each of the oils of lavender, thyme, lemon, nutmeg and orange flowers, three and a half parts of cloves and cinnamon and ten and a half parts of Peru Balsam. The solution made up to one litre, with 90 per cent alcohol. This was no emergency preparation, however, as it had to be allowed to stand for a few days and would still then need to be filtered. It would have to be made beforehand and kept ready for use.

Anyone with a snake bite should go to their nearest hospital accident and emergency department for prompt treatment. Venom antiserum for the European viper is available from specialist centres in Oxford, Liverpool and London, in addition to antivenom for certain other foreign snakes, spiders and scorpions.

Anaphylaxis – a real life-and-death emergency

Some people, when stung or after eating foodstuffs to which they are allergic, rapidly develop an allergic reaction called anaphylaxis, which is a medical emergency.[27] Patients liable to develop anaphylactic shock normally carry an Epipen (a self-injectable form of adrenalin) for this sort of emergency. They need to be injected with adrenalin immediately and should then be taken to hospital for further treatment.

In 2005, a child in the UK with a peanut allergy was bullied by some classmates. They smeared peanut butter on him, and the child developed anaphylactic shock but fortunately recovered after treatment with an injection of adrenalin.
A 15-year-old Canadian teenager was not so lucky. She died after kissing her boyfriend, who had earlier eaten a peanut butter sandwich. After going into anaphylactic shock, she was injected with adrenalin but could not be revived. She died several days later. About 100 people in the USA die each year as a result of food allergies, most of which are due to exposure to nuts.

A brief checklist of what to do in the case of suspected poisoning is given at the front of this book and again at the start of this chapter. If in any doubt, phone NHS Direct on 0845 46 47.

We now move into the next section of the book, poisons at work and in the home.

The Poison Problems: at Work and in the Home

Chapter 7

Poisonous Pigments and Dangerous Dyestuffs

IMAGINE LIFE WITHOUT COLOUR: it would be very dull indeed. Pigments and dyestuffs provide us with a means of using or copying some of the rainbow of colours found in nature. But, just like china in a china shop, they should be admired from afar and not touched.

Danger danger

Natural pigments come mainly from plants and minerals with very few coming from the animal kingdom. Plant pigments include chlorophyll for green; anthocyanins for reds, purples and blues; carotenoids for yellows, oranges and reds; and flavonoids for whites and creams. Plant pigments are not poisonous in themselves, although they may act as a signal that a plant is poisonous to those creatures that may wish to eat it (see Appendix III). In contrast, mineral pigments are often highly poisonous, as are many of today's synthetic pigments and dyestuffs.

Most dyestuffs are intensely coloured organic compounds. Mordants, such as the basic hydroxides of aluminium, chromium and iron, are used to combine with and fix the dyestuff onto the textile fibres. Mordants need to be used in those cases where textile fibres cannot be dyed directly. Pigments that are formed by the interaction of a dyestuff and a mordant are called lakes. These dyestuffs can be intensely poisonous, so the profession of 'dyeing' can be as dangerous as it sounds.

An added danger is colour blindness, as full colour is vital to make use of coloured signals as warnings of danger. Normal colour vision is trichromatic: all three primary colours – red, blue and green – can be perceived. But some people are dichromatic: they can only perceive two of these primary colours. There are a number of different types of colour blindness, which are usually inherited but sometimes acquired. A world in which colour is severely limited is very rare, but some people, called monochromats, see only black, white and shades of grey. Their world is like watching permanent black and white television.

Colour blindness is fairly common in men, but much rarer in women. Some defect of colour discrimination is present in about one in 12 men but only one in 250 women. Red-green colour blindness is the most common type, which is usually inherited, although it may, rarely, be acquired, caused by a disease of the retina. Anyone with red-green problems needs to take extra care at traffic lights. Deuteranopia is a condition where reds, yellows and greens are confused.

This chapter looks at colour danger signs, uses of colour (and sometimes mistakes) and the rainbow of colours and their respective dyes and pigments.

Medicine in colour

A number of pigments and dyestuffs have been used medicinally for many years, either as a colorant or as the medicinal product itself. Brilliant Green is an antiseptic and, in the past, was used in combination with Gentian Violet (also known as Crystal Violet). This mixture was called Bonney's Blue, named after the doctor who first reported its successful use, and was used for many decades for skin disinfection, as the treated area was then made readily visible.

However, some years ago Gentian Violet was shown to be carcinogenic[1] and so there was an understandable decline in its use, both alone and in the Bonney's Blue mixture. A number of other dyes also have antiseptic properties, including Malachite Green and Scarlet Red, a somewhat irritant antiseptic also known as Sudan IV. The use of the yellow dyes tartrazine and Sunset Yellow has diminished greatly over the years, following an increasing number of sensitivity reactions to them.

Multi-coloured mercury medicines

All medicaments have an official name, and many also have a common name. In the past, many of these common names were derived from the colour of the finished product, and many of these colourful medicines contained the highly poisonous mercury. Grey Powder, Blue Pills, Blue Unction, Yellow Ointment and Red Ointment all contained mercury. So too did Yellow, Red, Black and White Precipitate, these all being old-fashioned common names for various compounds of mercury.

Grey Powder and Blue Pills each contained about one third by weight of mercury. Grey Powder was given by mouth, and a dose of ten grains three times a day was given to treat smallpox. The *British Medical Journal* in 1910 reported that the results from its use were surprisingly good! Blue Pills were prescribed to treat syphilis, and were also said to be very effective in cases of cardiac dropsy, where digitalis alone had failed.

Yellow Ointment contained two per cent mercuric oxide (Yellow Precipitate) in Soft Paraffin. This was used for inflamed eyes, syphilitic sores and eczema. Blue Unction was a weaker form of this same mercurial ointment, but diluted with lard, which was a readily available form of fat, used to make most ointments in days gone by.

Until the mid 1960s, many medicines – particularly creams, ointments and lotions – were known by the Latin names that doctors, apothecaries and pharmacists had used in the past. This was a wonderful way to give the patient a fairly innocuous medicament, or even a placebo, with the grandiose Latin name hopefully filling the patient with great hopes for its efficacy. So we find that the impressive sounding 'Unguentum Rubrum' in plain English was simply Red Ointment.

This wonderful concoction contained two minerals of mercury, the oxide and sulphide, better known respectively as Red Precipitate and Vermilion. To make Red Ointment, both were mixed together with creosote and lard. Red Precipitate Ointment was much simpler to make: it was a mixture of one part Red Precipitate in nine parts of Paraffin Ointment. A hundred years ago, with no steroids and no antibiotics, this ointment, *Unguentum Hydrargyri Oxidi Rubrum*, was used for a wide variety of chronic skin problems.

The lotion of choice a century ago, used to bathe syphilitic sores and other skin lesions, was colloquially known as Black Wash. This doesn't sound nearly as impressive as *Lotio Hydrargyri Nigra*, to give it its official pharmacopoeial Latin name. This lotion contained Calomel, another mercury-containing mineral, mixed with glycerin and Solution of Lime.

Blue and red nightmares

We take colour for granted and don't question blue icing or yellow medicine. But when colourings start to accumulate in the body or change the colour of the urine or faeces, that's when we understandably panic.

Brilliant Blue FCF is a permitted colouring in foodstuffs, and while blue food is not common, blue icing is used on many a birthday cake.

Brilliant Blue FCF features in a special liquid feed that has caused consternation. A child was unable to eat normally and so was fed a special liquid diet through a tube. Over a period of time the child received a large quantity of one particular enteral feed, which contained Brilliant Blue FCF colouring. The child slowly started to turn blue, but tests showed this was not from cyanosis, the usual explanation. The doctors then realised the culprit and a change of feed solved the problem.

Magenta, a dyestuff used in Castellani's Paint, used to be used for certain skin complaints and as a food colouring. Today it is no longer considered safe, as cases of bladder cancer have been reported from those working in its manufacture. Magenta will reappear in just a moment.

When Dioralyte was first introduced in the 1980s as a treatment for the oral rehydration of children with sickness and diarrhoea, the original cherry-flavoured version caused problems.
When added to 200ml of boiled and cooled water, as per the instructions, there resulted a wonderful cherry-red coloured solution. The problem was that the same red colour reappeared in the child's nappy some time later, causing great anxiety and consternation to parents and medical staff alike, with fears that the red-stained nappies might be due to internal bleeding. Once this 'problem' was brought to their notice, the manufacturers very quickly removed the red dye from the product. All flavours of Dioralyte have been colourless ever since.

Some dyestuffs are used not only as colourants in food and medicines[2], but also as marker dyes in diagnostic procedures. Indigo Carmine is used in amniocentesis and in some urological procedures while Methylene Blue and Patent Blue V are used in investigations of the lymphatic system.

White lead paint poison

As long ago as Roman times, Pliny wrote about lead in his *Natural History*. He knew, even then, that it was poisonous if swallowed. The Romans only allowed convicts and slaves to work in their lead mines, because they knew of its poisonous nature.

However, White Lead, chemically known as basic lead carbonate, has been used for thousands of years as a white pigment in paints and cosmetics. Artists have greatly valued it as a pigment for its vibrancy and intensity, but because of its poisonous nature it has been banned from sale within the European Union since 1994, except under special conditions. White Lead paint is now only allowed on Grade I and Grade II* listed houses in the United

Kingdom for special historical reasons, and then only on the outside. Artists are now recommended to use a white paint made from titanium dioxide, which is so opaque that it is also used in correction fluid. Specially purified grades of titanium dioxide are even used as a white food colorant.

Apart from titanium dioxide, many other whites are available, including Bone White which is made from burned lamb bones, and Chinese White which is made from zinc oxide. Unfortunately, artists do not like the alternative whites on offer, including titanium dioxide, because they lack the vibrancy, sharpness and intensity of Lead White.

As an alternative to White Lead paint on the walls of buildings, lime washes have been used for centuries, but they need to be re-done each year. In some parts of the world, lime washing was once even believed to be the best precaution available against the plague.

Another poisonous pure white, antimony trioxide, is used today as an opacifier in paints and as a flame retardant in plastics because it is more effective than safer alternatives, in some applications.

Yellow peril

In some ways, yellow is a happy, sunny colour. In the world of dyestuffs and pigments, it often signifies extremely poisonous.

Chrome Yellow is the name given to the chemical lead chromate, while Primrose Chrome contains a mixture of this together with some lead sulphate. Yet another pigment, Cologne Yellow, is prepared by heating lead sulphate with potassium bichromate. All these chrome pigments are very poisonous.

In the past, cadmium sulphide was used as a pigment known as Cadmium Yellow, although depending on the amount of selenium and sulphur impurities present in the raw mineral, the colour could vary from yellow to orange to red to brown. It was once widely used in paints, artists' colours, printers' inks, vitreous enamels, rubbers and plastics but since the 1960s, when its poisonous nature became known, most of these uses have been phased out. Health and safety legislation now safeguards workers from excessive cadmium levels in the working environment.[3]

Another poisonous yellow pigment is zinc chromate which, when mixed with Prussian Blue, gives zinc greens, a range of particularly stable and light fast colours. Zinc chromates are also used in rust-inhibiting paints.[4]

Orpiment is a highly poisonous golden-yellow mineral, arsenic trisulphide, which has been used as a pigment since ancient times. Its modern name is simply a variant on the Latin for 'gold pigment', *auripigmentum.*

Naples Yellow is a light yellow pigment, originally held to be an Italian secret, which is manufactured from lead antimoniate. The mental illness and suicide of Vincent van Gogh may have been due, in part at least, to his frequent use of Naples Yellow, and to the lead and antimony it contains.

Gamboge is a bright yellow plant pigment derived from the resin of a tree that grows in Cambodia. This resin is a powerful purgative, causing severe griping if swallowed. In the past it was used as a medicine to expel tapeworms. Gamboge is not recommended for those artists who tend to suck their paint brushes.

A much safer yellow is luteolin, a yellow pigment extracted from the plant Dyer's Greenwood, which has been used since Roman times. The leaves and flowers of the plant contain the yellow pigment. This pigment was once used in combination with the blue dye from Woad to produce an excellent green, known as Kendal Green, named after the town in Cumbria where this combination dyeing of wool was developed.

Yet another yellow plant pigment comes from Wild Mignonette, otherwise known as Weld, which has also been traditionally used to dye fabric. Craft workers still use these plant-based dyes, but with modern synthetic fabrics, newer chemical dyestuffs must be used.

Bright but biting orange and red

There are a number of safe orange and red pigments and dyestuffs originating from minerals and plants, but many others are poisonous, including the ones below.

When Chrome Yellow, mentioned above, is put into an alkaline solution it changes colour and becomes Chrome Orange, which is still just as poisonous as Chrome Yellow.

A bright red pigment, Realgar, is the poisonous arsenic monosulphide, and is found naturally in association with the golden-yellow mineral, Orpiment.

Red Lead, extensively used in making corrosion-resistant paints, is a chemical called trilead tetroxide. Once upon a time, all paints contained lead, but now, due to its toxicity, all lead paints are governed by detailed regulations, regarding their preparation, storage and application.

Vermilion, an artist's pigment, occurs naturally as the ore Cinnabar, mercury sulphide. This same red pigment has been used for centuries in Chinese lacquer painting, as temple paint and in red ink, but this too has now been replaced in the interests of health and safety.

Another poisonous mercury-containing pigment is an iodide salt known as Iodine Scarlet, a deep red pigment which, interestingly, turns yellow when heated to 126°C.

The exotic-sounding Dragon's Blood is a resinous exudate from the fruit of some palm trees from Malaysia and Indonesia and is also obtained from the red sap from stems of the Dragon's-blood tree, which only grows on the island of Socotra in the Indian Ocean. Dragon's Blood is used for colouring varnishes and lacquers.

Solferino is the colour of rosaniline, a purplish red, so named as it was discovered soon after the Italian battle of Solferino in 1859. Rosaniline is the base of the fuchsin dyes, such as magenta, the hydrochloride salt of rosaniline. This was widely used in solution as a disinfectant for certain skin conditions in a preparation called Magenta Paint.

In crystalline form, Magenta is neither pink nor purple, but dark green, and only when it dissolves in water does it produce the typical purple-red colour. Magenta has also been called fuchsine, as the colour is similar to that of the flower Fuchsia, which was named after Leonard Fuchs (1501-66), a German botanist.

Mother Nature has fortunately given us many safe red pigments, such as Red Ochre. This mineral is also known as haematite or iron oxide, but most people know it by its more common name, rust. Calamine Lotion is a delicate shade of pink because of the presence of a minute quantity of this red pigment in the lotion.

The universal red, violet and blue pigments found in flowers, fruits and leaves, which are all water-soluble, are called anthocyanins. Madder is a plant that was used for centuries as a red dye. The orange-red pigment it contains is called Alizarin, which can be dissolved in alkaline solutions to give purple-red solutions, and can then be precipitated out as bright red, deep red, or violet lakes, by the use of metal salts as mordants.

Turkey Red is a fine, durable red that was obtained from the Madder plant in the past, but this colour can now be made synthetically.

Saffron is a species of crocus with purple flowers and brilliant orange-coloured stigmas, which are collected and dried, to be used as a (very expensive) natural dye. Safflower, otherwise known as Bastard Saffron, is a thistle-like plant whose dried petals are used both in cosmetics and also as a dye, which in acid solution gives a pinky-red colour. It is this particular colour that is the origin of 'red tape', as it is this dye that coloured the tape traditionally used by the legal profession to tie up their bundles of legal papers.

Cochineal is a red dye of animal origin made from the dried bodies of the females of a species of beetle, which come from Mexico and the West Indies. The deep red colour is due to the presence of carminic acid. The red lake pigment Carmine is obtained by extracting Cochineal with boiling water, and then precipitating the extract using alumina as the mordant. Cochineal is used widely to colour cosmetics, foodstuffs and medicines and is also used chemically as an indicator. Crimson is yet another rich red, used for thousands of years in Persia. It comes from the female bodies of another insect, called Kermes.

Other safe natural colourings include litmus, which is also used as an indicator, and is obtained from lichens after oxidation in the presence of ammonia. Litmus is red in acid conditions and changes to blue in alkaline conditions. Lycopene is the anti-cancer red carotenoid pigment found in tomatoes, rose hips and other berries, and Henna is a reddish orange pigment made from the leaves of a small oriental shrub of the loosestrife family, used for dyeing nails and hair and for skin decoration. Dyer's Mulberry and Dyer's Bugloss are both plants that have traditionally been used to produce pink and red pigments. Bixin is the principal pigment of a plant called Annatto. The violet/red crystals extracted from this shrub are used to colour foodstuffs with shades of red. Annatto will appear again in the chapter on Frightening Foodstuffs.

Passion for purple

Purple is an example of a colour obtained from both the plant and animal kingdoms, and another colour with both harmless uses and poisonous consequences.

Purple was the colour of the first synthetic dye, Mauveine. This was made accidentally by William Perkin in 1856. He was trying to make quinine, a feat of chemical synthesis not actually achieved until 1944, but ended up with a dark, sticky residue which, when dissolved in water, became a beautiful purple colour.
He originally wanted to call his mauve dye Tyrian Purple, but Tyrian Purple is a vat dye of great antiquity, which is made from Mediterranean shellfish called murex. This purple, in the past, was

reserved for Roman emperors and royalty. So Perkin's purple dye was eventually called Mauveine instead.

The Purple of Cassius, named after Andreas Cassius who died in 1673, comes from a colloidal gold reduced by a mixture of tin chlorides. This then results in a red-violet coloured solution due to the precipitation of finely divided gold on to the tin hydroxide in the solution. This exotic purple dye is still used in the manufacture of ruby-coloured glass.

Methyl Violet, also called Base Violet 1, is a violet-coloured aniline dye used to dye jute. It is also used as a bacteriological stain and as an indicator, but is perhaps best known as the purple dye used to colour methylated spirit. The dye is added deliberately to serve as a warning that it should not be swallowed, as it contains methyl alcohol, which is toxic and can also cause blindness.

Out of the blue

None of the well-known blue pigments are particularly poisonous, and some are even used in medicine.

Prussian Blue and Turnbull's Blue are cyanoferrates formed by the addition of an alkali cyanide to a solution of an iron salt. Prussian Blue is a very dark blue, given its name because it was discovered in Berlin, when located in Prussia. Today this dye is used to treat thallium poisoning, as we have already seen in the chapter on the treatment of poisoning. Although Mr Turnbull did not discover Turnbull's Blue, he gave the product his own name. He proudly named it Turnbull's Blue for posterity, but he was merely the chemist who manufactured this particular pigment in 18th-century Glasgow.

Azurite is a blue mineral which is, chemically speaking, a basic copper carbonate, the synthetic version of which is called Blue Verditer and is used as an artists' colour. Ultramarine is a vivid deep blue pigment, also much used by artists and originally made from the mineral Lapis Lazuli. Yet another blue mineral is cobalt oxide, used to make blue glass – the colour produced is called Cobalt Blue. Aquamarine is a greenish-blue semi-precious stone, which is actually a beryl-containing beryllium aluminium silicate.

Blue Vitriol is the old name for the beautiful blue hydrated copper sulphate. Many will remember growing crystals of Blue Vitriol in their schooldays. Because it is poisonous if swallowed, health and safety concerns in science lessons mean that today's schoolchildren have a far less 'hands on' experience, and growing crystals tends to be only a demonstration by the teacher. Today, Blue Vitriol is known as the mineral Chalcanthite.

Methylene blue is an important dyestuff used in dyeing fast fibres and in calico printing. It is also used as a bacteriological and microscopic stain and as a mild antiseptic.

While red-blooded animals, including humans, have a respiratory pigment containing iron called haemoglobin, other respiratory pigments have evolved in some species. Haemocyanin is the copper-containing respiratory pigment of molluscs and crustaceans, which quite literally makes them blue-blooded. Indigotin, better known as Woad, is a very ancient permanent dark purple-blue dye, long used as both body paint and as a fabric dyestuff, which comes from a plant of the same name. Delphinin is an anthocyanin pigment found in delphiniums and larkspur, the seeds of which are intensely poisonous. This pigment is found in many other blue and purple flowers.

Green with envy

Like blue, there are many harmless green pigments, but unlike blue, there are quite a few poisonous ones as well.

Many copper and chromium salts are richly coloured; copper salts are blue and green while chromium salts can be green, red or yellow. When copper is exposed to the air, a green colour forms on the surface. This happens because carbon dioxide in the air reacts with rainwater to produce carbonic acid. This weak acid then attacks the copper, producing the green copper carbonate, commonly known as verdigris. The word verdigris means 'green of Greece' in Latin, although why this name is used to describe the patina of copper is unknown. While verdigris is not very poisonous, the similar sounding viridian definitely is. Viridian is a chromium salt and a brilliant green pigment, much used by artists, despite its poisonous nature.

There are a number of other green dyestuffs and pigments that contain both copper and arsenic, and although toxic, they have been used as medicines in the past, but today we have much safer, more effective alternatives. Scheele's Green is copper arsenate, and in the past was used as a treatment for secondary anaemia and as an intestinal antiseptic.

Paris Green, Emerald Green and Schweinfurter Green are all basic copper ethanoate arsenates, which have been used in the past as insecticides for spraying fruit trees. But since these readily decompose to produce soluble arsenic compounds, which could poison anyone eating the fruit, their use is now severely restricted.

Lincoln Green is the name given to a brilliant green that, as its name suggests, was once made in Lincoln, and reputedly worn by Robin Hood and his merry men.

Malachite is another green, a native hydrated copper carbonate, used as an artist's colour. Another basic dyestuff that contains neither copper nor chromium, called Brilliant Green, was used medicinally as a bactericide.

Brown and out

Most brown dyes and pigments are of natural origin and are generally regarded as non-poisonous. Umber, named after the Italian province of Umbria, is a mineral composed of iron and manganese oxides. This brown pigment, on heating, produces burnt umber, which is a redder shade of brown. Similarly, raw sienna is a yellowish brown, while burnt sienna is a reddish brown, both named after the city of Sienna.

Sepia is a dark brown pigment made from the 'ink' produced by cuttlefish when they are frightened. Vandyke brown, named after the artist, is a deep brown which is made from a mixture of lamp-black and ochre, while bistre is a warm brown pigment made from the soot of beech wood.

Black as night

Black is the new black. If you walk into any clothes store, you will see that there any many different shades of basic black. Black pigments and dyestuffs can be prepared from a variety of natural and synthetic substances, and each black dye reacts differently to the fabric fibres, with no two dyeing exactly the same shade of black. The nigrosines are synthetic dyestuffs which are all used as black pigments.

Pitch, a bituminous or resinous dark-coloured substance of fossil plant origin, is carcinogenic. Lampblack is another carcinogenic pigment, made from the soot of a fuel-burning lamp. Lampblack used to be mixed with parchment size or fish glue to create Indian ink. Ink was also once prepared from oak-galls, which contain gallic acid, or from burnt cork or charcoal, which are both derived from plants. Japan black is a quick-drying black varnish made from asphaltum and drying oil while ebony black is named after the black wood of a tree found in Africa, India and South-East Asia.

Finally, our discussion of black has to include lead pencils which, of course, do not contain lead at all, but are made of graphite, a form of carbon, which, being grey and slightly shiny, looks just like lead. This is why in the past it used to be called black-lead or plumbago.

Chapter 8

Lethal Labours of Today & Yesterday

IMAGINE FEARING FOR YOUR health every day at work. It's unpleasant, but it's reality for many people: illnesses, injuries and disabilities can occur in, or because of, conditions in the workplace. These occupational diseases can be the result of work practices themselves, or due to exposure to chemical, physical or biological factors. Legislation has been introduced in an attempt to protect workers, but new hazards are continually appearing.

There are several distinct types of occupational disease: those caused by dust, chemical poisoning, skin diseases, radiation hazards, infectious diseases and occupational injuries. The problem with so many of these workplace hazards is that the true extent of the damage caused may not become apparent until many years later, and sometimes not until many years after the workers have retired.

Non-chemical work hazards can cause blindness, deafness and other problems. Some foundry workers develop cataracts in their eyes due to their prolonged exposure to intense heat in the workplace. They may also suffer from deafness from excessive noise involved in using such equipment. Carpal tunnel syndrome and similar disorders often develop because of repetitive actions or over-use of specific parts of the body; for example, road menders, miners and construction workers all tend to suffer from vibration white finger, the very painful circulatory condition that results from their years of using tools such as pneumatic drills, which also caused deafness too.

Fresh air, fibres and fumes

Dust diseases include pneumoconiosis, which is a fibrosis of the lungs caused by the inhalation of dust. Coal miners commonly suffered from pneumoconiosis as a result of inhaling coal dust. Other workers affected by dust diseases include quartz miners, china clay workers, foundry workers and metal grinders.

Asbestosis was, and still is, a similar work hazard, not only during asbestos mining from the earth, but also at every stage of its transport and processing into finished products. In addition, in both the construction and the demolition industries, workers have to handle asbestos during their work, so they too are in danger of breathing in the fibres and eventually contracting asbestosis many years later.

Yet another dust disease is caused by the inhalation of fungal spores from mouldy hay, straw or grain, and results in allergic alveolitis, also known as Farmer's Lung. This is an example of an occupational dust disease suffered by agricultural workers.

Many industrial processes involve the use of chemicals, which can cause poisoning of the workers by inhalation or skin absorption. The inhalation of fumes in the workplace can frequently result in lung damage, while the absorption of chemicals into the bloodstream, whether from the lungs or through the skin, can lead to widespread damage involving the liver, kidneys, bone marrow or other organs.

The damage caused by exposure to fumes during metal working varies depending on the toxic agent involved. Cadmium fumes cause kidney damage and fatty degeneration of the heart, while those from beryllium result in lung damage, and there are bone marrow problems from the fumes of lead or organic solvents. There are others that result in liver damage. Life was hard and very dangerous for foundry and mine workers in bygone days.

Allergy, radiation and infection

Many jobs involve physical contact with chemicals, and this can lead to contact dermatitis, due to development of an allergy or irritation caused by those chemicals in direct contact with the skin. If done properly, the regular use of barrier creams, gloves or protective overalls can dramatically decrease this type of risk.

Radiation is a very different type of hazard. The problems here are relevant to those working in the nuclear energy industry, as well as any place where work involves exposure to or handling of radioactive substances. The hazard of radiation also applies to anyone who works outdoors, such as agricultural workers, traffic wardens and market traders. They all need protection from the sun because of the increased risk of skin cancer that their work in the open air causes.

Some workers are even at risk of catching infectious diseases at work. In the woollen textile and carpet industries, workers are at risk of getting anthrax from handling raw wool before it is dyed and woven. Psittacosis, an endemic infection of birds, is a risk to pet shop owners and their employees, safari park staff and zoo keepers as well as the veterinary profession and their staff. Leptospirosis, better known as Weil's disease, can be caught from rats by sewer workers, miners, dock workers and others who may come into contact with vermin during their work. An example of a newer occupational hazard is the possibility of contracting viral hepatitis and AIDS by those who work with blood and blood products.

A death caused by anthrax occurred in November 2008. The man had been working with hide from West Africa, scraping off the hairs to provide a smooth layer, which was to be used in a bongo drum. He inhaled some anthrax spores and became ill, dying nine days after his admission to hospital. The Department of Environmental Health sealed off his house before giving it a very thorough cleaning. Another bongo player died in 2006 in Scotland. He was believed to have inhaled the lethal anthrax spores while playing his drum. A safer alternative would be to use plastic, instead of skin, for the drum membrane.

Percival Pott, chimney sweeps and cancer

The mining of coal has always been a very dusty job, and many miners ended up with what was commonly known as Black Lung. The doctors, of course, had fancy names for it: Coal Miners Phthisis, or Anthracosis. It was caused by inhalation of coal dust, and while it was not considered life threatening, it did cause catarrhal coughs.

As long ago as the eighteenth century, the surgeon Percival Pott (1714-88) identified Chimney Sweep's Cancer as an occupational hazard. This cancer of the scrotum was common in chimney sweeps, where contact with the soot in their clothes continually irritated the skin. Another name for exactly the same type of cancer was Gardener's Cancer, because many gardeners used soot to spread on gardens and allotments. In this case, the occupation was irrelevant as the cause was the soot, which acted as a chronic irritant around the scrotum because of the moisture and humidity of that area of the body.[1]

In 1913, the *British Medical Journal* published a report about gasworks pitch and cancer. It noted that men handling pitch, or engaged in making briquettes, were occasionally found to suffer from warty growths that could ulcerate and become the seat of an epithelial cancer. These men could also get particles of pitch in their eyes, which could induce severe inflammation of the conjunctiva and cornea, and may even lead to the complete loss of sight. By 1920 it was well known that cancer could be caused in mice simply by painting them with coal tar.[2,3]

Fibrosis and silicosis

Silicon is generally non-toxic, but silicates can become very dangerous when inhaled as silica dust. In the past, dust inhalation and lung disease were common among workers exposed to silica dust, such as sandstone hand grinders, print workers and those in the boot and shoe trades.

In the 1920s, doctors thought that the presence of silica in the lungs led to fibrosis and predisposed the workers to consumption, which we now call tuberculosis, an infection frequently affecting the chest. They also thought that the predisposition of coal miners, tin miners and gold miners to consumption might also be linked. How right they were: both tuberculosis and dust diseases that damage the lungs would later be recognised as occupational hazards in the mining industries.[4]

Silicosis is now a recognised occupational disorder caused by the inhalation of minute particles of silica, which are to be found today in the lungs of miners, stone-cutters and metal-grinders. This lung condition is one of a group of conditions known as pneumoconiosis, the symptoms of which are wheezing, coughing and shortness of breath. Silicosis can have a latency period of about ten years before the symptoms begin to show themselves.

In Guangdong province, China, an epidemic of silicosis has developed due to the appalling working conditions inside factories that supply the international jewellery trade. The workers, who may work 12-hour days with only one day off each month, labour in factories where windows are sealed and there is minimal ventilation. They don't even have face masks supplied to protect them from a fog of lethal particles. Those victims who have become ill and received compensation soon find that lawyers' fees and medical bills leave them destitute. Such industrial disease is just one part of the hidden price of China's rush to prosperity in the 21st century.[5]

Asbestosis – the delayed killer

Some silicates are fibrous, such as asbestos, a sort of magnesium silicate. Both the dust and fibres of asbestos have been found to be carcinogenic, causing lung damage by inhalation. Asbestosis has a latency period of up to 40 years, but once diagnosed, deterioration is rapid, often resulting in death within a year or so.

Asbestos was used for many years as a building material and in brake linings. However, extensive legislation and litigation in the Western world in the latter part of the twentieth century have now led to use of alternative materials, although asbestos is still widely used elsewhere in the world.

Companies have been bankrupted by lawsuits from asbestos-related claims for compensation from victims. In the UK, many employers and their insurers have delayed paying compensation by prolonging legal actions. Many asbestosis victims have died waiting.

Mesothelioma is a specific type of cancer of the pleura, the membrane that lines the chest cavity and covers the lungs. This cancer has about a 90 per cent increased incidence among those exposed to asbestos. It is even affecting the children of workers, although not until they are middle-aged themselves. They contracted this devastating illness from the dust that remained on their parents' clothes when they returned home from work. Once diagnosed with mesothelioma the average life expectancy is just 18 months.

Fatal solvent fumes

Carbon disulphide is a liquid that gives off foul-smelling fumes. The fumes are toxic by inhalation, by ingestion and even by absorption through the skin. The poisoning is cumulative, and a concentration of only one part in 3,000 parts of air is sufficient to produce a headache after only a few hours of exposure. The continued inhalation of small quantities of the vapour over many weeks or months can lead to chronic effects.

In the 1930s, the female workers in a rayon factory in the USA who were exposed to the fumes of carbon disulphide developed menstrual disturbances, and workers of both sexes were found, over a period of time, to have an increased mortality due to cardiovascular disease. Sexual impotence was also found to be a common symptom among the exposed male workers.[6]

Carbon tetrachloride is used in industry as a solvent and degreaser, and it was formerly used in some types of fire extinguishers and for dry cleaning. However, these uses have been phased out as carbon tetrachloride was found to be readily absorbed by inhalation as well as through the skin and was shown to be carcinogenic in animals. Safer solvents are now used in the workplace and for dry cleaning.

Benzene, an industrial solvent, has caused poisoning and death. During the 1930s there were nine cases of poisoning, including three deaths among the female workers in a Belgian mirror factory. The women were employed to varnish the mirrors after they were silvered. The second protective coat of varnish applied after silvering was a solution of gum-resin dissolved in benzene. After a time, the women were found to be wasting away, with severe aplastic anaemia. This turned out to be directly due to the effects of the benzene used as the solvent in the varnish they were using.[7]

Also in the 1930s, factory workers making mothballs and lavatory deodorant discs, which contained naphthalene, were liable to contract acne following exposure to the naphthalene fumes. Some workers even went on to develop systemic poisoning, starting with digestive disturbances, profuse perspiration, blood in the urine and finally death due to liver failure.[8]

Tri-nitro-toluene (TNT) is an explosive used in World War I. In 1916 the *British Medical Journal* reported that the Ministry of Munitions had had some 50 cases of fatal toxic jaundice out of the thousands of workers, mainly women, who were engaged in the making of munitions for the war. The symptoms of TNT poisoning included dermatitis, digestive problems, blood changes and jaundice.[9]

Purest is not the safest

Breathing pure oxygen below a depth of ten metres, divers can suffer convulsions and death. In the past, several divers died by drowning in this way before the cause of the problem was realised and dealt with. Today divers breathe air with an enriched oxygen content instead of pure oxygen.

Deep-sea divers, such as those working on the North Sea oil rigs, can suffer from nitrogen narcosis, known as the bends. This can occur when nitrogen, which makes up 79 per cent of air and so is contained in the tank of compressed air, is breathed in under high pressure. It dissolves in the blood and in the lipids in the circulation, but then becomes a problem if the diver returns to the surface too rapidly. If the decompression during the return to the surface after the dive is too fast, then nitrogen gas bubbles can form in the blood, producing gas emboli and decompression sickness, which can be very painful and life threatening. Treatment of this condition includes the use of a decompression chamber in which divers can be very slowly returned to normal atmospheric pressure, which in severe cases may take several days.

Gas at work

When seven chemical workers were exposed to chlorine gas in 1969, their exposure times lasted from as little as three minutes to as long as 45 minutes. The symptoms they suffered included chest pains, choking sensations, coughs, headaches and muscle pains. These effects lasted from three to nine days. There was respiratory failure in three of the workers and hypoxaemia in four of them, which was treated with oxygen therapy. Fortunately they all survived.[10]

In addition to dust diseases such as pneumoconiosis, coal miners were also at risk from explosive mixtures of air and methane gas, which they called firedamp. They used to take small finches such as canaries, in cages, down the mine with them, as an early warning system for the presence of gas. If the canary fell off its perch, they knew they had to get out of that part of the mine.

Methane gas is the chief constituent of natural gas derived from oil, which we use in our homes today for cooking and heating. In the past, methane was also called marsh gas. Produced during the decay of vegetable matter, methane can be a problem at landfill sites, unless precautions are taken to safely vent any gas produced by all that rotting rubbish.

Sulphur dioxide is a colourless, non-inflammable gas with a suffocating smell. Inhalation of high concentrations, as may happen within the workplace, can inhibit respiration and even cause death by asphyxia, while lower concentrations are irritant to the respiratory tract and to the eyes. When 230 workers at a copper smelting plant were exposed to high levels of sulphur dioxide about 30 years ago, 190 of them were later found to have liver damage and 168 also had blood abnormalities.[11]

Radioactive radium

The dangers of exposure to radiation were not fully appreciated for many years. Radium was discovered in 1898 by Marie Curie and her husband Pierre. The pure metal itself was not isolated until 1911 when Marie Curie and her assistant managed to isolate it by electrolysis. She later went on to discover two more elements and was awarded two Nobel Prizes for her work. She spent her life working with radioactive chemicals, and in 1934 Marie Curie died of leukaemia, caused by overexposure to the radioactive radiation. Her laboratory workbooks are still so radioactive that even today they need to be kept in a lead-lined safe.

Due to its radioactivity, radium glows in the dark, a property that was quickly utilised at the start of the twentieth century to make luminous paint. Workers used this luminous paint to paint the faces of clocks and watches. Unfortunately, many of these workers, who were mainly women and young girls, had a habit of licking their brushes in order to get a fine point. Contact with radium causes ulceration of the skin, so it did not take very long before many had developed cancerous growths, particularly on the mouth and jaw. Some workers became so contaminated that they even glowed in the dark themselves.

Phossy jaw and other problems

Phosphorus exists in three forms: white, red and black. The white form is the most common, and glows in the dark. It also spontaneously combusts in air and is a deadly poison.

Phosphorus initially became an occupational hazard in the match industry. Its dangers were first mentioned in the medical literature in 1838, and the first British case came to the attention of the medical profession in 1846.

In the match industry, workers who breathed in phosphorus vapour over a period of time developed 'phossy jaw', which slowly eats away the jaw bone. This happened to many workers in this industry in the nineteenth century, when phosphorus was used to make the match heads. Workers found that their teeth fell out as their jawbones crumbled, the pain was excruciating and their mouths ran with pus from the infected areas of their gums. Many children were affected too, as they were widely employed in the match industry at that time, purely because they were the cheapest labour available.

Legislation was eventually introduced, with the Children's Employment Commission of 1863 which led to the Factory Acts Extension Act in 1864. This latter act brought dangerous trades within the scope of the existing Factory Acts, and these laid down a minimum age of employment and hours of work. These changes to the legislation prevented any further use of children. So the match industry began to use more young women instead, who were the next cheapest source.

In 1898, there was a government report on the use of phosphorus in the manufacture of lucifer matches, the first type of phosphorus-containing match. The use of white phosphorus was eventually outlawed following a special international meeting – the Berne Convention – in 1906. A British law was duly passed in 1908 and the

lucifer was outlawed. Even so the new law did not come into effect until the end of 1910.
As usual Britain was somewhat late in the day with this legislation, as Finland had outlawed lucifers as long before as 1872, and most other European countries had done the same in the intervening years. But the manufacturers finally began to use a phosphorus sulphide instead of elemental phosphorus. The necrosis known as 'phossy jaw' completely disappeared as an occupational disease in the match industry.[12]

Inhaling metal fumes

Cadmium is used to prevent rust. It is used as a coating applied to metal, but during the welding process, fumes of cadmium are produced. These fumes can cause symptoms of delayed respiratory distress, which may not appear until some four to ten hours later. Long-term kidney damage has also been found. Cadmium tends to accumulate in the body, particularly in the liver and kidneys. Deaths have occurred due to the inhalation of cadmium fumes.

In 1971, a man died after brazing some metal using solder which contained about 25 per cent cadmium. Five years later another man died in similar circumstances. A few years after these deaths, an investigation of 27 coppersmiths, who had all been exposed to cadmium fumes during their work, revealed an association between cadmium exposure and liver impairment, kidney stones and restrictive airways disease.[13]

Manganese is another metal that can be an industrial hazard by inhalation. Workers breathing in the fumes from hot manganese can suffer from 'fume fever' which causes fatigue, anorexia and impotence.

In the 1930s, there was a report of four cases of chronic manganese poisoning in men who were all employed at the same manganese works. The symptoms appeared in one case after only eight months of work and in the other three men after 2½, 6½ and 8 years respectively. The men appeared to exhibit Parkinsonism. Their symptoms included involuntary laughing and crying, aggression, delusions and hallucinations, all of which have affected manganese miners in the past. This syndrome used to be called 'manganese madness', but with modern health and safety legislation, such cases are now rare.[14]

Zinc metal is irritant to the skin, and breathing in the fumes of hot zinc also causes fume fever with a sore throat, cough and sweating. Brassfounder's Ague was a disease which had similar symptoms, due to the zinc oxide fumes which are given off when the molten brass, an alloy of zinc and copper, is poured into moulds.

In the 1960s, a soldier died following inhalation of zinc chloride smoke in a confined space during a civil defence exercise. He had severe chest pains and became cyanosed. After 24 hours his condition worsened and he eventually died 11 days after his exposure to the fumes. The post-mortem showed severe lung damage and extensive necrosis of the affected tissues.[15]

Copper is a metal that is quite soft and easily worked at room temperature. As with cadmium and zinc, copper workers can also suffer from fume fever, when working with the hot metal or even with alloys containing copper. Copper dust can also be inhaled, giving rise to chest pain. In the 1970s, vineyard workers who used copper sulphate solution, which is blue in colour, in sprays were found to have blue spots on their lungs and many other lesions, which all contained particles of copper.[16]

Beryllium is a metal that mimics magnesium in the body and even displaces magnesium from certain key enzymes, causing them to malfunction. Beryllium itself has no use at all in the human body, and the lungs are particularly sensitive to poisoning from it. Beryllium is used in alloys, as a light structural material, and is also used in ceramics and in nuclear reactors. Beryllium, whether as fumes or dust, is an industrial hazard because it causes chronic lung inflammation and shortness of breath. These symptoms are of a serious condition called berylliosis, which can result from either brief heavy exposure or long-term, low-level exposure. This disabling condition is permanent and about one third of such patients die as a result, while others get lung cancer.

Nickel dust causes both nasal and lung cancer, and the gas nickel carbonyl can be lethal, even in tiny amounts. Nickel, even when used in stainless steel, can cause dermatitis, starting with the so-called 'nickel itch' caused by the acids in sweat reacting with the nickel in the stainless steel which is in contact with the skin. Watches, jewellery, spectacle frames and garment fasteners can all cause this problem, which can be solved very easily, by painting the offending metal surface with clear nail varnish, which then provides a barrier between the skin and the metal.

Vapours and amalgams

The hazards of handling mercury were known 2,000 years ago.

Mercury is a metal that is liquid at room temperature, and so can easily vaporise. The mercury vapour produced can be inhaled and cause poisoning via the lungs. The Romans would only use convicts and slaves to work in their mines, as they knew of the poisonous nature of the mercury within. In 1665 one of the first labour laws made it illegal for miners to work more than six hours a day in the mercury mines of what is now Yugoslavia.

In 1810 a troopship, HMS Triumph, was transporting flasks of mercury from Spain to London. During a storm one of the flasks broke open, releasing the liquid mercury. The mercury ran about the ship, lodging in nooks and crannies in the decks, and affected all 200 members of the crew, who breathed in mercury vapour for the remainder of the voyage. Three of the crew died, and so did all the cattle and birds that were also on board.

Barometer Maker's Disease had the same cause: breathing in mercury vapour, which in this case came from filling the barometers with liquid mercury. In the hat industry, in order to turn beaver and rabbit fur into felt, the pelts were dipped into a solution of mercury nitrate and then dried. Workers often suffered from 'hatter's shakes' and 'mercury madness', and it is from this industry that we derive the phrase 'mad as a hatter'.

Mercury can form amalgams with other metals. Some of these are solid, such as dental amalgam used in tooth fillings, which are a type of metal alloy. Others, which are mercury rich, are liquid, like pure mercury itself. The application of heat to these liquid amalgams results in the mercury vaporising, leaving the other metal behind applied to a surface, as in gilding.

In Birmingham in the early nineteenth century, workers gilded buttons using a gold/mercury amalgam, with one gram of gold being sufficient to gild 500 buttons. When heated, the mercury was given off as fumes, poisoning the workers, who were said to suffer from 'gilder's palsy'. This poisoning affected those living in houses nearby as well and only stopped after 1840, when a new process – electroplating – arrived.

Electroplating was more efficient and economical, as it would gild even more buttons for each gram of gold used. The Royal Navy, however, continued with the mercury amalgam method for another 100 years because the gilding produced by this method was more durable, so the naval gilders continued to suffer gilder's palsy.

In Russia, in the early nineteenth century, over 60 workmen died during the gilding of the dome of St Isaac's Cathedral in St

Petersburg, Russia. This massive task involved about 100kg of gold being applied to copper sheets, using the gold/mercury amalgam method. The finished golden dome was magnificent, but not without a disastrous effect on the workforce.

Until the middle of the nineteenth century, mirrors were always made using silver amalgam, so mirror-makers were another group who suffered from mercury poisoning, until the chemist Liebig improved their lot. He showed that, to make a mirror, a film of silver could be deposited on glass by chemical means, without the use of mercury.

Even in the mid twentieth century, mercury was an occupational hazard. For example, police detectives searching for fingerprints at crime scenes used a dusting powder made by grinding chalk and mercury together. Of course, many officers began to display the symptoms of chronic mercury poisoning. A safer alternative is now used.

In the past, mercury has had a wide range of uses in industry – in plating, felting, dyeing and tanning – where it has now been replaced by less harmful substances. Its long-time use in thermometers and barometers has also now largely been replaced, although it is still used in some electrical switchgear. Most batteries no longer contain mercury either, with the exception of hearing-aid batteries. With these, mercury is still used because this type of battery will continue working well until discharged, unlike other types of battery which tend to fade over a period of time.

Lead colic

Lead is another heavy metal, and a cumulative poison, for which our bodies have no practical use at all. In the past, lead was a major occupational hazard in the workplace, but it is now closely controlled by health and safety legislation.

Plumbers, who in the past worked with lead, tended to suffer from an excess of this substance, which caused poisoning, medically referred to as plumbism. The very name plumber is taken from the Latin word for lead, 'Plumbum', because since ancient times this soft and easily worked metal was used to make pipes and storage vessels for water and other liquids. Lead poisoning produced a characteristic blue line on the gums at the roots of the teeth, together with abdominal pain, constipation and nausea. Painter's Colic had these same symptoms, caused by the lead used to make paint in bygone days.

In the past, because lead was so widely used, there were many cases of lead poisoning in the workplace. In 1900, for example, there were 1,317 recorded deaths due to lead poisoning. Over the years there have been many reports of poisoning by inhalation of fumes containing lead from a wide variety of causes. Burning battery cases of lead/acid batteries, using an oxy-acetylene torch to cut lead-painted steel, print shop workers burning newsprint... These are just a few examples.
Even yarn workers in the textile industry developed lead poisoning from some of the dyestuffs used to dye their yarn. Pottery workers used lead glazes for their pottery, but the glaze arrived as a powder which had to be made up into a liquid glaze by stirring it into water. It was the dust released during the storage and manufacture into glaze which resulted in poisoning by inhalation of the lead-laden dust.[17] Even the children of pottery workers suffered from lead poisoning as a result of the powder clinging to their parents' clothes.

More metals

Those who work with cobalt salts tend to suffer from a form of dermatitis. Cobalt blue is a pigment widely used by artists and craft workers in porcelain, pottery, stained glass, enamelling, jewellery and tile making. Such workers are all at risk of developing dermatitis, and, in the longer term, of developing cardiomyopathy due to exposure to cobalt.

Chrome ulcers are an occupational skin problem for metal workers using this metal and will be mentioned again in Chapter 20. Destruction of the nasal septum is fortunately now rare in these workers, but the inhalation of dust containing chromium compounds may also cause severe lung damage and inflammation of the eyes. Stomach ulcers can also occur and there is a possibility that chromium compounds may be carcinogenic.

Platinum is a metal not essential to life, but as it is non-toxic it can be used in implants, which the body will tolerate. Certain platinum compounds, however, are poisonous and some are even used to treat cancer. People in regular contact with platinum compounds, such as jewellers, laboratory glassware workers and those involved in making electrical contacts or catalysts, may suffer from an allergic reaction, a condition called platinosis, exhibiting symptoms similar to asthma and the common cold.

Powerful occupational poisons

Nicotine is a toxic alkaloid derived from tobacco, which was used as an insecticide. It is a powerful poison, even when just applied to the skin. In 1933, the *British Medical Journal* reported the case of a girl who worked in an insecticide factory. She collapsed following the accidental spillage of a small quantity of a 95 per cent solution of nicotine onto the sleeve of her overall. Her life was saved by giving her emetics and by scrubbing her skin with soap and cold water, as hot water would have increased the skin absorption of the nicotine. In America at about the same time, a florist was also accidentally poisoned when a solution containing 40 per cent nicotine was spilt on her clothes. She suffered nausea, vomiting, sweating and difficulty in breathing, and it took her three weeks to completely recover.[18,19]

Paraphenylenediamine hair dyes and similar types of dye have been used not only to dye human hair, but also to dye fur in the fashion industry. A contact dermatitis, which may produce eczema, nervous symptoms, dizziness, weakness and even impairment of vision, were all symptoms suffered by those working with this type of dye.

A 21-year-old hairdresser's assistant died of liver failure in the 1930s following poisoning by hair dye, even though she had worn rubber gloves when applying the dye.[20] Fur workers also suffered from dermatitis and even those wealthy enough to purchase the fur products developed a fur dermatitis following contact with the dyed fur when they wore it.[21]

Ricin is an extremely toxic substance present in castor beans, from which castor oil is obtained. Even chewing a single raw castor bean is sufficient to cause poisoning. Ricin is a protein and so, fortunately, can be made harmless by heating. Castor beans are crushed to extract the oil, and the resulting seed cake, or pomace, is then treated with steam to destroy the poisonous ricin. However, sometimes the heat treatment is delayed or omitted. Ricin poisoning has even occurred in port workers who handled cargoes of castor pomace which had not been heat treated before dispatch.[22]

And finally, improvers and Baker's Itch

In the early twentieth century, chemicals such as potassium persulphate, called 'improvers', enabled millers to use a larger proportion of cheap wheats when making flour. They also allowed the bakers to introduce more water into the dough and therefore into the finished loaf of bread. It was claimed that the 'improvers' made the yeast work better by making the conditions in the dough more acid, enabling the British miller to compete with overseas competitors.

Baker's Eczema and Baker's Itch were two industrial medical conditions both attributed to the use of flour containing a small amount of 'improver', just 63 parts per million of potassium persulphate. In the 1920s such chemicals were in use in Great Britain and came to be a known occupational hazard to British bakers. This problem was not seen in either French or American bakers, as 'improvers' were illegal in France and America, and indeed these chemical hazards were unknown in this country before the introduction of 'improvers'.[23]

Poison at work threatens a wide range of people, from the industry workers mentioned in this chapter to the farmers in the next.

Chapter 9

Fatal Farming and Pestilential Problems

IN A PERFECT WORLD a pesticide would kill the pest without harming the host organism, be it plant or animal, and it would, ideally, then be de-activated and so rendered harmless.
But the world isn't perfect and many pesticides are very toxic to humans, particularly after repeated, long-term exposure to low levels or to a large single dose. Some countries have now banned certain pesticides because of the toxic effects suffered by their agricultural workers, or because of their persistence in the food chain.

Agent Orange and Co.

A defoliant is a special type of herbicide that causes leaves to fall off trees and bushes. When used in warfare it can deny cover to the enemy. Such chemicals were being developed at the end of World War I, but were never used in that conflict. The United States Air Force used defoliants on some jungle areas near Saigon in Vietnam in the 1960s. Agent Orange, the most widely used defoliant in Vietnam, was made up of a mixture of weed killers, known as 2,4D and 2,4,5-T, (2, 4-dichlorophenoxyacetic acid and 2,4,5-trichlorophenoxyacetic acid). These chemicals are synthetic auxins or plant hormones.

Dioxins first came to attention during the Vietnam War, as contaminants of the defoliant Agent Orange. They are highly poisonous materials, produced as by-products in the manufacture of commercial chemical products, including herbicides such as the polychlorinated biphenyls or PCBs, and also by the combustion of plastics. Dioxins were considered responsible for the development of chloracne, a persistent and severe form of acne, and have also been found to be potent teratogens and carcinogens.[1] The most toxic dioxin known is TCDD (2,3,7,8-tetrachlorodibenzo-p-dioxin).

In September 2004, Viktor Yushenko, the Ukrainian opposition leader, was poisoned with this most toxic of the dioxins, but fortunately survived, although before and after photographs in the press showed that he suffered a very bad case of chloracne. Tests found that Yushenko had a blood level of TCDD some 3,000 times higher than normal. It is believed that the poison was administered to the politician during a meal he had at the Kiev Dacha of Igor Smeshko, the head of the Ukrainian secret service. It was claimed that former KGB agents were probably involved in the incident, because of the forthcoming election. When the election was held in November, there was such an outcry at the irregularities found that the election had to be re-run on December 26th when Yushenko was declared the victor.

In December 2008, a food scare involving dioxins found in pork and pork meat products from Ireland made headlines. All such foodstuffs were removed from shops and supermarkets in a number of countries. The pigs on ten farms in the Republic and a further nine in Northern Ireland had been fed on contaminated pig feed. This pig meat had been mixed with that from a further 490 unaffected farms before this came to light. The contaminated feed was produced at a recycling plant where breadcrumbs used in the feed were heated with an 'inappropriate' oil and the contamination with dioxins occurred during this process. The Irish Food Board applied a special label to all such products made after December 7th that year to indicate that they were unaffected by this contamination with dioxins.

The agricultural and horticultural industries wish to optimise growth of their crops, be they plants or animals. To achieve this they need to use pesticides to kill off harmful organisms. A few pesticides are of natural origin, but most are now manufactured synthetic substances. As far as possible, the synthetic ones are designed to be effective against a specific pest, while avoiding unwanted toxic effects to other organisms.

The vast majority of pesticides and fungicides work either by direct contact or by a systemic action. However, in recent years biological pest control methods have been developed, and these are of particular use to those using organic growing methods.

Organophosphates

Contact pesticides work directly by hitting the pest, either from a spray, or as the pest crawls over a treated surface. Systemic pesticides are chemicals absorbed into the plant and transported throughout the plant in the sap. This type of pesticide is most useful against sap-sucking pests, like aphids, but far less useful for pests who chew their food, such as caterpillars, beetles and earwigs.

In a bid to improve their effectiveness, some pesticides contain combinations of both the contact and systemic types of chemicals.

Thorough spraying of all the plants requiring treatment, especially to the underside of leaves where pests tend to lurk in the shade, is essential with all pesticides.

The organophosphorus compounds, such as malathion, are an important group of pesticides that work by paralysing the nervous system by stopping the enzyme cholinesterase from working, which leads to a build-up of the neurotransmitter acetylcholine, with paralysis and then death. See Appendix IV for a more detailed explanation. Malathion, one of the least toxic, is a pesticide considered safe enough for ordinary garden use, being effective against houseflies, aphids and thrips, but it is only partially able to control red spider mite and scale insects. Indeed, it is so relatively safe that it is even used medicinally to kill the head lice and their eggs that infest so many schoolchildren.

Biological controls

Some greenhouse pests, such as whitefly and red spider mite, have already developed pesticide-resistant strains, so here biological control is a better alternative. The biological pest-control methods that are now being developed mean that pesticide use can be drastically reduced or even replaced altogether, which is of major importance to those who are using organic growth methods.

Natural predators as well as parasitic flies and wasps have been used, but these are often highly species specific. Other biological control methods include spraying the pest with fatal doses of hormones, or releasing quantities of sterile males of a pest species. This then results in a reduction of the breeding rate of the pest females. Another method is to spray plants with a biological pesticide, such as a bacterium or fungus that is toxic to a particular pest species.

There is also the somewhat controversial method of breeding disease-resistant varieties or species of crops by the use of genetic engineering techniques.

DDT and friends

Insecticides are designed specifically to kill a variety of insect pests, such as insects, mites, nematodes and molluscs. The earliest insecticides – derris, nicotine and pyrethrum – were all natural products, extracted from plants. A major advance came with the introduction of organochlorine insecticides such as DDT in the 1950s and 60s. DDT, short for dichlorodiphenyltrichloroethane, is a powerful and very persistent insecticide which was used to control mosquitoes in countries where malaria was a problem. Although not toxic to plants, the persistence of DDT in the food chain resulted in its accumulation in the bodies of higher animals, including humans.

Being fat soluble, toxicity from the organochlorines moved up the food chain with disastrous consequences, killing off many birds and small animals who had eaten insects sprayed with these insecticides, and then been eaten themselves, so accumulating the poison up the food chain. They were widely used for many years until concern about their safety led to restrictions in their use. Eventually a complete ban was imposed on their use in the Western world. However, they are still used in the developing world because they are so cheap, compared with the cost of newer, safer agents that are now used in the Western world.[2]

Chlorinated insecticides such as dieldrin and aldrin had high contact and stomach toxicity to most insects, but this type of pesticide also proved to be very persistent in the environment, also being stored in the bodies of birds and mammals. As a result, they too are now little used, except in the eradication of termites. The chlorinated hydrocarbon insecticides, such as chlordane and lindane, also proved to be toxic to higher animals and so are no longer used.[3]

Nowadays, the most important of the newer insecticides are the organophosphates and carbamates, which are used as liquids in sprays and dips, and also as dusts, granules and pellets. Like the organophosphates, the carbamate insecticides inhibit the enzyme cholinesterase, but they differ in that their action is generally less intense, and more rapidly reversible. They enter the central nervous system less readily than the organophosphates, with the result that severe effects on the brain are less common. These substances vary widely in their persistence, toxicity and selectivity of action.

Unfortunately overuse of some of these pesticides has led to the extinction of a number of species of insects in various parts of the world, while other species have now become resistant to their effects and continue to ravage crops.

Boron and fluoride insecticides

Boric acid was used in the past as an insecticide, for both ants and cockroaches. Borates were and are added to plant fertilisers because boron is vital for plant growth. However, both boric acid and its salts, the borates, are also poisonous to humans. A number of cases of boron poisoning appear elsewhere in this book, including victims who were babies, in Chapter 13.

Sodium fluoride has also been used as an insecticide, with such products once available in the US for ants and also for cockroaches. Unfortunately, they were also exceedingly poisonous to humans too, and so are no longer available. A case involving a child is included in Chapter 13 and two such poisonings are described in Chapter 14.

Safer killers

A safer natural alternative, pyrethrum, is a mixture of pyrethrins and cinerins obtained from chrysanthemum flowers. This powerful but non-persistent contact insecticide has a rapid knock-down effect.[4] It is non-toxic to mammals and so is widely used in the food industry and in the home, usually in combination with piperonyl butoxide, which acts as a synergist, making it even more effective. For houseflies, diazinon is a particularly useful insecticide, and rotenone, the active ingredient in derris, is a non-systemic type of natural insecticide which is widely used in both agriculture and horticulture.

Carbaryl is a carbamate insecticide which was used for many years as a lotion and as a shampoo for the treatment of both head lice and crab lice. It has also been widely used in veterinary practice, horticulture and agriculture. Following reports of tumours in experimental animals, the UK Department of Health advised, in late 1995, that carbaryl should be considered to be a potential human carcinogen. Consequently, its medicinal use for the treatment of head lice and crab lice since then has been restricted to prescription only for human use.

Killing larvae

Larvicides kill the pre-adult form of some pests such as caterpillars. Several organophosphates are used for this purpose, and are of particular use in tropical areas for the treatment of rivers where a number of pests cause illness in humans. Phoxim and pyraclofos are both used in the larvicidal treatment of rivers to control onchocerciasis, better known as river blindness.[5]

Temefos is another organophosphate, and it is effective against the larvae of mosquitoes, blackflies and other insects. It has been used in the control of dracontiasis (guinea worm infection), as it is effective against the crustacean host of the larvae of the guinea worm. Like phoxim and pyraclofos, mentioned above, temefos is also used in the larvicidal treatment of rivers in the control of river blindness.

Diflubenzuron is both an insecticide and larvicide, acting as a growth regulator by interfering with the formation of the cuticle by the larvae, and so inhibiting their development to the adult stage. It is used in agriculture and in the control of vectors, which transmit disease. It also has residual activity against mosquito larvae.

Slugs and snails

Molluscicides such as endon and metaldehyde are both used to kill slugs and snails. These agents are particularly useful to control the snail vector in schistosomiasis, also known as bilharzia, one of the most serious tropical diseases affecting humans.[5] Bilharzia was named after Theodor Bilharz, who in 1851 discovered and identified *Schistosoma* eggs in the corpses of those who had died of this tropical disease.

Today a single dose of praziquantel is sufficient to treat a person (the human vector) infected by this fluke parasite. Unfortunately, this drug is not effective against either the immature worms or the eggs, so there is a need to use molluscicides as well.

Metaldehyde is used in pellet form, which can be very dangerous to small children and to any animals that may find and eat the pellets. The symptoms of metaldehyde poisoning are somewhat delayed and include vomiting, diarrhoea, fever, drowsiness, convulsions and coma; kidney and liver damage may occur with death from respiratory failure following within 48 hours. Metaldehyde will appear again in a later chapter on household horrors.

Rats, mice and other rodents

Rodents are herbivorous or omnivorous mammals that have become pests all over the world, causing considerable economic and medical problems because they get into stored food and because they carry plague fleas. Rodents include squirrels, beavers, gophers, rats, mice, voles, hamsters, porcupines, guinea pigs and agoutis. The control of rodents is of vital importance to public health, and rodenticides are an essential control method, as the use of traps alone cannot control numbers sufficiently.

Aluminium and zinc phosphides are used as rodenticides in the form of pastes, as well as to fumigate grain. In the presence of moisture these pastes release a poisonous gas called phosphine, which accounts for the pesticidal activity. Phosphine has a garlic-like odour, which is repulsive to people and also to domestic animals, but apparently not to rats. However, it is exceedingly poisonous to poultry, and so this type of rodenticide should never be used anywhere near chicken runs or sheds.[7] It is also poisonous to humans and needs to be used with caution.

A few years ago, members of the ambulance, medical and nursing staff as well as parts of Arrowe Park Hospital, near Birkenhead on Merseyside, needed decontaminating because of phosphine gas. The gas was emanating from a dead body, which had been sent to the hospital. Having terminal cancer, the man had used a phosphide rat killer as a method of committing suicide.

In 2005, a horse sanctuary near Norwich, which had been careless and allowed some of its workers to inhale the fumes given off by aluminium phosphide tablets, was prosecuted with fines and court costs of £45,000. Breaching health and safety regulations regarding usage of the rat poison led to three employees suffering physical and mental injuries that may affect them for the rest of their lives.

Phosphorus is used in the manufacture of other rat and cockroach poisons, not only as the phosphide pastes already mentioned, but also as organophosphorus compounds.

In the past, red phosphorus was used, in the form of a 1 - 2 per cent paste, as both a cockroach poison and as a rodenticide.

White phosphorus has to be stored under water because it spontaneously catches fire with the oxygen in the air. Because of this, phosphorus should never be mixed with dry bait but should always be mixed with substances containing liquids, such as water, molasses or fat.

Due to its toxicity in humans, where as little as 15mg could be severely toxic and 50mg could be fatal, phosphorus, which was once a very popular rodenticide, is no longer recommended and now little used.

Other far more effective and less toxic rodenticides are now available. However, phosphoric acid is used widely in industry and indeed was used as an approved disinfectant in the outbreak of Foot and Mouth Disease in the United Kingdom a few years ago.

Sodium fluoroacetate is a highly effective rodenticide but is very toxic to humans and other animals, so it must be used with great caution by specially trained operatives. The toxic effects may be delayed for several hours after absorption by mouth or by inhalation. The symptoms start with nausea and vomiting, followed by feelings of apprehension with muscle twitching and heart irregularities, which lead on to convulsions, respiratory failure and coma. Death is usually due to ventricular fibrillation and the consequent heart failure.

Barium carbonate has also been used as a rat poison in the past. It is insoluble in water, but becomes highly toxic once swallowed because it dissolves in the acid in the stomach and can then be absorbed into the bloodstream. Death of the rodents then results from cardiac and respiratory failure.

Antu was another substance formerly used as a rodenticide, but it is now severely restricted in use, due to the presence of known carcinogenic impurities such as naphthylamine that are contained in it.

As with all poisonous substances, they can be used fatally, by accident or with cruel intentions. In the past, thallium-containing pastes and coated grain (such as Thalgrain) were marketed as rodenticides – and used to murder people too. One such example is Graham Young, whose tale is told in Chapter 2, as well as in the next chapter.

Bleeding rats, squill and strychnine

Warfarin is an anticoagulant that is widely used today. In medicine, it is given to patients who have had heart surgery, to prevent blood clots. It is also used by those who have had a deep-vein thrombosis (DVT), a pulmonary embolism or heart attack. Warfarin, in addition to its medicinal uses, is widely used as a rodenticide. The bait used contains a low level (about one part in 40) of the active agent. The rats are finally killed after eating about five doses. Because the bait has such a low strength, the risk of toxicity to man and domestic animals is not serious, so this type of poison can be safely left lying about on the floor.

Warfarin works by antagonising the action of vitamin K in the body. It does this by preventing prothrombin synthesis, which is vital in the blood-clotting process (described in Appendix V), and leads to internal bleeding in affected rodents. So the rats effectively become haemophiliacs, and bleed to death.

In humans, the antidote to warfarin poisoning is a large dose of vitamin K, to counteract the action of this anticoagulant. Some rats have now become resistant to warfarin, and so, for these creatures, the second-generation coumarin rodenticides are used. This type of rodenticide has only 100mg of the active agent in each kilogram of bait and so is not hazardous to humans.

These more concentrated forms are particularly hazardous, though: there have been a number of poisonings, including fatalities, with these more potent agents, and as a consequence their use is now somewhat restricted. Baits containing these powerful second-generation agents may only be prepared by trained personnel and must contain a marker dye to act as a visual warning of their presence.[8]

Red squill is another rodenticide, and it is both neurotoxic and cardiotoxic. It is very poisonous to rats and is incorporated as an ingredient of rat pastes. It acts on the central nervous system, and it is extremely irritating to the skin so should only be handled with rubber gloves. Its use as a poison is not considered acceptable to animals other than rodents, so it is included in the Animals (Cruel Poisons) Regulations, 1963, which prohibits its use in the United Kingdom except for killing rats.

Strychnine stimulates the nervous system by reducing the inhibition that normally controls it. It is very poisonous, with large doses producing convulsions. Strychnine used to be used to kill vermin, particularly the European mole. This poison is now also prohibited under the Animals (Cruel Poisons) Regulations, 1963, which banned its use in the United Kingdom. The use of strychnine to kill moles was prohibited from September 1st 2006. This is the result of an EU directive to ensure that all currently authorised pesticides meet modern human and environmental protection standards. Two products, which both contain aluminium phosphide, are now the approved alternatives to strychnine.

Pest-killing arsenic

Arsenic trioxide was used for many years in weedkillers, sheep-dips and as rodenticides.

In 1919, in Sussex there was a case of accidental poisoning caused by a tin of liquid weedkiller that contained arsenic. The leaking tin was placed beside a sack of sugar in the guards van of a railway train. During the journey, the sugar absorbed a quantity of the leaking weedkiller, although no one noticed at the time. It was not until the sugar was later sold, and had poisoned those who used it, that the cause was finally realised.

In 1921, a lady called Mrs Hanktelow from Beckenham was killed by taking a quantity of 'Eureka' weedkiller, which contained about 60 per cent arsenic trioxide. A report of poisoned apples, which also happened in the 1920s, was found to have been caused by a build-up of chemicals as an incrustation on the fruit, following their spraying with a mixture of the pesticides Bordeaux Mixture (copper sulphate and lime) together with Paris Green (an arsenical compound).[9]

Seed dressing comes undone

Mercury was used in seed dressings from the 1920s, with the intention of making the seed resistant to fungal disease. By the 1960s there were more than 150 different products on the market of the mercury type of seed dressing alone. Unfortunately, mercury is poisonous and this type of crop protection led to a number of mass poisonings in the Third World.

When seed dressings were given to Third World countries, instead of planting the grain to provide a crop the following year, as was intended, the poor and starving villagers used it immediately to make bread. The poisoning became apparent only after the bread had been baked and eaten. In northern Iraq, more than 5,000 people were poisoned in this way, and 280 of them died after eating bread made from mercury-treated grain in the 1970s.

The donating nations had supplied the seed dressing with the best of intentions, sending their gift via the World Health Organisation. Crops grown from the treated grain absorb very little of the mercury during growth and would have been perfectly safe to eat. However, as a result of the immediate use of the grain and the unfortunate and unintended poisonings, the use of mercury seed dressings is now much reduced.[10]

Failing fumigants

Ethylene dibromide and ethylene dichloride have both been used as insecticidal fumigants. But both these fumigants are readily absorbed through the skin, causing blistering as well as kidney and liver damage. Their use has been restricted because they were found to be carcinogenic in experimental animals. There was also evidence of persistence in both fruit and cereal crops that have undergone fumigation with these agents.[11]

Naphthalene has also been used in the past as a soil fumigant. But continuous exposure to the vapour of naphthalene led to cataract formation and then blindness in workers so this too is no longer used.

Worm-like species, the nematodes have their own pesticides, the nematodicides. Dibromochloropropane, an example of this type of pesticide, is also used as a soil fumigant. However, this substance can cause sterility in humans; both low sperm counts and evidence of testicular damage have been found in exposed workers. A substance used instead is ethylene chlorohydrins, which has a dual purpose, being utilised not only as an insecticide but also for forcing the early sprouting of potatoes.

Chloropicrin is a poison gas used to kill both insects and other parasites. It is used to fumigate and disinfect stored grain and soil. As it is a lachrymatory agent (it makes your eyes water), which is intensely irritating to the skin and mucous membranes, it has found an additional use as a warning gas when added to other even more toxic fumigants.

Mildew murder

Like pesticides, fungicides work either by direct contact or systemically. Fungal diseases such as mildew, blight and rust spread rapidly on plants once established, so treatment needs to be both fast and effective. Fungicides are now applied to both growing and stored crops as a preventative measure, either as a foliage spray or as a seed dressing.

The broadly toxic elements copper, mercury and sulphur were among the first fungicides used in agriculture. For example, the powder flowers of sulphur was used for many years by gardeners to dust over their plants to control fungal diseases, such as powdery mildew. Copper sulphate and mercury chloride were in use as long ago as the eighteenth century and lime sulphur was used to treat mildew from 1802 onwards. The well-known Bordeaux Mixture (a mixture of copper sulphate and lime dissolved in water) was also used as a fungicide for many years. These older contact type of fungicides have now largely been superseded by far more effective synthetic compounds, which work systemically.

As mentioned in the previous chapter, in the past vineyard workers who used copper sulphate sprays were found to have pathological lung changes, including blue spots on the surface of the lung, with lesions, nodules and other tissue damage. All the lesions were found to contain particles of copper. This lung damage and the resulting respiratory problems were due to progressive thickening and scarring of the tissues.[12]

Contact fungicides kill germinating fungal spores and so prevent further infection, but they have little effect on established fungal growths; in contrast, systemic fungicides, such as benomyl, effectively kill established growths of fungi in the plant tissues. Benomyl is active against a wide variety of fungal pathogens, especially powdery mildew. It is non-toxic to animal life and so is widely used by both professional growers and amateur gardeners alike. Some fungicides can be harmful to fish, so care must be taken near ponds, streams and rivers.

Frequent use of systemic fungicides can lead to the development of resistant strains, which can sometimes be overcome by the use of a different compound; however, a better alternative is to use biological control wherever possible. Organotin compounds can be extremely poisonous but are nonetheless used as fungicides and wood preservatives.

Those wretched weedkillers

A herbicide is toxic to plants, and is used to kill weeds and other unwanted vegetation. Non-specific herbicides such as sodium chlorate and paraquat quite literally kill all plants. Selective herbicides will kill only broad-leaved plants and so are useful when growing cereals and other grass-like, thin-leaved crops. Herbicides must be used with great care, as although they are designed to kill plants, some are also toxic to humans.

Concern about persistent toxicity has led to the development of some products that break down when they enter the soil. Other herbicides have been specifically designed as pre-emergent herbicides, to persist in the soil and to kill weeds when at their most vulnerable, as they germinate. Some other specialist herbicides, while being relatively non-toxic to animal life, are widely used by farmers as selective or post-emergent herbicides to kill off existing weeds as the crop grows. These special herbicides are used on many crops, such as maize, sugar cane and sorghum, so ensuring ease of harvesting of the maximum crop possible.

Both diquat and paraquat[13] inhibit photosynthesis non-selectively, acting as powerful contact herbicides, which are then inactivated on contact with the soil. However, if swallowed, they cause severe, often irreversible damage to the lungs, liver and kidneys. And this can be fatal: fashion icon Isabella Blow killed herself with paraquat in 2007, as detailed in Chapter 5.

The liquid concentrate of paraquat, one part in five strength, is now only supplied for agricultural use to approved users in the UK under the brand name Gramoxone. Preparations of paraquat are now formulated to contain an emetic or a laxative and some also contain a malodorous ingredient to deter ingestion. The available domestic garden product, 'Weedol', is only a one-in-40 solution of paraquat. This strength, though causing nausea and vomiting, as well as some respiratory changes when ingested, is not regarded as the lethal form.

Susan Barber's husband, Michael, came home early one day in May 1981 to find his wife in bed with another man. It was Richard Collins, his partner in the local darts team. Michael hit his wife and threw her lover out. The following day, Susan put half a teaspoonful of weedkiller into her husband's steak and kidney pie. Unknowing, he ate it, became ill and ended up in the hospital.

The doctors, not realising that he had been poisoned, suspected from the symptoms that he had pneumonia. Later his illness was diagnosed as a rare neurological condition called Goodpasture's syndrome. He was then transferred to Hammersmith Hospital in London, where he died. The cause of death was given as pneumonia and kidney failure. His 'grieving' widow was able to collect £15,000 from her husband's pension fund and she then set up home with her lover.

The pathologist who conducted the post-mortem had some lingering doubts about the cause of death and took the precaution of preserving various organs from the body. Some of these were sent to ICI, the manufacturer of a weedkiller called Gramoxone. Tests showed traces of the weedkiller in the preserved organs and the police began to make enquiries, which finally resulted in Susan Barber's arrest in April 1982. At the trial, Richard Collins, the lover, was charged with conspiracy to murder and received two years' imprisonment, while Susan Barber, the murderous widow, was jailed for life.

The triazine herbicides, such as simazine, a pre-emergence weedkiller, are also used as selective herbicides. Barban is a translocated herbicide used to kill wild oats without causing serious damage to wheat, barley and various legume crops. This carbamate derivative is relatively non-toxic in humans, but an allergic reaction to it can sometimes occur.

Benazolin is a post-emergent herbicide that is effective against chickweed and cleavers, and is relatively non-toxic. Glyphosate is another herbicide with which a large number of poisonings have occurred over the years. It is believed that its toxicity is largely due to a surfactant, polyoxyethyleneamine, which is included in the formulation of the proprietary product 'Roundup'.[14]

A final titbit

Dintro-o-cresol and dinitrophenol are interesting in that they were both formerly used as herbicides. Dintro-o-cresol, which claimed to be five times as potent as dinitrophenol, was also used as an insecticide. Their mode of action was to increase the rate of metabolism within the cells, and so, as well as their agricultural uses, they also came to be used medicinally in the 1930s as treatment for obesity. Unfortunately, dinitrophenol was very dangerous and fatalities occurred due to the induced heat stroke and swelling of the brain caused, which then led on to respiratory and cardiac failure.

Needless to say, there were many cases of poisoning, one of which is described in Chapter 14. Agricultural workers, who discarded their protective clothing when working in hot weather, were fatally poisoned, because these agents were absorbed through the unprotected skin, as well as by inhalation. There were eight fatalities among the many cases of poisoning, which occurred in a six-year period in the late 1940s. These dangerous substances are no longer in agricultural use.[15]

Chapter 10

Frightening Foodstuffs

MANY KINGS AND QUEENS employed royal tasters to sample their food for them, as this was such a favourite method of poisoning in the past. Simon, a monk at the court of King John, committed suicide in 1216 by deliberately drinking some wine that he knew to be poisoned, to try to get the King to drink it as well. He failed!

Boiled to death

King Henry VIII may be best known for his six wives, but he was also a bloodthirsty tyrant. He had more than 70,000 people executed during the 38 years he reigned – more than five every day.

In 1531, King Henry VIII was especially harsh. A man named John Roose was convicted of poisoning 17 people in the Bishop of Rochester's house; he had poisoned the broth served to both the Bishop's family and the poor of the parish and two people died as a result. The King wanted a particularly nasty punishment for John Roose, to deter others from food tampering. He passed a very special law: felons would be boiled to death and would be denied the last rites before their punishment. And so John Roose was boiled to death. This special law was in force for 16 years, and during this time a woman called Margaret Davey met the same fate. She was found guilty of tampering with foodstuffs at Smithfield, the famous meat market outside the London city walls at that time. Although Edward VI 'downgraded' the punishment for deliberate poisoning to the same as for murder, this law was not formally removed until 1863.

Unexpected additives

Food has often been adulterated so that cheaper produce can be passed off as being of a better quality than it really is. This was so rife in the early years of the nineteenth century that a German chemist, Friedrich Accum, who had lived in Britain for over 30 years, published his 'Treatise on the Adulteration of Food and Culinary Poisons' in London in 1820.

In one example, he revealed some Gloucestershire cheese to which some red lead had been added. This cheese was usually coloured with annatto, a natural colouring derived from a plant. A trader had used an additive to improve the colour of his poor quality annatto before selling it on to an unsuspecting farmer, who made the cheese. Unfortunately, the vermilion pigment used (which contains mercury) had itself already been adulterated with red lead by the druggist who supplied it. This druggist had thought it would only be used for house painting, and adulterated it to increase his own profit.

Herr Accum also exposed other doubtful practices, such as how flour was 'improved' when the harvest was poor, being mixed with potato flour, barley bran and other similar substances. This produced a rather dark coloured flour, which was then 'whitened' by the addition of alum. There was no limit to products being adulterated: milk was watered down, with chalk added to whiten it; copperas was added to beer, and capsicum to mustard; artificial tea was made from blackthorn leaves; and 'coffee' was made from horse beans, wheat or rye, which were first partially burnt and then ground down. Much of the wine at that time was made from spoiled cider.

In 1875, the Foods and Drugs Act said that 'no person shall mix any article of food with any ingredient or material to render it injurious to health'. But that didn't stop the watering down of milk and the continuation of other dubious practices. It would be many years later that legislation to protect the public from such food adulteration would be introduced and enforced. Even today there are dubious practices, such as adding water to chicken meat, changing sell-by dates, and selling pet food for human consumption.

Naturally occurring toxins can sometimes be found in foodstuffs. This may be due to an error in the preparation, or to a poisonous part (such as rhubarb leaves – read on below) being used instead of being discarded. And the spoilage of food may occur during storage, if either the temperature and/or humidity are unsuitable, which may lead to bacterial or fungal growth and spoilage. In the time of Herr Accum, most of the butter was rancid, the meat tainted and the fish stinking, because in those far-off days refrigeration had not yet been invented.[1,2]

Rhubarb, sorrel soup, rabbits, trees and ragwort

Rhubarb and custard is a traditional pudding in Britain. The young stems are boiled gently until tender, with a little sugar and ginger added to taste. The rhubarb leaves are always discarded as they contain poisonous oxalic acid crystals. Many accidental poisonings have occurred over the years, when small children have eaten bits of these leaves. Immediate first aid in such a case is to give the child a big drink of milk and then take him or her to the nearest hospital accident and emergency department. The calcium in the milk will combine with the oxalic acid to form calcium oxalate, an insoluble substance that cannot then be absorbed from the stomach.

Other plants also contain oxalic acid. Sorrel, a wildflower whose stems and leaves contain oxalic acid, is used, sparingly, in salads and soups. But, in 1989, there was a fatal case of oxalic acid poisoning caused by eating sorrel soup.[3]

In 1921, *The Lancet* carried an unusual report of belladonna poisoning, which was caused by eating a rabbit! The rabbit had eaten some deadly nightshade shortly before it was killed for the pot, and this remained in the meat. Fortunately, the poisoning was not too severe in those who ate the rabbit meat, but it did cause dilated pupils, a dry mouth, giddiness and a rapid pulse.[4]

If you take a holiday in Jamaica, then beware of a small fruit tree called the Akee. The seed covering, or arillus, of the unripe fruit of the akee contains a toxic substance called hypoglycin A, which is responsible for the so-called Jamaican vomiting sickness. The symptoms are acute and severe vomiting, low blood sugar, muscular weakness, central nervous system depression and convulsions leading to coma – such poisoning is frequently fatal.

Common ragwort, an abundant weed in Britain, is poisonous to livestock if eaten in quantity, and even a single ragwort leaf can damage a horse's liver. Dried ragwort is even more poisonous than the fresh plant, so great care must be taken during haymaking to ensure that no ragwort is contained in the hay bales, especially if it is intended to be fed to horses over the following winter. Ragwort should only be handled when wearing gloves, to avoid any risk of skin absorption. There is risk of chronic poisoning from both the common species here in Britain, and from the golden ragwort found in the USA. Both species have been used in the past in the herbal treatment of menstrual complaints. In Arizona, USA, in the 1970s, there were reports of liver damage and death following consumption of a herbal tea, called Gordolobos, made from yet another species of ragwort.[5,6] The widespread use of such herbal teas in the traditional medicine of the West Indies is a continuing problem, as chronic exposure to low doses of pyrrolizidine alkaloids, found in these plants, can lead to liver cirrhosis.

Poisonous plankton

In the Philippines, paralytic poisoning is caused by eating certain shellfish. And in Australia and the Pacific Islands there are more than

20,000 cases every year of food poisoning due to people eating coral reef fish. The cause of poisoning is the same for both shellfish and coral reef fish: the plankton these sea creatures eat produce saxitoxin, a nerve poison that does not affect them at all, but has dire effects on the humans who like eating seafood.

Toxic fungi

Aflatoxins are fungal poisons produced by the Aspergillus fungus, which grows on many vegetables, but particularly on peanuts. Aflatoxins cause liver damage and liver cancer in animals, and there is a similar link to liver cancer in humans. In 1978, a report in *The Lancet* told of two children who developed Reye's syndrome. This condition is usually associated with the use of aspirin in children, and will be described in detail in Chapter 18, Malevolent Medicines. These children were both found to have aflatoxin in their blood during the acute phase of the illness.[7] In October 2004, there was a health scare in Hungary, when it was discovered that supplies of the national spice, paprika – the mainstay of Hungarian cuisine, made by grinding down dried red peppers – were adulterated with some very inferior paprika from Latin America, which was heavily contaminated with aflatoxin.

Many outbreaks of ergotism, due to the ergot fungal infection of rye, have occurred throughout history, particularly in those parts of Europe and Russia, where rye bread is a staple part of the peasant diet. There was a major outbreak near the Urals in 1926-7, when over 11,000 people were affected. Russia suffered again in 1942-3 when Fusarium, another fungal infection, affected the grain in parts of Siberia resulting in thousands of people dying within two weeks of eating the affected bread. In 1979 *The Lancet* carried a report of an outbreak of ergotism, attributed to the ingestion of infected wild oats, in Wollo, Ethiopia.[8]

Poisoned porridge and other contaminated foods

Sometimes foodstuffs are found to contain a toxic substance in error or a permitted substance in a vastly increased amount.

In 1950, the *British Medical Journal* reported a case of poisoned porridge. Two people were taken ill within ten minutes of eating their bowls of porridge at breakfast one morning. The porridge oats used were found to be contaminated with potassium bichromate to the extent of one part in 400. How it got there was never discovered, but the symptoms included severe abdominal pain and vomiting. Fortunately, there was no lasting damage in this case, but there have been a number of fatal poisonings due to potassium bichromate in the past.[9]

During the Prohibition Era (1919-1933), when alcoholic drinks were banned in the USA, there was an epidemic of progressive paralysis in the south and mid-west of the country. It started with aching muscles, then numbness, followed by loss of sensation, weakness and eventually paralysis. Some 20,000 cases were diagnosed, affecting both the arms and legs of the patients. The cause was traced to drinking an imitation ginger extract. This was a medicinal product with a high alcoholic content that had escaped the Prohibition ban. Investigations eventually traced the problem to a contaminant in the product, which affected the nervous system. Very few deaths occurred, but most of the people affected only made a partial recovery after some months, and in many cases the paralysis seemed permanent.[10]

Potassium bromate is a chemical added to flour to improve or mature it, only in a very small quantity. However, in South Africa in the late 1960s, nearly a thousand people were poisoned by eating bread prepared from dough containing too much potassium bromate. On analysis the poisoned bread was found to contain almost 300 times too much.[11]

In early 1981, a mass poisoning began in Spain, particularly affecting those living in Madrid and the northwest regions of the country. This was traced to contaminated cooking oil, and by June 1981 the Spanish government had begun to remove it from the market. The last new case was found in September of that year. About 25,000 people were affected and more than 600 eventually died. Severe respiratory illness, with fever, rash and weak, painful muscles led on to long-term neuromuscular damage in about 20 per cent of those who survived.

The cooking oil was mainly rapeseed oil, but also contained a little aniline, as required by Spanish law for all imported rapeseed oil – to prevent it from being used for cooking. It had been imported into Spain for industrial use only, but someone decided to sell it and make a lot of extra profit. After treatment, which was supposed to remove the aniline, it was sold, house to house, and at street markets – and people bought it because it was very cheap. Too cheap! The cooking oil, when analysed, was found to contain rapeseed oil, other seed oil – including olive oil – but also liquefied pork fat, together with traces of aniline and fatty acid anilides.[12]

Sensitive to preservatives

In Europe, every food additive is given an E number. This includes not just colouring agents like tartrazine (E102), but preservatives, anti-oxidants, emulsifiers, stabilisers and thickeners, flavours and flavour enhancers, artificial sweeteners, nutrients and many others such as anti-foaming and anti-caking agents. The USA has a similar – though different – system of codes for food additives.

Chinese Restaurant Syndrome appeared in the 1970s with the increase in the number of Chinese restaurants at that time. Some customers suffered tingling and weakness of the face and upper body, and flushing of the skin. They had palpitations and felt anxious, nauseous and thirsty after eating their Chinese meal. These symptoms were partly due to a deficiency in vitamin B6 (pyridoxine), which was needed to safely eat the Chinese food: a flavour enhancer called monosodium glutamate is used in cooking Chinese food and pyridoxine is needed to metabolise and so remove the glutamate. The symptoms were due to the toxic effects caused by the glutamate.

The food industry uses preservatives to prevent spoilage of foodstuffs, but these can sometimes cause problems. Sodium nitrate and potassium nitrate are both used as preservatives in the food industry, particularly for meat products. Sodium nitrite is also used, but in much smaller quantities, as this is the dangerous one.

In 1997, there was a report of three people who became ill after eating sausages that had been preserved with a mixture of sodium nitrate and sodium nitrite, rather than sodium nitrate and potassium nitrate.[13] As a result of the incorrect sodium nitrite content, the sausage eaters became cyanosed (turned blue), and developed methaemoglobinaemia, a condition in which the haemoglobin in the blood is changed so that it can no longer carry oxygen.

Over 70 years ago, in 1936, three members of a Middlesborough family died within two hours of consuming some sodium nitrite, taken in error for salt (sodium chloride). These were the first known fatalities due to nitrite poisoning in Great Britain at that time.[14]

Some people seem to be particularly sensitive to sodium nitrite when it is used as a preservative in food or drink. One person developed headaches after eating cured meat products, and also when given a weak solution of the suspected preservative in water to drink.[15] Another had recurrent attacks of muscle pain when he drank beer that contained sodium nitrite as a preservative.[16] Sodium nitrite is not allowed at all in foodstuffs intended for small babies due to the high risk of methaemoglobinaemia.

In Australia in the 1970s, because of the hot weather, a 36-year-old man accidentally took sodium nitrite tablets (1g daily) for several weeks, thinking that they were salt tablets. One particularly hot day, after taking double the usual dose, he became ill. He became obviously cyanosed, was perspiring and was quite distressed but recovered after receiving treatment, which included the administration of oxygen.[17]

Sulphur dioxide and various sulphites are also used as food preservatives. Sodium, potassium and calcium sulphites are all used because, when in aqueous solution, they form sulphur dioxide, which kills bacteria. And because it is also an anti-oxidant, it prevents food from going brown. It is widely used with both fruit and vegetables because it preserves the natural colour. When grapes are crushed, the juice will start to ferment naturally, due to wild yeasts present on the skin of the fruit. This natural fermentation can be prevented by as little as 10 parts per million of sulphur dioxide. When making wine, specially cultured yeasts, which can survive sulphur dioxide, can then be added to produce wine of the desired quality. Just before bottling the wine, the sulphite is added to prevent any further fermentation. This sulphite will react with other components in the wine, dissipating as it matures during storage. However, young wines, lacking the all-important storage stage, often contain noticeable amounts of sulphur dioxide.

Salt substitutes

Sodium chloride is the chemical name for ordinary salt, which we add to our food during cooking, or at the table. Because a low sodium diet is helpful for those with high blood pressure[18], many chemicals have been marketed as salt substitutes over the last century. In America, there were a number of poisonings, including some fatalities, from the use of table salt substitutes containing lithium chloride, by people on just such a low sodium diet. The early symptoms included drowsiness, weakness, loss of appetite and nausea; later victims experienced tremors and blurring of vision, followed by confusion, before falling into a coma and death. If the salt substitute was discontinued at an early stage of poisoning, then the symptoms faded away over the next three to four days.[19]

The other chemical widely used as a salt substitute was potassium chloride. This substance was also used medicinally at that time in the 1930s and 1940s as a diuretic, to prevent kidney stones. A fatal case of poisoning happened in Canada in 1940, connected to this very chemical. A man had taken between 30g and 35g of potassium chlorate over a period of three days in mistake for potassium chloride, which he used instead of salt. The poor man died five days after the last dose.[20] Potassium chlorate was widely used at that time as a mouthwash and gargle and also in toothpowders, but was not supposed to be swallowed.

Sweeter than sugar

Artificial sweeteners such as saccharin, cyclamate and aspartame are used by manufacturers as a cheap alternative to sugar. Saccharin is 300 times sweeter than sugar and aspartame is about 200 times as sweet, while cyclamate is about 30 times as sweet as sugar. Many people use these sweeteners as a means of reducing calorie intake while weight watching.

Saccharin was found by chance, a happy accident, discovered by a chemist at Johns Hopkins University in the USA in 1879. Graduate student, Constantin Fahlberg, was attempting to make some toluene derivatives and got one of them on his hands. He went to lunch and noticed that his bread tasted sweet, then realised he had not washed his hands before lunch and that the sweet taste must have come from the chemical he had made that morning. It was widely used until safety testing in 1977, when it was banned in the USA for a time due to cancer concerns, but only for extreme and unrealistic doses. There was such a public outcry that it was allowed again while further testing was carried out, and only in the year 2000 were official concerns dropped.

A similar problem occurred with cyclamate, a sweetener preferred by many since it did not have the bitter aftertaste of saccharin. Cyclamic acid and its calcium and sodium salts were intense sweetening agents used in soft drinks, foods and sweetening tablets. They too were banned for a time, but now have official backing once again.

Aspartame is another sweetening agent, which is widely used in the food industry and can be bought by the public to use in place of sugar. However, it must be used with care by those who suffer from PKU (mentioned in Chapter 5) because one of its breakdown products is phenylalanine.

A sticky honey situation

Since 2006, bees have been a cause of concern, because of a phenomenon called colony collapse disorder, which has massively reduced the bee population in the USA. No cause has yet been found, but production of honey there has halved and more honey has been imported, but not without problems. A similar decline has recently occurred among European bees due to the spread of the parasitic mites – Varroa – which decimate hives.

Imported honey has been found, in many cases, to have come from China, but by devious routes, involving repackaging and re-exporting, to hide its true origin. This is referred to as 'honey laundering' and is an attempt to avoid health and safety checks, import fees and other tariffs imposed to prevent 'dumping' of tainted foodstuffs on to world markets, at below the cost of production.

Chinese honey has been and still is tainted, because back in 1997 bees were nearly wiped out by a bacterial epidemic. The beekeepers, instead of burning the hives to remove the infection, chose to treat them with chloramphenicol, an antibiotic that can cause aplastic anaemia, a serious blood disorder. Although this practice was outlawed, some honey is still tainted, not only with chloramphenicol but also with other antibiotics, such as ciprofloxacin.

Adam's ale

A safe and adequate water supply is vital for a healthy life. Typhoid and cholera are just two of the diseases that may result from foul water supplies. The antibacterial properties of silver were known thousands of years ago, and silver has been used since ancient times to sterilise water. The Greek historian, Heroditus (485-425 BC), reported that the Persian King Cyrus the Great, who died in about 530BC, used silver for this purpose. Drinking water and swimming pools can be made safe by as little as 10 parts per billion of silver – although this is too expensive to put into action these days.

Any one visiting a local swimming pool knows that chlorine is another substance used to disinfect the water. It was first used in 1897 during a typhoid epidemic in Maidstone, Kent. It then became the main means of disinfecting drinking water both in Britain and throughout the developed world, preventing further outbreaks of water-born diseases such as typhoid, cholera and meningitis.

In Peru in 1991, the authorities stopped using chlorine to purify drinking water, following a campaign by environmentalists about the health threat of organochlorine compounds. The result was a massive outbreak of cholera: over a million cases were reported and 10,000 people died. Needless to say, the authorities started to chlorinate the water again as soon as they realised their mistake, although it took some time to bring the disease under control. This is a prime example of how risks and benefits must both be considered. In this case the benefits of chlorinated water to prevent cholera far outweighed the risks that worried the environmentalists.

Unfortunately, even in the developed world, water contamination can get the better of us. Babies have been poisoned by contaminated water mainly in the rural areas of Britain, such as Lincolnshire and East Anglia. This poisoning is usually a result of the baby being given feeds made with water from wells that have become contaminated, with nitrates, for example, in groundwater run-off from the fertilisers used on the fields.[21]

In 1988, a lorry-load of aluminium sulphate was emptied, by mistake, into a reservoir near Camelford, in Cornwall. This chemical contaminated the local water supply, causing diarrhoea, mouth ulcers and blisters, malaise, joint symptoms and memory defects in the local population of some 20,000 people who drank the water. The memory problems did not appear until some months later. Even six months later, aluminium deposits were still found in the bones of those affected.[22] A report issued in 2005, following a four-year study into Britain's worst water poisoning incident, found evidence of long-term health problems, including asthma and arthritis, suffered by the victims of 1988, some of whom have since died.

Even bottled water has problems

Bottled water is marketed as pure, from a spring or similar. But even this form of water can be dangerous.

In the early 1990s, miniscule traces of benzene were discovered in Perrier water, forcing the French company to withdraw 160 million bottles worldwide. Analysis of a number of the drinks showed the benzene was caused by a reaction of the preservative, sodium benzoate, and the antioxidant, vitamin C. Then in 2005 and 2006, tests on some 230 drinks on sale in Britain and France again showed high levels of benzene – eight times the level permitted in drinking water. Benzene is an aggressive carcinogen and may lead to leukaemia and other cancers of the blood. There is currently no known safe level for benzene. Part of the problem is that benzene is present in the air already, and we breathe in more than 200 micrograms of it every day.

In 1995, Scottish biochemistry lecturer, Paul Agutter, was tried in the High Court in Edinburgh for attempting to murder his wife. On a Sunday evening in August 1994, he gave his wife, Alexandra, a gin and tonic to which he had added poison. He used atropine, the poison found in the plant Deadly Nightshade. His wife noticed that it didn't taste right and stopped drinking after a few sips. But later that evening, she became ill and was admitted to hospital. She was treated for acute poisoning and fortunately made a complete recovery. The remaining undrunk gin and tonic was analysed and was found to contain a large amount of atropine. Agutter tried to blame the tonic water, which had come from Safeway.

The supermarket was forced to issue a nationwide recall on some 55,000 two-litre bottles of tonic water after seven more cases of poisoning came to light. He had posed as a victim, but the supermarket CCTV showed Agutter placing more poisoned bottles on the shelf in an Edinburgh store, and he was then arrested. He wanted to kill his wife so that he could then be free to marry his lover. But this lady deserted him as soon as he was charged, and although his wife stood by him throughout his trial, she divorced him as soon as he was imprisoned. Found guilty, he was sentenced to serve 12 years in jail.

Contaminated foodstuffs

Insecticides of the organophosphate type are very good at killing insects, but they can cause great problems if they contaminate foodstuffs, which happened in India in July 1997.

A group of 60 fit and healthy young men, aged 20-30 years old, ate their midday meal in a works canteen. The meal included chapatis, cooked vegetables, beans and halva. Over the next three hours, all 60 became ill, suffering nausea, vomiting and abdominal pains. They were treated at the local healthcare centre, but four of the men needed hospital treatment and one of them died a few days later. The doctors initially assumed it was caused by some sort of food poisoning, such as botulism. Blood tests, however, revealed the presence of the insecticide malathion, and this was traced to the chapatis, which were found to be heavily contaminated.[23]

A few years ago, a nationwide food alert came about because of a very unusual contaminant: a large-animal tranquilliser.

In 2003, a Cumbrian manufacturer of game pies and sausages unwittingly bought some deer meat that was contaminated with Immobilon, a strong tranquilliser used by veterinary surgeons to sedate large animals such as lions and elephants. Immobilon was many times stronger than the more usual sedatives, and many deaths, both accidental and deliberate, resulted from its use, before it was banned in the 1990s. But in 2003 it was still being used, illegally, to kill red deer, and this is how it got into the pies and sausages. A nationwide alert had to be issued by the Food Standards Agency warning people not to eat the affected food.

Alcohol poisoning

Alcohol presents growing problems and few people realise how dangerous it can be. Even ordinary beer, wines and spirits can kill. The current craze for binge drinking can be particularly dangerous. People who are drunk should be placed in the recovery position and watched carefully for several hours in case they vomit or their condition deteriorates, in which case they should be taken to hospital. If vomiting is excessive, it can lead to a massive drop in the blood glucose level, which may need intravenous fluids to correct. And apart from the risk of vomiting and the dehydration it can cause, there is also the risk of choking to death on the vomit too.

While alcohol is a poison itself, it can also be adulterated by the addition of other poisonous substances, which may be cheaper or easier to obtain. In America in the early 1950s, some 23 people were poisoned by drinking Korean saké which contained 16 per cent methanol. Five of them died of respiratory arrest after several hours of coma, and another one died four hours after admission to hospital. The post-mortems showed swelling of both the brain and lungs, with inflammation of the stomach and fatty livers. The other 17 saké drinkers were all treated, and unfortunately one more died, but the rest recovered.[24]

Unappetising arsenic

There have been many poisonings of food with arsenic over the years (as you've likely already discovered with the numerous arsenic anecdotes sprinkled throughout this book), for an amazing variety of reasons. One of these is because of its use as a pesticide. Apple orchards used to be sprayed with the pesticides Bordeaux Mixture (copper sulphate and slaked lime), and Paris Green (copper arsenate). Unfortunately, poisonous incrustations were left on the fruit as it ripened, only to poison the unwary.

Cocoa was also poisoned in the 1930s, when it was treated with impure potassium carbonate contaminated with arsenic. The potassium carbonate was used to make cocoa soluble, but the new ‘chocolate with arsenic’ flavour wasn’t the plan![25]

Interestingly, in very small doses, arsenic seems to boost the metabolism and increase the formation of red blood cells. Vichy water contains about two parts per million of arsenic, which may account for its supposed tonic effect.

A famous case of mass arsenic poisoning in Britain happened in 1900: over 6,000 people were affected and 70 died. The cause was found to be contamination of the invert sugar used to brew beer. Invert sugar is a mixture of glucose and fructose which has been treated for use in brewing or confectionery. This particular invert sugar had become contaminated with arsenic during its manufacture. The poisoned beer was found to contain enough arsenic that a night’s drinking contained a dangerously high dose. At first some of the victims were thought to be suffering from the effects of too much alcohol, but one doctor realised that their symptoms could also be caused by arsenic poisoning.

Even today, arsenic poisoning can occur. In the summer of 2008, Cornish health officials had to warn people not to eat a sea plant called marsh samphire (glasswort). The plant had been picked illegally on mudflats in the Hayle Estuary in Cornwall and was sold for use as a garnish on seafood dishes in local restaurants. These mudflats were known to be contaminated with high levels of arsenic from past tin mining in the area. Even a small quantity of the plant could cause problems with swallowing and salivation.

Various third world countries continue to be at risk of arsenic poisoning. The surface water in countries such as India, Bangladesh and Thailand is often heavily contaminated with sewage and should not be used. This polluted water causes diarrhoea, which kills some 20,000 people each year in Bangladesh alone. In the 1960s, a UN project encouraged and even paid for tube wells to be dug in such countries to provide clean water both for people to drink and for the irrigation of rice crops. Although done with the best of intentions, the tube wells were sunk down to the depth of arsenic-containing rock strata. This was not known at the time, but the water from the wells then contained high levels of arsenic, leached from the underlying rocks. Over a period of time, this has resulted in the slow poisoning of many millions of people, including about 40 million in West Bengal and Bangladesh. This arsenic poisoning causes, among other things, disfiguring leprosy-like skin eruptions, particularly on the

hands and feet, which make walking and working both difficult and painful. After many years of continual exposure, cancerous growths also develop.
Once the problem and its cause were realised, the Indian government began to issue chlorination tablets to add to the water. These react with the arsenic, oxidising it to an insoluble form which then combines with iron, also present in the water, making the water safe to drink.

Recently it has been discovered that the powdered dried roots of the water hyacinth plant can rapidly reduce the arsenic levels of contaminated water from 200 micrograms per litre to below the World Health Organization (WHO) guideline of ten micrograms per litre within an hour. It is hoped that a filtration system incorporating the powdered roots can be developed to purify drinking water and irrigation supplies.[26] The water hyacinth plant thrives in arsenic-rich landscapes, such as Bangladesh, and has until now been regarded as a troublesome weed in water courses. Further research into this interesting possible solution continues.

Heavy as lead

Many countries have taken action to reduce lead exposure from environmental sources, by limiting or banning its use in food, paint, petrol and other sources. In Great Britain, the amount of lead present in food is carefully regulated and must be less than one part per million. If the food is for babies and small children, it must contain less than 0.2 parts per million of lead. In 1995, WHO reduced the recommended upper limit for lead in water supplies to ten parts per billion.

Even ordinary household dust can contain lead, although the amount will depend on how near the home is to major roads or industrial sources. The soil contains a certain level of lead, both from its natural mineral content and from airborne particulates deposited on it.

Vegetables and fruit grown near busy roads or in urban areas may contain excess lead, which has been absorbed by the plants. This is now less of a problem, since lead-free petrol has been introduced, and many more diesel-powered cars have become available. But though picking blackberries from hedgerows at the roadside may seem like a good idea, no matter how carefully the fruit has been washed to remove the surface dust, there will still be lead content: lead from the soil, and dissolved in the ground water, is taken up by the roots of plants.

When preparing food, leafy produce should always have the outer leaves removed, other produce should be peeled and then all should be well washed before eating. Most lead passes through the body without being digested and any that is absorbed tends to be deposited in bone, hair and teeth.

In the past, many people accumulated lead, largely as lead phosphate in their bones, from their lead-glazed cooking utensils. The risk was found to be particularly high where acid foods, such as beers and wines, or pickles and preserves, were stored for a long period of time in lead-glazed containers. This could happen when a large batch of a homemade produce was made, and was then stored away for future use. Lead poisoning has occurred with home-brewed cider, moonshine liquor, apple juice and barley water, all of which were stored in earthenware jars and resulted in poisoning due to lead from the glaze being leached into the drink during storage.

Old pewterware dishes, mugs and plates, which were in everyday use centuries ago, have been found to have a very high lead content, as have lead crystal decanters, so any food or drinks stored in them pose a threat. Even today, new cooking, drinking and storage containers have been found to release lead into the food placed in them. The glazes used in the manufacture of pottery and earthenware, especially the handmade variety, may contain lead. Enamelled cookware from Asia, and particularly China, may also have a very high lead content.

Lead was used in the early days of the canning industry to solder the seams of cans. This lead affected the food and resulted in poisoning when the canned food was eaten. In 1848, Sir John Franklin led an expedition to find the fabled North-West sea passage to the Orient. The ship became stuck in pack ice and all 129 members of the expedition were dependent on tinned food, regardless of its lead contamination, for their survival.

In 1988, the bodies of some members from this ill-fated expedition were found. Tests showed high lead levels in their tissues, consistent with acute lead poisoning, which, no doubt, dulled their senses and hastened their demise.

More than 60 years after Franklin's fateful expedition, in World War I (1914-18), trench neuritis was the name given to symptoms suffered by the troops. Trench neuritis was really a sign of lead poisoning, caused by the contaminated canned food the soldiers were eating in the trenches.

Lead-lined containers were used by the Romans when making wine, which must have resulted in it being heavily laced with lead. Even cider makers in Devon used lead pipes to draw their cider from the cask to the pump. The cider, being more acidic than beer, dissolved some of the lead from the pipes, and the resulting drink gave the cider drinkers lead colic. In 1922, some 93 people were affected by lead poisoning of beer in Isleworth in West London, but this was caused by the lead glaze enamel on the iron-brewing vessels.[27]

Lead acetate, known as Sugar of Lead, has been used to sweeten wine for thousands of years. This substance caused a form of colic in those who drank too much, with stomach cramps, constipation, tiredness and anaemia, ultimately leading to insanity and a lingering death. The practice of adding lead to wine to 'improve its flavour and keep quality' continued well into the nineteenth century, a piece of lead shot being deliberately added to each bottle, where it would slowly dissolve.

In the past, many babies and young children accumulated lead simply by chewing painted surfaces. All paint once contained lead, and this included painted nursery furniture, the children's painted toys and, of course, lead soldiers too. All the household woodwork, if painted, was a hazard, and still is, if removed by sanding or a blowtorch. We will return to this topic in the next chapter.

Despite legislation, lead salts continue to be deliberately used to adulterate food. In 1969 there was even a case of both lead, and chromium, poisoning caused by the use of chilli powder that had been deliberately adulterated with lead chromate. Medical literature is littered with hundreds of such cases of lead poisoning.

Problems in Japan

Mercury is a heavy metal, and a cumulative poison, widely distributed in nature. Some fish, plants and plankton have developed the ability to concentrate it, without it harming them, as in the example below. But this concentration moves up the food chain, leading to increasing levels of poisoning of the fish in contaminated areas, which may contain up to ten times as much mercury as the normal level found in fish elsewhere.

During the 1950s, one of the worst examples of environmental mercury pollution occurred in Minamata Bay, Japan. The local chemical company had been discharging mercury compounds into the bay for over 30 years, at a rate of about 100 tons a year. Over the years the mercury had built up, in the form of organomercury compounds, made by the micro-organisms that lived in the sediment of the bay. These then poisoned the fish, which the local people had always eaten. Over the years the villagers slowly became crippled by 'Minamata disease'. This was caused by organomercury poisoning of the central nervous system. Some of those affected eventually died as a result.[28]

In recent years, the Food and Drug Administration (FDA) in the USA has been concerned about the high mercury levels found in tuna sushi, so beloved of New Yorkers. Most of the sushi tested was from bluefin tuna, which generally has higher mercury levels than other species such as yellowfin and albacore tuna. As an occasional treat, it's fine, but eaten every day it could cause neurological symptoms and increase the risk of cardiovascular disease too.

Another metal that has plagued Japan is cadmium, a metal with a number of uses, including batteries. Cadmium is found associated with zinc-containing minerals, and the ore is extracted by mining. But there is always the problem of waste, with spoil tips, which can become environmental hazards.

In Japan, long-term exposure to cadmium resulted in the development of a 'new' disease they call Itai-Itai, meaning 'It hurts'. Coming to light over 25 years ago, the cause was traced to eating rice grown on contaminated land. The rice fields had been watered from a source polluted by waste, containing cadmium, from a zinc mine nearby. Itai-Itai was the result of chronic cadmium exposure, which had weakened bones and joints, making movement both difficult and painful. The rice, which the sufferers had been eating, contained ten times the cadmium level of normal rice. An additional factor to the disease was found to be the low vitamin D levels of the sufferers.[29] vitamin D is essential for the normal turnover of calcium in the bones.

Thallium for tea

Thallium and its salts have been used to kill rodents and insects for many years. Thallium was the poison of choice for a number of murderers, as we shall see in a couple of Australian examples. Being tasteless and colourless, thallium salts are easy to disguise in foodstuffs by those with murder in mind.

Caroline Grills lived in Sydney, Australia. She married Mr Grills in the early years of the twentieth century. Nearly 40 years later, when she was 63 years old, she nursed her stepmother-in-law, Mary Ann Mickelson, through her final illness.

A number of elderly relatives all died in fairly quick succession at about the same time: 87-year-old Christina Mickelson died in 1947; then a family friend, Angeline Thomas, who was also in her eighties, died. However, when the rather younger 60-year-old, John Lundberg, died the following year after his hair had fallen out, and then a short while later, when his relative and Caroline's stepmother-in-law Mary Ann Mickelson died, with similar symptoms, people, not surprisingly, grew suspicious.

The mysterious illness had also begun to affect John Lundberg's widow and her daughter. Their hair started falling out, they both felt very tired and they both had difficulty moving about. Eventually, a suspicious relative contacted the police. The common factor in all four deaths was Caroline, always there helping to care for the victims, and always making them endless cups of tea. The tea was sent for analysis. Thallium was found, and its discovery was just in time to save the lives of Mrs Lundberg and her daughter, although Mrs Lundberg lost her sight as a result of the poisoning. Caroline Grills was tried for the murders and found guilty. She was sentenced to life imprisonment.

In another Australian case of poisoning a few years later, no doubt copying Caroline Grills' 'modus operandi', a woman in New South Wales used thallium to kill off two husbands. This lady used a rat poison, a paste containing two per cent thallium sulphate, to poison her second husband, Mr Fletcher, in 1953. She gave him several doses mixed in his food, causing his hair to fall out, giving him pains in his hands and feet and making him feel nervous and cry. The cause of his condition was not diagnosed, and after 11 days of increasingly severe illness, he died. At the post-mortem, thallium was found in his body tissues. The police then remembered that four years previously the widow's first husband, Mr Butler, had died of similar undiagnosed symptoms. They exhumed Butler's body and found thallium in his tissues as well. Like Caroline Grills, the woman was tried, found guilty and sentenced to life imprisonment.

Thallium is no longer used as a rat poison, but accidental poisoning with thallium has still occurred. In Guyana in 1987, hundreds of people were poisoned and 44 died after drinking contaminated milk. This milk had been produced by some cows that had eaten molasses laced with thallium sulphate. The molasses was poisoned bait intended to kill rats, but unfortunately the cows found it first. They liked the taste and ate it, with fatal consequences to the dairy customers.[30]

Galvanised into action

Fifty years ago, the Ministry of Health issued advice that galvanised iron vessels should not be used for the soaking, cooking or storing of acid fruits.[31] Such containers were in common use until new plastics became available for household purposes in the 1960s. Today, nearly all bowls and buckets are made of polypropylene (polythene) and are safe to use for foodstuffs of every type.

Galvanisation is a process where a layer of zinc is deposited onto steel to protect it from corrosion. The zinc layer on the surface is quickly oxidised by oxygen in the air, and it is this zinc oxide layer that prevents further corrosion of the steel from the air or from water – but not from acidic foods. Corrosion can occur if acidic foodstuffs are left in contact with the galvanised surface, and this can cause zinc poisoning when the food is eaten.

Storage of acidic fruit juices in galvanised containers has caused many mass poisonings, resulting in vomiting, stomach cramps, diarrhoea and fever. Out of eight outbreaks of food poisoning attributed to zinc in the past, seven were found to be due to the use of a galvanised container in which fruit had been soaked, cooked or stored after cooking. The acidity of the fruit, or its juice, had dissolved the zinc, leading to symptoms. Only a short time elapsed between eating the poisoned fruit and the onset of symptoms, which luckily were mild, only nausea and vomiting. Fortunately, the symptoms were followed by a rapid recovery period.

Infamous botulism

Food poisoning is usually a brief, self-limiting form of gastro-enteritis, although more serious outbreaks of food-borne illness have been associated with certain bacteria, such as Salmonella and Escherichia coli, which both produce exotoxins.

Botulism is a very serious form of food poisoning that comes about from the exotoxin produced by the bacterium *Clostridium botulinum.* This is the same toxin as in the anti-wrinkle treatment Botox. Botulism has been responsible for many deaths over the years, from undercooked meat and particularly from canned meat that has not been sterilised properly.

This form of food poisoning was first described in 1895 by Van

Emermengem. It was discovered that the bacterium responsible is an anaerobe, meaning it can live and grow without oxygen. The botulinum toxin it produces is a virulent poison, but it can easily be destroyed by heat, as normally happens during cooking, or during the sterilisation process which follows the canning of the food.

Unfortunately, in the early days of the food canning industry, failures occurred for a variety of reasons. Some were due to incomplete sterilisation, some to poor sealing of tin lids and some to poor soldering of the seam down the side of the can. In the United Kingdom during the years 1882 to 1919, there were no less than 51 outbreaks of food poisoning, and all but one of these cases involved canned meats. In the years 1919 to 1922, there were a further 14 outbreaks in Great Britain, of which ten were caused by tins of canned meat which had been imported from South America. Other outbreaks were traced to potted meat prepared from wild duck in Loch Maree, Scotland, in 1922, and 'vegetable brawn' in London in 1935.[32]

Botulism was also common in the United States and Canada, and there were 84 outbreaks in the years 1906 to 1920, resulting in a total of 206 deaths.[33] These were all traced to tins of fruit or vegetables, where botulinum spores naturally present in the soil were also on the fruit and vegetables. Adequate sterilisation was needed as the spores are highly resistant to heat. If any spores did survive the sterilisation, they would not produce sufficient changes to cause rejection of the tin, in the short time until it was inspected on the production line. However, with the passage of time, eventually such tins could become 'blown' due to germination of the spores and subsequent growth of the bacteria. The contents would then have an offensive odour when the tin was opened.

And finally, typhoid

In 1963 there was a major outbreak of typhoid in Aberdeen, and smaller outbreaks elsewhere in the UK, including the South Shields area on Tyneside and Bedford. These outbreaks resulted in hundreds of people being hospitalised, some for as long as three months, and many deaths. The cause was eventually traced to some South American corned beef. The large tins (the size of catering packs), which each contained six pounds in weight of corned beef, had been produced and canned in Argentina. They were used in shops where customers could buy a few slices at a time of an array of different cooked meats.

The tins of corned beef had been contaminated because of poor sterilisation in the factories in Argentina, and this problem was compounded by the use of the local river (which contained sewage) as the place to cool the tins after sterilisation. Some of the tins had weak seams, which burst during the sterilisation process. The cooling of the tins in the river water thus allowed contamination of the meat by the typhoid bacteria from the sewage.

The infection spread because the slicing machines in the shops and supermarkets became contaminated when slicing the corned beef, passing the bacteria on to other cooked meats that were sliced on the same machine. This cross-contamination caused many more cases of typhoid. Because some patients had not eaten any corned beef at all – only another sliced cooked meat – the investigation of the outbreaks was initially very confusing.

Chapter 11

Household Horrors

NAPOLEON BONAPARTE WAS ONE of Britain's most feared and hated enemies. When he was eventually captured, it was clear that no simple prison could hold the great military leader. Instead he was exiled to the remote island of St Helena. And there, in 1821, he was killed. Not by his guards, or by an avenging British military, but by his wallpaper.

Arsenic again

Arsenic is one of the most widely known deadly poisons, yet in the nineteenth century it was an everyday ingredient of many popular domestic products – including dyes, paints, fabrics, wallpapers and flypapers. Only in the 1860s did the medical profession cotton on to the harmful effects of arsenic in domestic products, and it took almost another 100 years for the full impact of arsenic around the home to be realised and start being properly assessed.

The cause of Napoleon's demise was only formally identified in the late twentieth century when modern analytical techniques were used to identify high levels of arsenic in a hair sample and in the dye used for the green pattern on the wallpaper, which was found to contain an arsenic-based pigment.

But Napoleon's fate was by no means unique. Accidental arsenic poisoning in the nineteenth century was easily possible. Chemical dyes were the height of fashion and wonderfully bright colours such as Scheele's Green and Paris Green (both copper arsenates) were used to print wallpaper. With no proper heating, houses were frequently damp and this meant the wallpaper became mouldy. The organisms that caused this mould thrived in these surroundings, living off the plaster, the wallpaper paste (a mix of flour and water) and even the wallpaper itself to help them multiply. In the process, they produced trimethylarsine gas, only in tiny amounts, but when people breathed it in over a period of time, this led to chronic arsenic poisoning.

As late as 1912, a report in the *British Medical Journal* mentioned the use of arsenic in a carpet dye, which had caused a poisoning in Germany. Although by then such use was illegal[1], carpets containing these dyes would remain on many a floor, perhaps for decades. Patients complained of chronic diarrhoea, which stopped when they left their home, only to come back when they returned. Again, these cases of arsenic poisoning were probably caused by the gas produced by mould growth in the carpet, as had happened with the wallpaper. It was not until the 1930s that the mould theory was finally proved to be the correct explanation.

Because wall coverings and carpets often stayed in place for so long, such problems could lie dormant for years. In the 1950s, Clare Booth Luce, the American Ambassador in Rome, was diagnosed with chronic arsenical poisoning. Thinking this might be an evil plot by communist agents, the CIA investigated. After much careful searching, eventually the cause was found – arsenic dust from the flaking paint of her bedroom ceiling in the United States Embassy. This dust had been falling over her clothing, the furniture and her food for several years.[2] And in the room above, the CIA found the culprit: a washing machine. Its vibrations had caused the release of the poisonous paint and dust.

Even though arsenic's use around the home was limited at the beginning of the twentieth century, it was still commonly found in other products. [3] Many liquid weedkillers at that time contained high arsenic content. Some manufacturers added a dye to colour the weedkiller as a safety measure, but there was no legal requirement to do so. Many wood preservatives also contained the poisonous sodium arsenite, but these could not be coloured as the dye would then have stained the wood being treated.

Hammond's Vermin Remedy, available in the early years of the twentieth century, was regarded as a poison because of the large quantity of arsenic it contained. Purchasers could only obtain this particular product from pharmacies, where all sales were strictly recorded.

One of the most notorious uses of arsenic was in flypaper. A number of murders were committed using arsenic extracted from flypaper. It was so very easy to do – by simply soaking the flypaper in water, the arsenic dissolved into the liquid. Such was the concern after the Seddon case in 1912 (described in more detail in Chapter 2) that the law was eventually changed. All flypapers containing arsenic were brought within the Poisons Rules, but not until 26th February, 1925.

As recently as the 1970s, there was a report of 11 cases of arsenic poisoning from drinking well water in America. The water was contaminated by grasshopper bait, which had been buried in the soil near the well. The arsenic in the bait dissolved in the groundwater and then seeped into the well.[4] Even today, millions of people in the Third World are suffering from arsenic poisoning, as a result of contaminated drinking water, as mentioned in the previous chapter.

Lead or dead?

Lead is another major culprit in accidental poisoning. Although serious lead poisoning is now rare in the UK there is still a lot of lead around the home. Since Roman times, lead has been used for water pipes, because it is so soft and easily worked, even at room temperature. In the past, water from peaty areas of the UK, which is acidic, dissolved the lead in water pipes and storage tanks and caused frequent cases of lead poisoning. When this problem was identified, householders were told to run the tap for several minutes before using any water for drinking or cooking. This was so water that had been lying in the pipes for a while would be flushed away, as it would have a much higher lead content. This acidic water needed to be treated with lime at the waterworks to harden it, as hard water didn't dissolve the lead.

Even today, where copper piping is present, hot tap water should never be used for food preparation or drinking. This is because the hot water can dissolve some of the lead from the solder that joins the copper pipes together throughout the system. It is some years since the use of lead solder was abandoned in modern central heating and plumbing systems. However, lead solder is still common in many older homes.

Metallic lead is not usually a hazard, although in the past children have been poisoned by swallowing lead shot, or even fishing or curtain weights, which also used to be made of lead. Recently a children's comic was prosecuted by trading standards because yellow pencils given away free with the magazine contained three times the permitted level of lead in the paint on them.

Paints and varnishes used to contain a lot of lead, but EU legislation now requires that lead paints must carry a warning: paint with only one per cent lead has to carry a hazard warning and such is the danger that even those paints containing as little as half a per cent lead must carry the warning 'Not to be applied on surfaces likely to be sucked or chewed by children'.

Old, flaking paint can contain lead, or may expose previous layers containing high levels. Care must be taken when removing any old paint, which should never be removed by dry sanding, rubbing down or burning off with a blow-lamp as these methods will release lead dust, which could contaminate food and drink or be inhaled. Paint manufacturers have produced safety information booklets about these dangers and the way to treat them. They recommend that a liquid paint stripper be used, followed by wet sanding.[5]

Danger – strippers at work

The downside of liquid paint strippers is that they are very toxic themselves. A substance known as methylene chloride is widely used as an ingredient of liquid and gel paint strippers. Exposure to the fumes produced by these products has caused poisoning, with symptoms such as skin rash, swelling, mental impairment, temporary diabetes, mellitus and even death.[6] It is essential that these products are only used in well-ventilated areas.

In normal circumstances, accidental and even intentional contact with paint strippers can have dire results. One medic reported his own experience after accidental exposure to the substance, listing symptoms including vomiting, nausea, anorexia, lassitude, urinary frequency and mental impairment lasting up to 24 hours.[7]

Hard to swallow

One of the most incredible facts about domestic dangers is the number of poisons that were intended for human consumption. In Victorian times a product called 'Condy's Disinfecting Liquid' was sold in pharmacies. It was 'flavoured' with lavender oil to prevent it being drunk by mistake. It also contained potassium permanganate, an oxidising agent. Drinking this liquid would have been highly dangerous, leading to nausea and vomiting, corrosion of the alimentary tract and oedema, with both liver and kidney damage and eventually death, up to a month later. Amazingly, this product was used, well diluted with water, as a gargle for sore throats as well as for bathing wounds.

Other similar products such as Condy's Red Fluid (containing sodium permanganate) and Condy's Green Fluid (containing sodium manganate) were widely sold as cheap disinfectants in the early twentieth century.

Potassium chlorate was also used in toothpaste at the start of the twentieth century. It claimed to treat spongy gums and prevent the build-up of tartar. It was also used as a general antiseptic and was included in other dental products, such as mouthwashes and gargles. Kidney and liver damage were a known hazard of overuse.[8] Potassium chlorate will appear yet again in the chapter about the Noxious Nursery.

Stannous fluoride, like sodium fluoride, has been used as an aqueous solution for application to the teeth to prevent decay. Toothpastes containing stannous fluoride are also used, although staining of the teeth can occur.

In 1976, in the USA, a three-year-old child died about three hours after drinking a solution of stannous fluoride, which had been intended only as a mouthwash to allow its application to the child's teeth. Unfortunately, the child swallowed the solution instead of spitting it out. Within five minutes the child vomited, then developed a convulsive seizure and later went into shock; cardiac and respiratory arrest followed and the unfortunate child died.[9]

Boron the deadly

Boric acid has been used for centuries for medicinal, cosmetic and household purposes. Borax, the sodium salt of boric acid, has been similarly used, and has also been used as an astringent and an emulsifier in creams. If applied to raw and weeping skin there can be sufficient absorption of boric acid to cause boron poisoning. In the UK, the concentration of boric acid was restricted to five per cent in talcum powder and only 0.5 per cent in products for oral hygiene.

These talcum powder products have now been completely banned from use on children under three years old as in the past infants have died of boron poisoning (see the chapter on the Noxious Nursery).

The last prescription product for oral hygiene containing boron was Bocasan, containing sodium perborate. As sachets of powder to be mixed with water for use as a mouthwash, this product carried warnings that it was not to be used for longer than seven days, due to the risk of boron poisoning, and that it was not suitable for those with renal impairment or for children under five. This product too has now been discontinued.

Deaths have also resulted in the past from absorption following the washing out of body cavities with solutions of boric acid. In the UK the use of boric acid in cosmetics and toiletries is now severely restricted.

Descaling products

Formic acid and both its sodium and calcium salts are safely used, in very dilute solutions, as food preservatives. Formic acid was widely used for many years in various industries, including textile dyeing and leather tanning. It has also found a household use in descaling products, removing the build-up of lime scale from domestic kettles. Strong solutions, containing up to 60 per cent formic acid, have been used in some of these products. Chemically, formic acid is similar to acetic acid, but it is far more irritating and pungent.

A number of poisonings have resulted from people drinking descaling products containing formic acid. Such poisoning is usually fatal, but not until several days later. Those who survive are permanently scarred internally, due to the acid's corrosive effects. Major complications include damage to the throat, oesophagus and stomach. Some survivors sustain so much damage to their mouth, throat and oesophagus that they need an operation to create an opening from the outside directly into the stomach to allow them to be fed through a tube.

Other symptoms of this type of poisoning include problems with breathing, circulation, blood clotting and kidney failure. In 1980, there was a report of three patients who had each swallowed kettle descaler, containing between 40 and 55 per cent formic acid. All three eventually died, although not until between five to 14 days after admission to hospital.[10] Today, descaling products contain citric acid, which is far less harmful.

Unnatural gas

Gas is another domestic killer. Until the arrival of natural gas in the 1970s, all gas supplies in the UK were produced from coal. Gas mantles were used in incandescent lamps, to light the home, before the advent of electricity. Coal gas was used in these lights and also later in ovens and in gas fires.

Coal gas was a mixture of a number of gases, including carbon monoxide. Carbon monoxide can also be formed during the incomplete combustion of natural gas and some other fuels. It is highly toxic when inhaled, especially by babies, small children and the elderly. Even today this gas is still a leading cause of death in house fires.

In the past, when coal gas was still in use, many suicides were due to people quite literally 'putting their head in the oven'. They turned the gas supply on, without lighting the oven, and breathed in the gas, soon afterwards dying from carbon monoxide poisoning.

Some such cases appear in Chapter 16.

Gas played a key part in one of the most notorious mass murders of the mid-twentieth century. John Reginald Halliday Christie, born in Yorkshire in 1898 and hanged in 1953, confessed to the murder, by strangulation, of six women including Mrs Evans, the wife of Timothy Evans, who had lived in the same house as Christie.
Evans had been convicted and hanged for the murder of his baby daughter in 1950. Evans had also been charged with the murder of his wife at the same time, but the case never came to court. Following a special inquiry by the Home Office and several debates in Parliament, no definite conclusion was reached. However, the view that Evans was innocent and that Christie had killed both mother and child increasingly came to the fore.
Carbon monoxide was found in the blood of three of Christie's victims. At his trial, he explained that he got the women partially drunk, then encouraged them to sit in a deck chair with a sun canopy above it. A rubber pipe carried coal gas from the domestic supply to the canopy and made the women unconscious. Christie then strangled and raped them.

Although it isn't poisonous, natural gas still has the potential to be deadly because of this carbon monoxide risk. Many deaths occur each year, caused by poor maintenance of appliances and lack of ventilation. This can lead to incomplete combustion of the natural methane gas in appliances and central heating boilers, producing deadly carbon monoxide.

Natural gas is both colourless and odourless, which means that it could be dangerously undetectable, so any leakage could lead to an explosion without warning. For this reason, a particularly smelly substance, a chemical called a mercaptan, is added to natural gas to act as a warning. Methyl mercaptan is said to be the world's smelliest molecule, with the nose able to detect as little as 0.02 micrograms per litre of air.

Dead clean

For many years, toilet cleaning blocks and mothballs have used a substance called naphthalene. Children who swallowed mothballs, thinking they were sweets, developed haemolytic anaemia, in which the red blood cells are destroyed.[11] Fortunately, the children recovered from their 'mothball anaemia' after a blood transfusion. There was even a report of blood problems in a newborn baby because of such poisoning. The child's mother gave in to cravings for the smell of naphthalene when she was 28 weeks pregnant.[12] Some women get the strangest cravings, but shouldn't always give in: it was found that a dose as low as 2g of naphthalene can be fatal to a small child.

Household bleach, available in a variety of strengths, is sodium hypochlorite solution. If swallowed, bleach reacts with our stomach acid with dire results, releasing fumes of hypochlorous acid and causing irritation and corrosion of the mucous membranes of the mouth, the oesophagus and the stomach with, needless to say, much pain and vomiting. The acid fumes cause coughing and choking, and may also cause severe respiratory tract irritation and even pulmonary oedema.

An 18-year-old swimming pool attendant who added 16 per cent sodium hypochlorite solution to swimming pool water every day found that her fingernails began to drop off after several weeks. They grew back normally when she stopped using the bleach, but the problem returned the following year when she started using it again.[13] Another young girl who had the strange habit of sucking socks which had been bleached suffered episodes of vomiting, abdominal pain and bronchopneumonia over some years. This was eventually traced to her curious habit, although it took a lot of detective work to discover the bizarre cause in this case.[14]

It is vital to remember that different types of cleaning products, such as toilet cleansers and bleaches, should not be mixed or used together, as this can produce poisonous chlorine gas.

Ammonia is yet another chemical that was used for over a century for general cleaning purposes. It should be handled with great care, as exposure to the concentrated ammonia vapour may cause injury to the eyes, and inflammation of the respiratory tract or even spasm of the throat with resultant asphyxia. It should *never* be mixed with bleach, as this produces chloramine gas, which is also very toxic.

Many cases of poisoning due to exposure to chlorine and chloramine gases have resulted from the accidental mixing of incompatible cleaning fluids.

Always read the label, which lists the active ingredients and explains the dos and don'ts of that specific product .

More killers in the home

If arsenic is well known as a deadly substance, then cyanide is positively world famous. But did that stop it being used as a household cleaning product? Not a bit. In 1934, the Ministry of Health gave the following advice on a common method of eradicating bedbugs: 'In ridding a house of bed-bugs, it cannot be too strongly emphasised that fumigation by hydrogen cyanide is a dangerous process and should be undertaken only by responsible persons with full knowledge of the nature and properties of the gas, and who are skilled in the use of gas masks and oxygen breathing apparatus. In order to minimise risk, a lachrymatory gas (which makes your eyes water) may be mixed with the hydrogen cyanide to act as an indicator. Failure to detect the lachrymatory gas, however, can of itself be accepted neither as indicating the absence of hydrogen cyanide nor that ventilation is complete.' In other words, 'You have been warned.'

Nitrobenzene was widely used in the manufacture of polishes, perfumes and soaps in times gone by. Nitrobenzene is a highly poisonous volatile liquid and as little as 1g can be fatal. Poisoning can result by drinking the liquid, inhalation of the fumes or absorption through the skin. The toxic effects are usually delayed for several hours and the symptoms include nausea, prostration, burning headache, cyanosis, haemolytic anaemia, vomiting and convulsions, leading to coma and death. Bootblack containing nitrobenzene caused a case of poisoning in 1904[15], and a further case was reported in 1912, when some nitrobenzene was placed on cotton wool to ease, of all things, an aching tooth. [16]

Phenol, or carbolic acid as granny would have called it, was widely used as a disinfectant in the home, and in hospitals, as soap and a liquid. Bars of carbolic soap were also placed in drawers and cupboards of clothing to prevent moth attacks. But phenol is a highly corrosive substance, especially in its concentrated form, and can cause nasty burns to the skin. Any such burns should be treated by application of glycerol.

In the 1920s, a man suffered extensive burns to his hip, thigh and scrotum after a bottle of phenol broke in his pocket during a train journey. He eventually died of his burns.[17] Some years later a hospital patient died after liquid phenol, which is very concentrated, was applied by mistake, instead of a much weaker diluted solution, as a dressing of an arm wound.[18] And a woman who had regularly used phenol for cleaning purposes, without wearing protective gloves, went to the doctor because her urine had turned lime green. Investigation showed that this strange colouration was caused by a phenol derivative. No doubt she wore gloves after that.[19]

Turpentine was used in Victorian households for many cleaning purposes and as an ingredient of homemade furniture polish. Today it is still used as a thinner in paints and varnishes and as a solvent. In the past, people did not realise that it could be very poisonous. In the 1930s, a man died after drinking about 170ml of turpentine. At the post-mortem the lining of his stomach was found to be in shreds, lying in small pieces in the gastric cavity. The wall of the stomach itself felt like leather.[20] In homes where the floors were regularly cleaned with turpentine, the occupants sometimes developed poisoning by inhalation and absorption through the skin, which resulted in blood abnormalities and petechiae (red/purple flat pinhead spots) all over the body and in the mouth.[21]

Carbon tetrachloride was used as a 'dry cleaning fluid' both commercially and in the home for many years. It is only in the last few decades that its poisonous nature, especially to the liver and kidneys, has led to its replacement by less toxic substances. Both the liquid and its vapour are poisonous. Should the vapour come into contact with a flame the highly poisonous phosgene gas is produced.

Half a century ago a little girl, only four years old, who had been seen awake in bed at 9.20 p.m., was found to be in a comatose state beneath the bedclothes when checked on at 11 p.m. There was an empty bottle of 'Thawpit' dry cleaning fluid beside her in the bed. The skin of her face was found to be severely blistered, and she died shortly afterwards. The post-mortem showed congestion of the lungs, with kidney, spleen and heart damage plus fatty changes to the liver. The cause of her death was found to be inhalation of the vapour.[22]

Other cases of poisoning at the time included a naval officer who drank some carbon tetrachloride in mistake for gin. He refused to have any medical treatment and subsequently died five days later.[23] Another man, aged 53, drank about 50ml of carbon tetrachloride while under the influence of alcohol. His kidneys were badly damaged as a result and he died some 24 days later.[24] A child of three who swallowed about a thimbleful of carbon tetrachloride fortunately recovered after treatment.[25] Someone had stupidly stored the carbon tetrachloride in an old orange juice bottle, from which the child drank.

Paraffin, also known as kerosene, has long been used as a fuel oil for stoves and lamps, as a solvent in insecticide sprays and as a cleaner and degreaser both in the home and in industry. It was sold in a number of different grades, dependent on the use. Inhaling or drinking it leads to stomach upset, cough, heart problems and depression of the central nervous system. The main problem is the development of 'kerosene pneumonia', which comes either from breathing paraffin in or absorbing it through the bloodstream.

Many cases of paraffin poisoning in children have been recorded over the years, such as a 12-year-old boy, whose clothes became soaked in kerosene, and who developed epidermal necrolysis, a severe type of rash where the top layers of skin peel off, leaving large, red raw areas, which are then very susceptible to infection.[26]

It may seem unbelievable today, but in 1915 the *British Medical Journal* carried a report of the successful treatment of a case of diphtheria, in which kerosene was administered, with the taste masked by the use of sarsaparilla. A dose was given every four hours for three days. At that time, doctors wondered if kerosene would become another useful drug, but fortunately for us all, newer and more effective medicines were developed instead.[27]

Other poisons found in the home included metallic oxalates, used in laundry blue bags and in polishes, and oxalic acid, used to remove ink stains and iron mould, as well as for cleaning leather and removing the colour from calico printing. Salts of lemon, also known as Salt of Sorrel and chemically as potassium tetraoxalate, was used to remove rust and ink spots and also as an ingredient of metal polishes. Modern cleaning products are a great deal safer than those used by our grandparents.

An angel of death

As we have seen, Victorian households contained many hazardous chemicals used as cleaning products, so it's hardly surprising that some were used for more sinister means. Catherine Wilson was a nurse who befriended sick people with some wealth, and then tried to get them to change their will in her favour. Then she set about feeding them with a variety of poisons. For a while she lived with a Mr Dixon, but he began to drink heavily so she poisoned him too.

In 1862 she was caring for Sarah Carnell, who had already rewritten her will in favour of Catherine. One day the nurse brought her patient 'a soothing draught'. Mrs Carnell took a mouthful, but immediately spat it out, saying it had burnt her mouth. She called for her husband, who saw that there was a hole burnt in the carpet where the fluid had landed. Nurse Catherine fled but was arrested several days later. The mixture she had given to Mrs Carnell contained enough sulphuric acid to kill as many as 50 people. She was charged with attempted murder, but was cleared as her defence had been that the pharmacist had given her the wrong bottle. No one could be sure, so the charge was dropped. On her release, however, she was promptly re-arrested. Word about this case had spread and resulted in post-mortems being carried out on some of her other patients. Seven bodies were exhumed and a variety of poisons found in them. So Nurse Catherine was tried yet again and this time was found guilty. She was hanged outside the Old Bailey in October 1862, and it is said that some 20,000 people attended to watch.

And finally ...

The unfortunate tale of Sidney Smith, a British clergyman who became the Canon of St Paul's Cathedral in London. He was a renowned essayist and wit of his day. He died in 1845 after mistakenly drinking some ink. When his wife told him what he had done, his last words were claimed to be, 'Bring me all the blotting paper in the house'.

Chapter 12

Murderous Make-Up

COSMETICS TODAY ARE A far cry from those used in ancient times. Application of today's powders and potions will no longer poison you, let alone kill you, as they once could. The simplest cosmetics are powders, dusted on to the skin. Many others are pastes or creams. Clays, ochre and charcoal, together with various oils, waxes and greases, have all been common ingredients of face and body paints for centuries. Any substance applied to the skin to improve the appearance, beautify or make a person more attractive, can be called a cosmetic.

Henna and tadpole hair dye

Egypt has given us the earliest evidence of the use of cosmetics. Tombs contained not only mummies, but also many burial gifts – comforts and luxuries, including cosmetics, provided to accompany the person buried into the next world.

These gifts included kohl, made from stibnite, a poisonous black mineral containing antimony, used to decorate the eyes and paint the eyebrows. Various eyeshadows of mineral origin were found, and the plant dye henna, still used today, was used not only as hair dye, but also for colouring the palms of the hands and soles of the feet and decorating the finger and toenails. Ancient Egyptian ladies even used mud from the River Nile as a facemask.

Today's perfumes are alcohol-based, but in Ancient Egypt the art of distillation had yet to be discovered, so perfumes were always oil based. One sacred perfume used in the embalming process was made from myrrh, cinnamon, cyprinum, juniper, honey and raisins steeped in wine. Many fragrant oils were used as body rubs and skin softeners,and to add to bathwater. The Ancient Egyptians even used crushed, dried tadpoles in oil as a hair dye.

Walnut hair restorer and powdered pumice tooth cleaner

By comparison, the women of Ancient Greece tended to use few cosmetics, and then only to colour the lips and cheeks. But we also know that they used hair dyes of both mineral or plant origin to change their hair colour.

The Romans were also very keen on hair dyes, as well as on bleaches, using recipes culled from far-flung outposts of their empire, such as Gaul and Germany. The green outer covering of walnuts, for example, was used to restore colour to greying hair. Another recipe included black wine, raw crow's egg and putrefied leeches.

As a result of all this dyeing and bleaching, the Romans found that their hair became very coarse and brassy, and so, to improve its lustre and texture, they also used conditioning creams. Many of these preparations contained strange ingredients; for example, one contained pepper and hellebore together with rats' heads and droppings. If all else failed, the Romans resorted to wearing wigs.

The removal of unwanted body hair was also very important in the Roman society. This was done by covering the area with a paste made from a mixture of lime and orpiment, that poisonous arsenic sulphide mentioned in Chapter 7. Or more safely – but more painfully – by rubbing the skin with powdered pumice. The pumice stone in your bathroom is a piece of volcanic lava that has cooled and solidified. The Romans would have had a plentiful supply of pumice from various volcanoes such as Mount Vesuvius, south of Rome, and Mount Etna, on the island of Sicily.

The Romans, like the Egyptians, also used facemasks, only theirs contained not Nile mud but powdered plants mixed with honey or lanolin, and were applied at night and then washed off the next morning using scented water. They also used powdered pumice in vinegar to clean their teeth – rather rough on the enamel though.

Kohl, carmine and woad

On their faces, Arabic women used white batikha, a powder made from ground shells mixed with borax, rice, lemon and eggs, and kohl eye make-up made of soot, lead ore, burnt copper, rosewater and sandalwood.

Indian women also used kohl – made from the poisonous stibnite, or the somewhat safer lampblack – mixed with oil for their eye make-up. They also used the poisonous vermilion to stain their lips, and the safer henna, not only to stain the soles of their feet like the Egyptians, but also to paint intricate designs on their hands and arms.

The Chinese and Japanese have used a lot of cosmetics in the past. To produce the traditional white face and bright red lips of the geisha they used a heavy layer of rice powder to dust the face and carmine to paint the lips, cheeks and nostrils.

In Britain, the ancient tribes used woad, a blue plant dye, both as a body paint and for tattooing. The Romans brought their own cosmetics with them when they invaded Britain and some of their cosmetics were duly adopted by the Celts. After the Romans left Britain in the fourth century, the idea of personal adornment came to be frowned upon, mainly due to the spread of Christianity.

And so, for several centuries, the Anglo-Saxons and later the Normans used very few cosmetics, although they did use a wide range of herbal lotions and mixtures for medicinal purposes. Then, during the Middle Ages, knights returning from the Crusades brought with them the knowledge of many new things, including cosmetics and perfumes, that they had seen both on their travels and while living in the Middle East.

Slowly, over the centuries, the fear of the Church and its teachings that vanity was evil began to give way to an increased use of cosmetics. The herbal preparations which had until then been used only medicinally now began to be used cosmetically as face creams and body lotions.

Elizabeth I – The Virgin Queen

Elizabeth Tudor used many cosmetics. A white face was considered to be very fashionable in her time and continued to be so for several centuries, and so lemon juice and rosewater were both used to bleach the skin. A pale complexion and white face demonstrated that you did not need to work; exposure to the sun while working in the fields produced the ruddy complexion of the farm labourer.

A variety of ingredients were used as face powders to produce the required fashionable whiteness. Ceruse (their name for white lead), a mixture of sulphur and borax powdered with finely ground alabaster or even perfumed starch set with a film of egg white were all used as face powders. Unfortunately once the egg white dried, it tended to crack. And the ceruse also turned grey as the day progressed.

Thick white face paint was much needed at that time to hide the ravages of smallpox, which in those days most people caught at some time in their lives.

The cheeks were rouged using the same white lead, but this substance was now coloured with red ochre or red mercuric sulphide. The lips were coloured by a salve made from cochineal mixed with gum arabic, fig milk and egg white.

Needless to say, some people died of lead poisoning just from their use of face powder or rouge.

Belladonna drops were put in the eyes to dilate the pupils and so gave ladies that wide-eyed and innocent look. Ladies in search of a husband may well have appeared more alluring as a result, but in fact using these eye drops meant that they could not focus properly and so had only a hazy view of their suitors, who may well have been pockmarked, old or just plain ugly. The eyes were lined with kohl, probably made yet again from the poisonous stibnite.

The Queen had naturally red hair, which in the past had been regarded as associated with witches and the devil. But now red or even golden hair became the absolute height of fashion. Red hair was achieved with a henna dye, while for golden hair turmeric, rhubarb steeped in wine or a rinse of saffron water was recommended. Today, people worry about a receding hairline, but not in Tudor times. It was the fashion then, as in medieval times, to pluck the eyebrows completely away and to pluck along the hairline to give greater length to the face.

One of the Queen's beauty lotions contained eggs, including the shells, with burnt alum, sugar, borax and poppy seeds, all mixed together with water. Queen Elizabeth never married and never had any children, but those ladies of the court who became pregnant while using these lead and mercury cosmetics were prone to miscarriage or their children to birth defects as a consequence.

In 1724, an Act of Parliament was passed regarding the ingredients used in cosmetics. As a result many cosmetics became less harmful and safer to use. However, many people still preferred to use the old lead-based cosmetics as they gave much better coverage of pockmarks and other blemishes.

Even two centuries later in the 1920s, cases of lead poisoning due to the use of face enamel with a very high lead content were still being reported.[1] In 1936, a serious case of lead poisoning was reported in the *British Medical Journal* concerning an actor whose theatrical grease paint was found to contain nearly 40 per cent lead, present as lead oxide.[2]

Pink disease and calomel cream

Acrodynia, commonly called 'pink disease', was not just a problem for babies and small children, as mentioned earlier. It also resulted from adults taking calomel powders (calomel is mercury chloride) as a purgative. It was called 'pink disease' because of the unnaturally bright pink colour of the cheeks, nose, buttocks, hands and feet from the mercury poisoning.

In South America, adults regarded the bright pink colouration in a positive light, and so calomel powders and creams, marketed as beauty aids, were still on sale until quite recently. They were used by many people in Mexico as well as in some of the southern states of the USA. One such product called Crema de Belleza Manning, which contained up to ten per cent calomel, caused mercury poisoning to such an extent in some people that they required clinical treatment. This cream was widely advertised by its manufacturers for 'skin cleansing and prevention of acne'!

Walnut juice and other wonderful hair dyes

Henna has been used for thousands of years as a hair dye. About a hundred years ago, it was also used in conjunction with indigo. A paste of one part henna and two parts indigo was left on the hair for varying lengths of time, according to the shade required: one hour for light brown, one and a half hours for darker brown, etc. It was a safe and effective natural hair dye, but some people are never satisfied and always want something better.

Potassium permanganate is an oxidising agent that occurs as tiny blackish-green crystals. A single crystal dropped into a glass of water dissolves to produce a beautiful purple-coloured fluid. But if you get it on your skin or on your clothes it will stain them brown. Potassium permanganate has been used in the past to dye white hair a chestnut brown colour and to produce an artificial tan. But, even 80 years ago, these uses were considered dangerous and inadvisable. Potassium permanganate is still used today, but for its antiseptic and astringent rather than cosmetic effects. It is used in a very dilute aqueous solution of one part in 10,000 for bathing areas of weeping eczema, to prevent infection of the weeping areas and also to help it dry up.

Silver nitrate is another chemical that stains the skin. This was the main problem with its use as a hair dye. In the late 19th and early 20th centuries, silver nitrate lotions of up to four per cent strength were used to dye eyebrows and eyelashes, and when combined with potassium sulphate could achieve different shades of brown or black. Combined with pyrogallic acid and ammonia these lotions could dye grey hair to jet black in a single application. But another problem with silver nitrate was the expense, so hairdressers looked for a cheaper alternative. They found it in a chemical called paraphenylenediamine.

Mrs Potter's Walnut Juice Hair Dye was one of a number of these cheaper products that contained this substance as the active ingredient. There was not a drop of walnut juice in the product – the brand name was chosen simply as it alluded to the traditional use of walnut bark, which had been used since ancient times to dye wool a dark brown colour.

Paraphenylenediamine was present as an ingredient in hair dyes for at least a century. It was well known to doctors for causing dermatological and other problems even a hundred years ago. The use of soap, in an attempt to wash out the dye, aggravated the dermatitis that this type of hair dye caused. Many people were allergic to it.

Some ladies were so badly affected by the poisoning from the use of this type of dye that the only sensible course of treatment, although somewhat drastic, was to shave the head. The symptoms they experienced following use of these hair dyes included, initially, swelling of the head, neck, tongue, eyelids and face, followed by skin eruptions, eczema, nausea, nervous symptoms, sleeplessness, dizziness, weakness and even impairment of vision, sometimes leading to permanent blindness.

In 1922, Inecto Ltd issued a warning on the use of one of its products, which contained paraphenylenediamine and resorcin. However, this warning was insufficient to protect the company in a legal action brought forward following severe adverse reactions. Damages of £200 were awarded, a large sum of money in those days. It was claimed at that time that many hairdressers were using these hair dyes containing the problem ingredient paraphenylenediamine because they were so much cheaper than those containing silver nitrate.[3]

Many hair dyes were very smelly to use, but in the 1920s a new odourless type, the copper pyro hair dyes, were introduced. With these, copper chloride and pyrogallol gave light brown, and the addition of ferric chloride gave dark brown. Or, by varying the quantities of the different ingredients used, even a natural looking black was possible.

Amidol hair dye, which contained different chemicals, was considered to be the best black hair dye available at the time and did not stain the skin, as silver nitrate did. Shorter application times with this dye gave various shades of brown, and several applications would be required to achieve a true black.

Lead acetate and sodium thiosulphate were also used as hair dyes and were included in products marketed as hair colour restorers, such as Grecian 2000 and Restoria, which are still on sale to this day. Apart from the henna mentioned at the beginning of this section, all these hair dyes are poisonous to a greater or lesser extent.

The importance of a patch test

The poisonous nature of hair dyes was well known even in the early decades of the 20th century, and deaths were reported from their use. In *The Lancet* in 1934, there was a report about a case of fatal systemic poisoning of a young hairdressers' assistant. The young lady was only 21 years old and death was found to be due to liver damage caused by contact during her work with the chemicals in hair dyes and permanent wave solutions.[4]

In spite of the known risks, paraphenylenediamine continued to be used as a hair dye because it was so easily applied, the colour was very permanent and it was capable of producing a range of natural looking shades. The hair dyed with it could also be permanently waved without difficulty, whereas this was not the case with those products containing metallic dyes, such as the silver nitrate and copper pyro dyes.

In 1937, the Medical Officer for the Ministry of Health issued a report in which it was stated that these substances should never be used without a previous patch test being done to determine any abnormal susceptibility to the ingredients. It is now recognised that about four per cent of apparently normal, healthy subjects are sensitive to paraphenylenediamine and that one per cent are acutely sensitive.

There have been a number of studies that have linked certain cancers to the use of hair dyes. Such findings have always been vigorously refuted by the manufacturers. Another risk was accidental ingestion, which caused vomiting, massive swelling, gastritis, increase in blood pressure, dizziness, tremor, convulsions, coma, cardiac and respiratory difficulties, sometimes needing treatment with an emergency tracheotomy. When used as hair dyes, the concentrations of the chemical phenylenediamine and its derivatives are now restricted by the Cosmetic Products Regulations, 1978.

Many hair dyes still contain phenylenediamine even today, but in much lower concentrations than used in the past, together with the addition of many other ingredients.

Hair dyes can now be safely used, providing that a patch test has been successfully completed before use.

Women still suffer adverse reactions to hair dyes, usually because they could not be bothered to take the time to perform the patch test before dyeing their hair.

Deathly depilatories

Depilatories, or hair removers, have been used for many thousands of years for purification, cleansing or cosmetic reasons. A variety of substances, mainly poisonous, have been used for this purpose. As already mentioned, the Romans were very keen on the removal of body hair, and they favoured the use of orpiment – that is, arsenic sulphide.

In more modern times, other less toxic sulphides have been used as hair removers, including the smelly sulphides of barium, sodium and calcium. Barium sulphide was used as a fine powder, which was first mixed with wheat starch, then made into a cream with water, and spread on the part to be treated. It had to be left for five to ten minutes before being removed with a blunt knife or spatula. This cream could not be used too frequently, because it caused dermatitis.

A patented German preparation available in the 1930s contained strontium sulphide mixed with corn starch, talc, dextrin, nerolin and essential oil. The dextrin in this preparation was claimed to protect the skin and hair follicles and prevent the bad odour associated with sulphides during use. [5]

In the 1950s, it was suggested that the X-ray method of depilation[6], in use for the 40 previous years, was the method of choice. In the 1950s, X-rays were very popular and were used in a number of imaginative ways, until the dangers of excessive exposure were recognised, and precautions regarding their use were introduced. There were even special machines in most shoe shops at that time, so that children's feet could be X-rayed and viewed through an eyepiece to see if the shoes allowed sufficient 'growing room'.

Thallium acetate used to be recommended as a hair remover. Various hair removal creams containing thallium were once on sale as cosmetics, but the hair loss was really a symptom of thallium poisoning, caused by absorption of thallium through the skin. Thallium acetate is thankfully no longer used for such purposes.

Instead of thallium, thioglycollates are now the favoured

ingredient for depilatory creams, and they are also used in the active lotion of home permanent wave and hair straightening kits. When used in hair straightening, the thioglycollate is subsequently neutralised by potassium bromate, or a similar but less toxic substance.

Contact dermatitis caused by thioglycollates can occur, with symptoms of swelling, redness, itching, and papular rash, which will all magically disappear when use of the offending substance is discontinued. As with hair dyes:

When permanent waving, patch testing before use is advised. Rubber gloves should always be worn when using these lotions.

This is because barrier creams do not provide acceptable protection. Anyone who finds that they are sensitive to thioglycollates should take great care never to come into contact with them again.[7]

Lipstick and eyeliner

Ladies are frequently surprised to learn that most lipstick is inadvertently swallowed and the more frequently it is applied, the more is swallowed. Lipstick is essentially a stick composed of waxes and oils blended together with dyes and perfumes, all of which must, of course, be perfectly safe when ingested. However, years ago some lipstick imported from Asia was found to contain the poisonous lead carbonate.[8]

Kohl is the traditional eyeliner used in the Middle East, India and Asia. This black substance is traditionally made from a poisonous mineral called stibnite – antimony sulphide. There have been many, many cases throughout the twentieth century of lead poisoning that have been ultimately traced to various cosmetics.

Even today, there have been reports of lead sulphide in eye make-up, which has caused poisoning.[9] Manufactured in India, this eye make-up was imported into Britain. Although such products containing lead are banned in the United Kingdom, eye make-up, such as kohls and surmas, may be imported privately, and some of these contain up to 83 per cent lead.

Today kohl tends to be mainly used for ornamental cosmetic purposes, but traditionally it was used as both a means of spiritual protection and a protection against conjunctivitis. This traditional protective use is unfortunately still continued, even today in the UK, by some misguided Asian parents on their children. They persist in using poisonous kohl and surma products, which they have imported privately from their homeland, when they could use Kajal, which is virtually lead free, and so much safer.

The scourge of skin colour

In addition to changing hair colour, cosmetics can also be used to change skin colour. The desire to lighten or darken skin colour has led to the production of a wide range of cosmetic products. Such changes are not without problems though, and these should be considered carefully before embarking on what may be irreversible and life changing.

A photographic developer called hydroquinone, which is still in use today, was first used in the early years of the 20th century. Those using it soon noticed that, after a period of time, it removed the pigment from skin. Entrepreneurs were quick to spot a new market and soon many preparations of creams, ointments and lotions containing hydroquinone were marketed for the treatment of pigmentation disorders. Many are still available today, particularly in Africa and South America. The use of hydroquinone in cosmetics and toiletries in the UK is now carefully controlled by the Cosmetic Product Regulations.

Hydroquinone reduces skin pigmentation in all races by interfering with the enzyme system involved within the skin. A radiographer who was exposed to a great deal of hydroquinone developed 'darkroom hepatitis' as a result of continuously using the film developer in the darkroom to develop X-rays. And he wasn't even trying to improve his own looks![10]

Coloured people who use skin lightening products probably initially purchased them for cosmetic reasons. Unfortunately, after about six months' use, those who use them regularly start to develop a blue-black hyperpigmentation called ochronosis. This effect occurs even when products containing a sunscreen are used. These people then fall into the 'skin lightener trap' as they then use other hydroquinone preparations to try to remove the disfigurement that has resulted. But the damage has already been done and they are now caught in a vicious circle. Most of them will only experience a partial improvement of their condition, even with the most careful management.

Ammoniated mercury was also included as the active ingredient in some skin lightening creams used in various parts of Africa. This resulted not only in mercury poisoning, but also damage to the kidneys.[11]

Depigmentation can occur naturally in some medical conditions, such as vitiligo, an autoimmune disease most noticeable in dark-skinned races. The late pop star Michael Jackson suffered from vitiligo, although many thought he was a victim of skin-lightening products. He was diagnosed with the condition in 1986 and admitted to it during an interview with Oprah Winfrey in February 1993. Special cosmetic applications can be used to camouflage the affected area, even on the darkest skinned patient. The British Red Cross has a specialist service to teach patients how to use camouflage make-up successfully.

Faking a tan

For the lighter skinned who wish to appear suntanned, products containing colourants are available. The orange food colourant canthaxanthin is widely used today in fish farming. It is given to salmon and trout in their food to colour their flesh, as this improves the price that the fish can be sold for at the fish market. In the past this food colourant was sold for humans as Orobronze capsules, which were taken by mouth as a means of producing an artificial suntan.

The main problem with the use of this 'cosmetic' product was that it stained not just the skin, but the underlying flesh as well. This led to problems for doctors – and surgeons in particular – because if they needed to operate on a patient who was taking these capsules, the flesh appeared to be an abnormal colour and masked the normal colouration that would guide the surgeon as to the state of the patient while on the operating table.

Canthaxanthin was also used medicinally for a short time in the management of the medical condition erythropoietic protoporphyria, a metabolic disorder where the skin needs special protection from sunburn. Unfortunately, this use led to problems with the vision of such patients.[12] There was even a report about someone who took canthaxanthin to produce an artificial tan and who subsequently died of aplastic anaemia as a consequence.[13]

Recently a new tanning product, Melanotan, has been introduced. This is injected daily for the first seven to ten days until the desired level of colour is obtained, and then this colour is maintained by weekly top-up injections. The product stimulates the melanocytes, the cells that produce melanin, to produce even more melanin than normal, thus affecting a tanned appearance. But no long-term studies have yet been done to determine the product's long-term side effects, or if it will increase the risk of malignant melanomas, so it has not officially been licensed for use.

A different approach to achieving an impressive suntan was to use 5-methoxypsoralen, an ingredient of bergamot oil that was included in a number of highly popular suntan preparations. This same substance was used medicinally, taken by mouth by sufferers of psoriasis and vitiligo in a treatment introduced about 30 years ago.

This treatment involved exposure of sufferers' skin to ultraviolet A light. Concerns about the possible risk of phototoxicity were expressed, particularly about a type of photosensitivity resulting from this use of 5-methoxypsoralen, which became known as Berloque dermatitis.[14] Apart from the short-term risks of sunburn, this treatment fell from use some years ago because of the long-term risks of skin cancer and accelerated skin ageing.

Para-aminobenzoic acid (PABA) was used as a sunscreen for many years, usually as a five per cent solution. It effectively absorbs ultraviolet light throughout the harmful UVB range while absorbing little or no UVA light. The UVB rays are harmful as they can cause mutations leading to skin cancer. However, photosensitivity reactions to PABA have occurred, as have hypersensitivity reactions to other chemically related drugs. Adverse skin reactions, such as vitiligo, have also been reported following its use.[15] PABA can also stain clothing rather badly, which is a less than helpful problem with its use.

Safer sunscreen products have been developed in recent years, such as zinc oxide and titanium dioxide. Zinc oxide reflects ultraviolet radiation and so is included in many sunscreens, including the sticks used by sportsmen to apply to their cheeks and nose. Titanium dioxide also reflects ultraviolet light and so is used in sunscreen preparations to prevent sunburn. It is also contained in some face powders and other cosmetics. Having a similar action to zinc oxide, it is also used in creams to relieve itching and in the treatment of certain skin problems, such as nappy rash.

Poisonous perfume and a noxious necklace

Nitrobenzene was known a century ago as Oil of Mirbane or Artificial Oil of Almonds. It was widely used in cheap perfume and soaps. This pale yellow liquid, which smells somewhat like almonds and has a sweet taste, could cause poisoning by absorption through the skin or by inhalation. It is very toxic, and the ingestion of as little as 1g can be fatal. But, unfortunately, the symptoms of poisoning are usually delayed for up to 12 hours, by which time treatment may be too late. Starting with nausea, a burning headache, prostration and intense cyanosis, with vomiting and gasping for breath, the symptoms lead on to convulsions, coma and death in a few hours after they appear. It is interesting to note that strychnine, another deadly poison, was the recommended antidote, together with the use of stimulants, in the treatment of nitrobenzene poisoning in the 1920s.

In the 1960s, several medical journals carried reports about a woman who wore a necklace that included some 'beads' that were actually the seeds of the castor plant. She experienced a severe allergic response when one of the seeds was broken, exposing the neurotoxin ricin within. She had suffered two earlier mild attacks of itching on her neck, which were also later believed to be due to this poisonous necklace. Ricin is the poison that killed Georgi Markov, detailed in Chapter 3.

Creating the body beautiful

Cosmetic surgery has come a long way in recent decades, with botox and fillers to smooth out wrinkles, as well as more serious surgery to treat and repair disfigurements, both accidental like burns, and congenital like cleft lip and palate. Silicone implants are the modern way of increasing bust size. But even back in the 1920s and earlier in the twentieth century, some women were persuading doctors to help them improve their looks.

Before the silicones were invented, doctors used a special paraffin wax for similar purposes. It had a melting point about ten degrees above body temperature, so the operating theatre had to be kept very warm to prevent the molten paraffin wax from setting in the syringe before it was injected. Thus were deformities of the nose, ears and face improved a century ago. Liquid paraffin was also used for this purpose, but because it was liquid, various lumps and bumps could result in the wrong places.

In the 1970s, a case of liquid paraffin being inserted in the wrong place happened to the face of a 49-year-old Asian woman. She developed painful swelling and redness of the skin some seven months after she received injections of liquid paraffin around the eyes and in both cheeks for cosmetic purposes. Although her condition did improve after some surgery and steroid medication, it was felt that complete surgical removal of all the affected tissue was the only effective treatment.[16] A few years later another patient, who had received injections of liquid paraffin into her breasts some 60 years previously, developed florid chronic inflammatory local reactions with hard cysts forming.[17]

Since the introduction of silicone breast implants in the 1960s, there have been a number of reports of problems. Women undergo breast augmentation or reconstructive surgery following mastectomy or for cosmetic purpose, but there is always some risk of migration of the silicone, with cyst formation and other complications. In the past, there were many anecdotal reports of connective tissue disorders concerning silicone implants. And, in the 1990s, litigation by patients affected by autoimmune disease and cancer, which they claimed resulted from leakage from silicone implants, led the manufacturer Dow-Corning to file for bankruptcy.

Even today there are problems, particularly in the USA where there are many so-called cosmetic surgeons who are simply back-room quacks offering cut-price cosmetic surgery, sometimes even from make-shift clinics set up in hotels rooms. They tend to use industrial-grade liquid silicone, which is an inexpensive material often used in adhesives and paint.

The toxic effects of liquid silicone have been known since 1979, when the FDA published an article warning of the 'serious and sometimes fatal complications' of silicone injections, including 'swelling, discolouration, cyst formation and migration of the silicone particles to the brain, lungs or heart'. The silicone migrates, creating granulomas, which are benign, inflammatory tumours. These can wreak havoc by appearing anywhere, on the face or wherever else the silicone was injected.

Even the breast implants comprised of silicone encased in plastic have caused problems as some of these have also leaked, allowing the liquid silicone to escape into the surrounding tissues. In 1992, the FDA called for a moratorium in the US on the use of breast implants. However, despite all the problems and attendant publicity, large epidemiological studies and a review by the British Medical Devices Agency have failed to show any association between silicone breast implants and connective tissue disorders.[18]

Anti-wrinkle creams

Some cosmetics containing oestrogens, the female sex hormones, have been marketed in the last century. In 1938, an American medical journal carried a report about a face cream that contained oestradiol, sold commercially as an anti-wrinkle cream. The use of this type of product led to claims of a number of adverse effects, such as precocious puberty in children who used them, and gynaecomastia (development of breasts) in men, while elderly ladies became very distressed when they developed post-menopausal bleeding. The report included details of how the anti-wrinkle cream was tested by being applied daily to the skin of some experimental animals. Testers found that it was absorbed through the skin into the bloodstream of the animals, which resulted in definite internal effects.[19] Needless to say, the use of such hormonal creams is very carefully regulated today.

Poisonous preservation

Whether packaged in pots, jars or tubes, as soon as the lid is first removed any cosmetic may become contaminated, from the air we breathe, or more likely from contact with our fingertips. Microbial contamination can readily spoil cosmetic preparations, particularly creams and lotions. All cosmetics need preservatives to prevent deterioration due to chemical changes such as oxidation. If they are prepared without a preservative such contamination could cause some very nasty infections.

Today in the UK, we have legislation that regulates cosmetic products and the ingredients used to make them. The use of colouring agents in cosmetics is also controlled. Any cosmetics containing lanolin must be labelled 'contains lanolin' because so many people were allergic to it in the past. Many of these hypersensitivity reactions were in fact due to impurities, such as detergent, used in processing, and naturally occurring fatty alcohols. When lanolin is suitably treated to remove these impurities the incidence of hypersensitivity is reduced to almost zero.[20] It is reassuring to know that, today, lanolin causes far fewer problems – it is now far more highly purified than it was in the past.

A final sobering thought

In November 2008, researchers at Imperial College in London suggested that women exposed to hairspray during the early stages of pregnancy are more than twice as likely to give birth to a son with a serious genital defect, known as hypospadias. This unfortunate condition results in the urinary opening being on the underside of the baby boy's penis. Certain chemicals in the hairspray aerosol, known as phthalates, may be the cause. The first three months of pregnancy are always the most crucial for the baby's development.

Chapter 13

The Noxious Nursery

IT IS A SAD and unfortunate fact that every year there are many cases of poisoning of babies and young children, some of which are fatal. Whether accidental or deliberate, every fatality is a tragedy.

Poisoning of babies and children can occur in many ways. For example, they may be given an incorrect dose of medicine for their size and weight. Or, the poisoning may even happen before birth if the mother receives medication that can pass to the unborn child by way of the placenta and umbilical cord. Or, the mother's medication may pass to a breastfed child in the breast milk. All doctors, midwives and pharmacists have information about the toxicity of specific drugs to the foetus in pregnancy, and to the baby in breastfeeding. However, there are also adverse and allergic reactions which cannot be anticipated. There may even be some poisonings caused by a parent or carer giving a child a noxious substance deliberately.

There are many cases each year of accidental poisoning of children. Most babies put things in their mouths as a way of exploring new objects, and so may eat things they shouldn't. It is also very likely that small children will mistake tablets, especially the sugar-coated ones, for sweets, or liquid medicines for drinks. Medicines come in all shades of the rainbow, whether liquid or solid, and should be kept locked away, out of sight.

Food intolerances or hypersensitivities can drastically affect growth and development, as already mentioned in the Malfunctioning Metabolism part of Chapter 5, and such problems can be regarded as yet another form of poisoning. Today many children appear to be hyperactive, disruptive and lacking in concentration. But a change of diet has been found to work wonders for some. This change of diet excludes sweets, artificial colourings and refined fatty foods, but includes fish oils (as a source of omega-3 fatty acids), together with fresh fruit and vegetables, and adequate vitamins and minerals. Any food intolerances these children may have also need to be investigated. Any foods a child is intolerant to must also be excluded from that child's diet, for intolerance is also a form of poisoning. This type of diet has been found to have had a calming effect, and children became far better behaved as a result.

Poisonous treatment

When medical treatment leads to poisoning, there is normally a detailed investigation, with careful consideration of the cause of the problem and then the use of an alternative, safer treatment thereafter. In the past, doses of medicines given to children were usually calculated as a simple fraction of the adult dose. But we now know that this was not the best method, as babies and small children frequently differ greatly from adults in their response to drugs. Because their body systems are immature, they use, break down and excrete drugs at different rates from adults.

Today, paediatric doses are carefully calculated based on the child's body weight, the estimated body surface area and age. The first edition of the *British National Formulary for Children* was published in 2005 to provide the most accurate and up-to-date information about medicines, their uses and the most appropriate dosage for babies and children. This reference source is revised annually, with the many changes made between editions published on its website, www.bnfc.org.

The boiled lobster look

In the past, boric (also called boracic) acid was used as an antibacterial and antifungal agent, in solutions and in dusting powders. It was also impregnated in dressings, such as boracic lint. Although it has now been superseded by more effective and far less toxic disinfectants, it was very widely used for many years, until it was finally realised that if applied to raw, weeping or abraded skin (such as with nappy rash), it would be absorbed through the broken skin and could then cause boron poisoning.

It is highly poisonous: as little as a teaspoonful of boric acid crystals can be fatal. Boric acid solution looks exactly like water – they are both colourless solutions – and that has been the cause of many a poisoning in the past. In 1927, there was a report of the fatal poisoning of a number of babies, each born weighing about seven pounds, who had been fed varying amounts of boric acid solution, given in error for water.[1]

In 1945, an American medical journal carried a report of a baby with severe infantile eczema who was fatally poisoned following the application of large amounts of ten per cent boric acid ointment to extensive areas of broken skin. Sufficient boric acid was absorbed to cause irreparable damage to the baby's central nervous system, causing convulsions, blindness and deafness. A significant feature was the 'boiled lobster' appearance of the skin, which is characteristic of boric acid poisoning.[2]

A few years later, another report came from Canada about six babies ranging from six to 11 days old. They had all died after being given one feed of milk, which had, unfortunately, been diluted with a 2.5 per cent solution of boric acid, in error for sterile water. The first death occurred after a period of 19 hours, and, one by one, all the other babies died, the last one not until five and a half days later.[3]

Borax is the sodium salt of boric acid, sodium borate. So it is not surprising that similar warnings apply to its use. Indeed, this substance is considered to be so hazardous if administered to babies that the Royal Pharmaceutical Society of Great Britain has advised pharmacists for many years now that they should no longer supply glycerin of borax or honey of borax, even with an appropriate warning, to anyone, of any age. These now obsolete preparations were used as throat paints for the treatment of sore throat. In 1928, there was a fatal poisoning of a two-week-old baby who was given a dummy dipped in a mixture of honey of borax and glycerin of borax. Both the *British Medical Journal* and *The Lancet* carried reports of this particular poisoning.[4]

Keep your camphor and castor oil straight

Camphor has been used medicinally for hundreds of years. But, as with many things, it can be dangerous in careless hands. For example, it is exceedingly dangerous to apply camphor (or menthol, which has the same effect) to the nostrils of babies. A report from as long ago as 1913 tells of a nine-month-old baby who immediately collapsed when camphor ointment was applied to its nostrils.[5] Camphorated oil applied too liberally can result in absorption directly through the skin, entering the bloodstream and causing fatal camphor poisoning. Physicians of yesteryear were of the opinion that camphorated oil was a powerful stimulant to the respiration and circulation, when in fact it has the opposite action and the apparent stimulatory changes that they observed were weak, delayed and probably resulted from biochemical changes in the liver.

Poisoning with camphorated oil in children has also occurred due to someone mistaking it for castor oil, which is a purgative given to treat constipation. The person administered it by mouth, causing nausea, vomiting, colic, dizziness, delirium, convulsions and paralysis, leading to difficulty breathing.

Fortunately, death from this type of poisoning is rare, although deaths have been recorded from taking as little as 1g, while some people have survived taking as much as 5g. In 1950, a child swallowed and then also aspirated some camphorated oil, which caused asphyxia. Fortunately, this was successfully treated with atropine and oxygen and the child survived.[6]
In 1973, the doctors at Newcastle General Hospital carried out a review of 175 admissions of children to the hospital for poisoning, and found that no less than ten of these cases were due to children drinking camphorated oil.[7]

In the 1980s, the Royal Pharmaceutical Society of Great Britain advised pharmacists that they should no longer supply camphorated oil because of the poisoning dangers. But even today caring and well-meaning grandparents still occasionally ask pharmacists for camphorated oil, which they recall perhaps from their own childhood, little realising that this poisonous preparation is now obsolete.

Paints and unguents making things worse

Magenta paint, also called Castellani's paint, was a disinfectant used in a number of skin complaints. But, some 20 years ago, concerns were raised about it causing cancer, and this has now severely restricted its use. Magenta paint contains magenta, boric acid, phenol and resorcinol in solution. Application of this paint to a six-week-old baby, some 30 years ago, resulted in the child turning blue due to cyanosis, because absorption of the paint prevented the blood from carrying oxygen around the baby's body.[8]

A few years later, phenol was found in the urine of four out of 16 babies with seborrhoeic eczema who were treated with magenta paint. The paint was applied twice a day for 48 hours, and the presence of phenol in the urine indicated that there must have been significant absorption through the skin, causing some degree of phenol poisoning.[9] As the magenta paint also contained boric acid, it is probable that these babies had boron poisoning too. Magenta paint might have helped their eczema, but today we can readily treat eczema without such drastic side effects.

Salicylates, such as aspirin, were used for many years as painkillers. They have been taken by mouth or used externally, as salicylic acid, in ointments for a range of painful conditions, including some skin complaints. In the 1930s, a seven-year-old child who was suffering from psoriasis was treated with salicylic acid ointment. The ointment, which was applied only to the patches of psoriasis, resulted in death some 40 hours later, probably due to absorption of salicylate through the skin.[10]

The fatal intoxication of another child, aged ten, who was being treated for rheumatic endocarditis, was reported in 1934. In this case, after five days of treatment involving the administration of both injections and rectal doses of sodium salicylate, it is hardly surprising that the symptoms of salicylate intoxication were seen, together with kidney failure. Despite intensive treatment this poor child died some three days later. A post-mortem revealed that the child had died due to kidney lesions and fatty degeneration of the liver, which had both developed as a result of the salicylate poisoning.[11]

Picric acid (trinitrophenol) also has disinfectant properties and was used in the past to treat burns. However, it is toxic, due to absorption through the skin, and is now rarely used in medicine. But, in 1912, it was still very much in use. A ten-week-old baby who had suffered severe burns, caused by boiling water, to one of its feet was treated with a dusting powder containing picric acid. Vomiting and severe diarrhoea, followed by a fever, developed, with the pulse varying from 100 to 150 beats per minute. A general all-over redness, or erythema, developed and the child died 22 days after the accident as a direct result of picric acid poisoning.[12] By 1939, *The Lancet* was reporting that hospital doctors had seen so many cases of this toxic erythematous rash that some of them were now hesitating to use picric acid, even as a first aid treatment for children.

Mercury in grey and pink

Despite its known poisonous nature, mercury was used extensively as a medicine for centuries. It was used as an antiseptic, an antibacterial agent and even a mild purgative for children, in the form of grey powder. Grey powder contained one-third part each of mercury, dextrose and chalk, and was still in use in the 1920s. This mixture could be made into tablets or diluted with lactose (milk sugar) and supplied as powders for small babies and children. Another formulation, widely used for babies with constipation, contained grey powder mixed with rhubarb, ginger and sodium bicarbonate.

As well as this type of purgative preparation, teething powders and skin preparations such as dusting powders and ointments also contained mercury. In babies and young children their use led to the development of the condition acrodynia. Otherwise known as ‘pink disease', acrodynia was actually a manifestation of the signs of mercury poisoning, characterised by bright pink cheeks, which unfortunately made the baby look very healthy!

A fatal case of acrodynia was reported in 1965: a six-month-old baby was given a total of nine out-of-date mercurial teething powders, each of which contained 26 per cent calomel – the old name for mercurous chloride. This report, published by the British Medical Association, was made over a decade after calomel teething powders were banned from sale in 1954.[13] But there were no product expiry dates on medicines that long ago. The results of this report and survey by the BMA were considered to be conclusive evidence that 'pink disease' was caused by poisoning with mercury. Calomel had been the 'active' ingredient of most teething powders on sale since their introduction as long ago as 1812, according to the claims made by one manufacturer.

It is only in the last few decades of the twentieth century that medicines began to be labelled with expiry dates, but today all medicines have them. Everyone keeps a few basic remedies at home for future use. Once in a while, it is a good idea to check the expiry dates on these and replace anything that is out of date. You should also always check both the expiry date and the required dose when about to use a medicine. Old out-of-date medicines, and any prescribed medicines no longer in use, should be returned to your local pharmacy for safe disposal. Never flush medicines down the toilet: your water company strongly disapproves of such contamination, as it may pass through the local water treatment works unchanged, escaping into local water courses, where it could then cause environmental damage.

In a book called *Secret Remedies* published by the *British Medical Journal* in 1909, calomel was found in a number of analysed children's medicines. Stedman's Teething Powders, the similar sounding but different Steedman's Soothing Powders and Pritchard's Teething and Fever Powders all contained calomel, and the last product contained the poisonous antimony oxide as well. Another product mentioned in the same book, Fennings Children's Cooling Powders, did not contain mercury – but contained potassium chlorate, another poison, instead.

In the 1960s, a four-year-old boy was treated for a skin complaint with ammoniated mercury ointment, which slowly poisoned him. The treatment continued for two months, during which time he developed increasingly progressive difficulty walking, and other symptoms. He was treated for mercury poisoning with dimercaprol, mentioned in the treatment chapter. He eventually recovered completely, except for absent Achilles reflexes.[14]

There have been many reports of damage to the unborn child caused by mercury. The mother's exposure to methylmercury, an organomercury compound, once commonly used in pesticides and seed dressings, while pregnant was found to delay the neurological development of affected children up to the age of seven years.[15] Fortunately, these compounds are no longer used as much, even in the developing world, as newer less toxic products are now available, as mentioned in Chapter 9.

Further fatalities from misguided treatments

Thallium acetate used to be recommended as a hair remover, but the hair loss was really a symptom of thallium poisoning. A single dose, taken by mouth, was used to remove hair from children who had ringworm of the scalp. Ringworm is a common fungal infection, which is easily treated nowadays by any one of several anti-fungal creams and ointments. Shaving the head is definitely no longer part of the treatment – and neither is thallium.

Many children were poisoned by the thallium acetate treatment for ringworm of the scalp. A four-year-old boy died after an overdose of this substance, having been given a succession of doses instead of the required single dose. The *Pharmaceutical Journal* reported this case in 1949, with the comment that even a single dose produced signs of poisoning in about 40 per cent of cases.[16]

It seems unbelievable that about 75 years ago this poison was advocated to be administered to children, and specifically only to those who had not yet reached puberty, in a dosage of eight milligrams per kilogram of bodyweight, based on the weight of the child. A number of poisonings due to thallium have been described in earlier chapters and yet more are described in the next.

Owing to the many fatalities, including 14 out of 16 children treated in Granada, Spain, caused by this horrendous treatment, it is hardly surprising that the use of thallium acetate was largely abandoned by the late 1930s. However, it was obviously still in use to a sufficient extent – some two decades later it required additional warnings, to the effect that it should not be used more than once in any three-month period due to its cumulative effect.

Santonin was traditionally used as a treatment for worms. It is obtained from the dried flower heads of a species of Artemisia and is very toxic, particularly so if it is given in an oily solution, such as with castor oil.

One case of santonin poisoning occurred in 1919 precisely because of this reason: a five-year-old girl was given a dose of santonin in castor oil as a treatment for worms. A large number of threadworms and a few roundworms were expelled, but the child became very ill and weak, and looked likely to die. After three weeks, she had made a good recovery but was now completely blind and remained so for a further three months.[17]

Santonin poisoning mainly affects vision: white objects appear first violet and then later yellow. This effect may continue for several hours. Larger doses may also cause headache, nausea, vertigo, vomiting, diarrhoea and convulsions, followed by coma. Many fatalities have occurred over the years; it was estimated that as little as 60mg could cause severe poisoning and that two such doses could be fatal.

Neem, also called Margosa or Azadirachta, is a gum-secreting tree that grows in the tropics of Eurasia and North Africa. The seeds contain an essential oil, known as neem oil or margosa oil, which was used in these areas as a hair tonic and for skin diseases. It has also been used in India as a remedy taken by mouth for minor ailments. This has caused severe poisoning in many children, resulting in Reye's Syndrome (see the section on aspirin in the chapter about Malevolent Medicines).[18] Hair preparations – both lotions and shampoos containing neem oil – are in use in the UK today because they are effective at treating head lice infestations, killing both the adult lice and the unhatched eggs. But they should not be used on broken skin, to avoid any absorption.

Naphthalene was used medicinally in the early twentieth century. In 1911, the *Pharmaceutical Journal* reported the fatal poisoning of a six-year-old child who had been given seven doses of naphthalene by mouth as a treatment for worms.[19] Some years later in 1917, the *British Medical Journal* carried a novel suggestion for its use in preventing body lice. It was suggested that vests made of butter muslin should be dipped in a solution of naphthalene and sulphur dissolved in benzol. Benzol is better known today as petrol! It would certainly be extremely dangerous for any child encased in such a vest – if he were not overcome by the petrol fumes, he would be liable to ignite at any minute from the coals in the grate.[20]

Ophthalmia neonatorum is an eye infection of the newborn, caused by gonococcal infection from the mother. Gonorrhoea was considered to be so prevalent, even 50 years ago, that treatment of all newborn babies for this condition was considered necessary. Eye drops containing one per cent silver nitrate were used for many years to treat this condition as silver is a very effective antibacterial agent. But many babies treated with silver nitrate eye drops then developed another eye infection, conjunctivitis. A large study published in 1949 showed that silver nitrate was less effective than other antibacterial agents and its routine use soon fell into disuse.[21]

However, in America in 1975 a baby had a silver nitrate stick, which contained 75 per cent silver nitrate, applied to its eyes instead of the prescribed one per cent solution. After one hour, there was a thick purulent secretion, and the eyelids were red and swollen. The corneas had a blue-grey bedewed appearance, with areas of opacity. Fortunately, after copious washing and application of antibiotic and other eye drops there was a marked improvement. After a week, steroid eye drops were started and the residual damage was limited to slight corneal opacity.[22]

Atropine, from the plant Deadly Nightshade, acts on the eye to expand the pupil. A child, aged five years, suffered from atropine poisoning when only two drops of atropine one per cent eye drops were put into each eye. The child became restless, started singing and shouting and finally developed convulsions and a red rash all over the body, with dry mouth and dilatation of the pupils. Treatment with an injection of morphine was fortunately followed by a full recovery.[23] Another report of mental disturbance in a boy aged ten followed the instillation of five drops of atropine eye drops one per cent into each eye.[24]

Sodium nitrite is today used medicinally as an injection in the treatment of cyanide poisoning. In 1917, a baby was poisoned when given sodium nitrite in error, due to the label having been misread as sodium citrate, which was the intended medication.[25] Such errors are frighteningly easy to make, which is why nurses always check with each other before administering medicines.

Nicotine, obtained from tobacco, has been used medicinally in the past. In the 1930s, a five-year-old boy was given an enema consisting of 60ml of strong tobacco juice in one litre of water, administered as a treatment for threadworms. It seems the dose given was too much for this little boy. He started to vomit continually immediately after he was given the medication, and then he became unconscious. Three doses of sodium benzoate and copious enemas to eliminate the nicotine were needed to effect a recovery.[26] Much safer alternatives are now used.

Potassium chlorate was widely used in mouthwashes, gargles and toothpastes in the first half of the twentieth century. In America during 1939, potassium chlorate was used to treat a child with bad mouth ulcers. The child was given one teaspoonful of a saturated solution of potassium chlorate every three hours. After a week of this treatment, the child began to show signs of brain damage. It was calculated that a total of 7-8g of the drug had been taken by this time. The treatment was then stopped and the child began to improve, although at that time it was thought unlikely that the child would ever recover sufficiently to be entirely normal.[27]

Mistakes with morphine

Morphine is used today mainly in the treatment of severe pain. In the nineteenth century, laudanum (just another name for an alcoholic solution, or tincture, of opium, which contains about ten per cent morphine plus a number of other alkaloids) was used by all and sundry, and many became addicted to it (see Misuse and Abuse, Chapter 19). At that time, laudanum was used for pain relief, for sickness and diarrhoea and was even included as an ingredient in a number of so-called 'soothing syrups' for babies until the law was changed in 1908.

Before this there were many number of products containing laudanum used to soothe the fractious baby. Godfrey's Cordial was a lethal mixture of alcohol and opium, Mrs Winston's Soothing Syrup, Morrell's Teething Syrup, Grandma's Secret and Mother's Treasure all contained opiates of one sort or another and Kopp's Baby Friend was probably the worst morphine-based soother, as it killed at least 11 youngsters in 1905 and 1906.

Even with medication containing morphine, mistakes have been made. The *British Medical Journal* in 1912 reported a case where a baby was given a morphine suppository in mistake for a glycerin one. Fortunately this four-month-old baby recovered after treatment for morphine poisoning.[28]

In January 2004, a six-week-old baby was fed with a liquid soap solution called Infacare. This was given in error for a medicine with a similar sounding name, Infacol, used to treat wind and colic in babies. The baby had been in hospital because of a lung infection, which was successfully treated. The baby was then discharged and sent home.

The parents were given a bottle of the incorrect product by a nurse, together with an oral syringe with which to administer it, as they left the hospital. The mother was not aware of the error until her baby had to be re-admitted to hospital because of severe vomiting after she gave the child a dose of the soap solution. Presumably neither parent bothered to read the manufacturer's label on the container, another common cause of mistakes with medicines.

Poisoned before birth

Some paediatric poisonings happen either before the baby is even born, or after birth by way of the breast milk fed to the baby. In either case, the poisoning results from something the mother took which then affects the baby.

Both opium and morphine can cross the placenta to affect the unborn child and can also pass into breast milk, and so may continue to poison the baby after it is born as well. If the mother is a drug addict, then the baby is born with a problem. Many babies, even today, are born in need of treatment for withdrawal symptoms, because their mothers are addicted to an opioid drug, and so the baby has also developed an addiction to the drug 'in utero' during the pregnancy.

Alcohol is yet another substance that can cause problems for babies. It too can cross the placenta and pass into breast milk. Foetal alcohol syndrome is the name given to a characteristic pattern of abnormalities recognised in babies born to alcoholic women. Alcohol has been shown to adversely affect the levels of the hormones involved in breast-milk production. If the mother is an alcoholic, the baby drinks alcohol too. It was only about 20 years ago that alcohol was removed from a number of gripe mixtures, traditionally given to babies with colic.

The selective serotonin re-uptake inhibitors, better known as the SSRIs, are a type of antidepressant that came into use in the late 1980s. The best known of this type is Prozac (fluoxetine). SSRIs are generally regarded as less toxic in overdose than older types of antidepressants such as the tricyclics. However, babies can still be at risk if they are breast-fed by a mother taking an SSRI.

Symptoms of colic were reported in one six-week-old, breast-fed baby whose mother was taking the usual 20mg dose of fluoxetine each day. The baby's colic disappeared completely when the breast milk was replaced by a bottle feed.[29] The manufacturers of antidepressants generally advise that breast feeding be avoided by mothers taking these drugs.

Some ergot alkaloids are used in the treatment of migraine, and if taken by women who are breast feeding they too will pass into the breast milk. Cases of such poisoning have occurred, in which nursing mothers gave their babies the symptoms of ergotism, such as vomiting, diarrhoea, weak pulse and unstable blood pressure.[30]

Lead is very soft, easily shaped and has been used to make many things. This may explain why, before the age of plastics, it was chosen for the curious production of nipple shields. A mother used the lead nipple shields while breastfeeding, causing lead poisoning to her baby. The use of these or even the application of lead acetate ointment to the nipples by breastfeeding mothers could result in sufficient lead in the breast milk to cause encephalitis in the baby. This particular type of poisoning only came to light because it was also associated with a child developing rickets, the vitamin D deficiency disease that affects the bones.[31] The lead poisoning had affected the baby's use of calcium. Needless to say, the doctors who reported this case, in *The Lancet* in 1949, suggested that, in view of the dangers to babies, lead nipple shields should no longer be used.

Some medicinal preparations made in India have been found to contain lead, including a baby tonic that was on sale there in the 1970s. More recently, eye preparations, such as kohl and surma from India, which have been imported into the UK for use on Asian children living here, have been found to contain up to 83 per cent lead. Although the sale of these products is prohibited in the UK, that does not prevent well-meaning relatives, still living in various Asian countries, from sending these poisonous cosmetics privately, in the post, for use on children here. Further details of poisonous cosmetics are given in the previous chapter.

Chloroquine has been used for many years for both the prevention and treatment of malaria. About 40 years ago, in America, a woman was prescribed chloroquine to prevent malaria and took it during four out of seven of her pregnancies. This resulted in complications in the children born, including mental retardation and neonatal convulsions.[32] Another child whose mother was treated with chloroquine suffered ear damage, and eye problems were found in two sisters, probably due to their mother taking choroquine tablets daily for more than three years, which covered the periods of both pregnancies.[33,34] For a woman to prevent getting malaria while pregnant, if living in a part of the world where this disease is endemic, without taking anti-malarial drugs, must be very difficult indeed. The anti-malarial drug has also caused fatal poisoning. In the mid 1970s, the deaths were reported of six children, ranging in age from 11 months to four years, who had each swallowed up to five tablets containing chloroquine.[35]

A 17-year-old mother suffering from depression drank a small quantity of a liquid; it was only about 30ml in volume, but it contained about 400mg of arsenic. This happened during the final three months of her pregnancy. Four days later, she gave birth to a premature baby who died of respiratory distress syndrome some 11 hours after the birth. Post-mortem examination of the baby showed high concentrations of arsenic in the dead baby's liver, brain and kidneys, having passed through the placenta from the mother.[36]

Thiomersal is a mercury-containing substance that was used as a preservative in the past. Its use in medicinal products, including injections and vaccines, has already been mentioned in the chapter about Potent Potions. During the 1970s, a study found that there was an increased rate of malformation in 56 children born to mothers who had been exposed to thiomersal and possibly other drugs at some time during the first four months of their pregnancy.[37] At about the same time, there was a report concerning 13 children who each had an umbilical hernia. Treated with a tincture of thiomersal applied to the umbilicus, ten died as a result of poisoning from the concentration of mercury in the tissues.[38]

Some children will eat anything

Some paediatric poisoning problems include those cases where poisoning resulted from children eating something they should not have done, such as something noxious that was left within reach of the child; frequently, sugar-coated tablets that the child thought were sweets.

Quinine was used for centuries to treat malaria, and today it is used to treat night cramps. Through the years, there have been many fatalities caused by quinine poisoning. In 1931, the fatal poisoning of a 30-month-old child was reported. The child had eaten 26 sugar-coated tablets containing a total of more than 8g of quinine and could not be saved. In 1972, a three-year-old child became completely blind within 15 hours of ingesting a quantity of quinine. Treatment with bupivicaine was given and, after 12 hours, central vision was normal but peripheral vision was poor. A month later, there were signs that the damage to the child's eyes was permanent.[39]

Quinine can also cross the placenta, as was shown when a baby was born and developed jitteriness shortly after his birth. This was attributed to quinine withdrawal symptoms because the mother had been drinking a lot of tonic water for the last four months of pregnancy.[40] Tonic water contains a small amount of quinine (the cause of its bitter taste). Although the amount that is excreted in breast milk is considered too small to affect a breastfeeding baby, it can clearly affect the baby before birth.

The dangers of tonics and tablets

Strychnine was used for many years in very small doses as a tonic. Tonics containing iron, quinine and strychnine were official preparations in the British Pharmacopoeia in the first half of the twentieth century, and were available as a syrup or as Easton's Tablets. Each of these tablets contained 100 micrograms of strychnine, that is, one tenth of a milligram.

A one-year-old baby ate between ten and 20 Easton's Tablets and died, and a two-year-old child also died of strychnine poisoning after swallowing about ten tablets of a type which each contained 2mg of strychnine – the fatal dose for an adult is probably between 50 and 100mg.[41] Until the mid 1980s, Metatone, a well-known tonic, contained a small quantity of strychnine as one of its ingredients. This product has long since been reformulated, without the strychnine, in line with modern thinking on tonics.

Iron tablets are particularly poisonous to children. In 1950, the *British Medical Journal* carried a report of six cases of poisoning in children, including two fatalities. They had each swallowed up to 40 sugar-coated ferrous sulphate tablets.[42] The previous year, the same journal had reported the poisoning of a baby who had taken some pills which contained not only ferrous sulphate, but copper sulphate and manganese sulphate as well. The number swallowed was unknown, but after a period of apparent recovery, there then followed the sudden death of the infant. The post-mortem found that there was destruction of the stomach lining, which had led to lethal liver damage.[43] At about the same time, an American medical journal similarly told of the death of a 17-month-old little girl, who had swallowed some iron tablets thinking they were sweets.[44]

In 1961, *The Lancet* reported the death of a small child who had swallowed about 3g of ferrous sulphate in tablet form, noting that the poisoning had followed three distinct phases. First, there was vomiting, possibly within 30 minutes of taking the tablets; also observed in this initial phase were pallor, shock, blood in both vomit and stools, with circulatory collapse and coma. The second phase showed some clinical improvement, but this was followed by liver, metabolic and brain injury with renewed restlessness, collapse, convulsions, coma, jaundice, fever, leucocytosis and abnormal brain function. During the third phase, in those patients who had survived that far, there was gastritis and from three to six weeks later, the symptoms of gastro-intestinal obstruction, but the liver and brain then appeared to recover.[45]

In 1977, a toxicology journal told of a 16-month-old child who died of acute liver failure, some five days after eating a quantity of ferrous sulphate.[46] In treating such poisoning in children, where as little as one gram (five 200mg ferrous sulphate tablets) can be fatal, the speed of treatment is paramount. Any baby or child thought to have eaten iron tablets of any type should be rushed to the nearest accident and emergency department as soon as possible.

Iron tablets are so common but so harmful that **all iron preparations** now carry a special warning label that states:

'Keep out of the reach and the sight of children.'

Trouble with tricyclics

Tricyclics antidepressants have been in use since the late 1950s. Ever since then, there have been many reports of children being poisoned by them.

In 1967, the *British Medical Journal* told of a 14-month-old boy who ate 100 tricyclics tablets. He developed status epilepticus, low blood pressure and a high fever. The child only made a partial recovery and was left with permanent brain damage. Unfortunately there was a delay of about 15 hours before treatment was started, which may have been responsible for this outcome.[47] In every case of poisoning early treatment is absolutely vital.

In 1979, the *British Medical Journal* told of the death of a four-year-old child who drank a quantity of imipramine syrup. This particular tricyclic antidepressant was widely used for many years as a drug therapy for bed wetting because one of its side effects is urinary retention. It had been prescribed to this child for this purpose.[48] Today, such therapy is not considered appropriate for children under seven years of age, and then only when other alternative measures, such as an enuresis alarm, have been tried and failed.

The following year another four-year-old child, in America, suddenly collapsed for no apparent reason after a morning of normal activity. She was taken, in a comatose state, to her local doctor. She then had a grand mal type epileptic seizure, which progressed to status epilepticus, as she was being transferred to hospital. The frantic parents had not suspected poisoning from an overdose, but once at the hospital, part of an antidepressant capsule was recovered from the child's oesophagus. This was quickly identified as a tricyclic antidepressant, which then allowed the most appropriate treatment to be used. The child recovered without further mishap.[49]

More nursery nightmares

Colchicine is a naturally occurring alkaloid used in the treatment of gout, which is obtained from the corm and seeds of the meadow saffron. In France in 1977, three babies aged from 18 to 26 months recovered after ingesting some colchicine, the quantity taken being estimated as ranging from 0.1 to 1mg per kg of body weight. Their recovery was due to them all receiving very early stomach wash-out. A fourth child, who did not receive treatment until some five hours after ingestion, subsequently died.[50]

Disopyramide is an anti-arrhythmic drug used to treat irregular heartbeat of adults. A two-year-old boy swallowed 600mg (just six capsules) of this medicine. His blood pressure fell and his heartbeat became irregular, he went on to develop convulsions and he died 28 hours later.[51]

Hyoscine is used today in travel sickness tablets for adults and children. In the past it was used to treat insomnia and also as a tranquilliser. In those far-off days, in 1910, a five-year-old boy ate several hyoscine tablets and was unconscious by the time the doctor arrived. However, an injection of pilocarpine much improved the boy's condition within just ten minutes and by the next morning he had recovered completely.[52] Hyoscine has already appeared in Chapter 3.

A verdict of accidental death was recorded following the fatality of a four-year-old boy in 1959. He drank some brake fluid, which contains polyethylene glycol and so tastes very sweet. Despite being taken to hospital immediately for emergency treatment, he was dead a few hours later. He had been playing with an older friend on a bombed site in central London. Having climbed into a lorry parked there but not locked, he found the tin of lethal brake fluid in the cab. In those days, emergency treatment was much simpler than a decade or so later, as the next case illustrates.

The chemically similar ethylene glycol is an ingredient of antifreeze solutions, and it also tastes very sweet. A two-year-old child who drank about 100ml of ethylene glycol in the 1970s was found to be unconscious the following morning, despite some earlier vomiting. Fortunately he eventually recovered after being given treatment which included the intravenous infusion of a number of drugs plus peritoneal dialysis over a period of eight days.[53]

Diethylene glycol is another similar substance, also used in antifreeze solutions. But, in 1995, the *British Medical Journal* published a report of this substance causing kidney failure in Bangladesh. The diethylene glycol in question was used no doubt as an inexpensive sweetener in the manufacture of paracetamol elixir, a children's pain and fever relief medicine. Three years later, an American medical journal reported that there was an epidemic of similar cases in the Third World.[54]

Metaldehyde was used as a solid fuel, called Meta, in the 1920s and 30s. Meta burnt with a non-luminous carbon-free flame. This fuel was supplied in the form of tablets, and many cases of poisoning of small children occurred. In 1929, the *British Medical Journal* reported two cases of acute poisoning of the central nervous system in children who had eaten Meta tablets, having mistaken them for sweets. They both fortunately survived following treatment, which included both a stomach wash-out and a purge.

In 1935, there was a report of a further case involving a baby, aged 20 months, who also recovered, but a fatal case in 1939, concerning a two-and-a-half-year-old child who swallowed only one tablet, led to a call for the public to be warned of Meta's dangerous qualities and for the State to introduce measures for the safer distribution of this fuel.[55]

In 1968, *The Practitioner* carried a report of a 20-month-old baby who swallowed an unknown quantity of metaldehyde. The baby was treated within half an hour by gastric lavage and an intravenous infusion of dextrose in saline, and seemed to recover. On the third day, however, there was a relapse. The intravenous infusion was restarted and with further treatment, the symptoms finally resolved by the fifth day.[56]

Metaldehyde is still in use today, but now it is used in the garden as a slug and snail killer. It is coloured bright blue to make it easy to see, and it is hoped that no children will eat blue food.

Before the invention of plastics and man-made fibres, all clothing was of natural origin: cotton, silk, wool or leather. Woollen clothing was liable to attack by the larvae of the wool moth, and so mothballs were left in drawers or hung in wardrobes amongst the clothes to prevent this damage. Naphthalene was the traditional ingredient of old-fashioned mothballs, but it is very poisonous if eaten. There have been many cases of small children eating mothballs, having mistaken them for sweets.

One little boy, aged three, began to pass dark-coloured urine the day after he swallowed at least half a mothball. The next day he was profoundly anaemic, because the naphthalene in mothballs destroys red blood cells. However, after a blood transfusion he made a complete recovery.[57] This case of 'mothball anaemia' occurred in Britain in 1949, and the same year in America there were four more cases involving children. Each case resulted in the same 'mothball' haemolytic anaemia, but all the children recovered when given blood transfusions.[58]

In the past, phosphorus was used as rat poison, as discussed in the chapter on Fatal Farming. In 1976, an American medical journal reported three cases of acute phosphorus poisoning in children after they accidentally ate unknown amounts of rat poisons. Two children, one aged two and the other aged three, died within five hours of eating the poison. Both children's hearts stopped, they both stopped breathing and they passed 'smoking stools'. Another older child, aged seven, was luckier and recovered within three days.[59]

In Victorian times, it was thought that phosphorus would improve mental ability, because it had been shown that brain tissue contained a lot of it. Unfortunately, the scientists of the day did not realise the importance of phosphorus in the working of every cell of the human body and particularly in enzymes, energy-producing compounds and nerve tissue. As a result many phosphorus-containing medicines were devised and used, even though they were useless, and some people took to eating match heads, which contained phosphorus, as a means of obtaining it, or sometimes as a means of committing suicide. Many young children were accidentally killed by sucking match heads. Some babies even died in this way, although this may have been murder, rather than accidents.

Formic acid was used for many years in products to help descale kettles. Despite its irritating and pungent nature, children have swallowed this type of product on many occasions. In 1968, *The Practitioner* carried a report of a two-year-old boy who slowly recovered – but never fully – after swallowing an unknown quantity of a kettle descaler, which contained 60 per cent of formic acid. The most serious complications that resulted were a severe laryngeal stridor (noisy breathing), a tracheostomy to allow him to breathe and a gastrostomy to allow him to be fed. These permanent interventions were vital due to the damage done to the little boy's throat. The after-effects of this poisoning would be with him for the rest of his life.[60]

Murders by the dozen

While some poisonings are accidental, some are deliberate, even to small children. Unfortunately there are many cases of poisoning caused by parents or carers who deliberately administer noxious substances to children. It would seem that childcare is too difficult for some.

Lydia Sherman was born in Burlington, USA, in 1824. She was orphaned at the age of nine and then brought up by an uncle. In 1840, she moved to New Brunswick, New Jersey, where she met a widower, Edward Struck, who had six children. They married and moved to Manhattan, where Struck joined the police force. Together they had seven more children. When they had been married for some 18 years, by which time his children had all grown up and left home, Edward was discharged from the police force and became depressed as a result. Lydia decided to put him out of his misery, so she poisoned him with arsenic.

She then found that she was unable, singlehandedly, to support the remaining younger children still at home, and so she poisoned baby William, four-year-old Edward and six-year-old Martha Ann all in one day. When 14-year-old George became chronically ill she laced his tea with arsenic and when 12-year-old Ann Eliza had a number of chills and fevers one winter, she was poisoned too. Lydia's eldest daughter, also called Lydia, died of natural causes two months after the last of her brothers and sisters were poisoned; she was a young woman of 18 by that time.

The widowed and now childless Lydia was next hired by a shopkeeper in Stratford, Connecticut, to care for his invalid mother. She was then recommended as a housekeeper to a wealthy farmer called Dennis Hurlburt. He first hired her and then quickly married her. But it was only a few months later that he too was dead, having been poisoned, and leaving Lydia a legacy of $10,000.

Next, a widower called Horatio Sherman began visiting, as he wanted to hire Lydia as a housekeeper, and also needed her as a nurse for his baby since his wife had recently died. He proposed to her and they married, but Horatio not only drank a lot, he also abused her and spent her inheritance from Dennis as well. So, Lydia poisoned the baby, Frankie, followed by Horatio's 16-year-old daughter, Ada, before putting some arsenic in her husband's brandy bottle. Lydia finally went on trial in 1872, having poisoned a total of three husbands and at least seven children. She was found guilty of second-degree murder and spent the rest of her life in prison.

At about the same time as Lydia's rampage in America, in England another 'mother' was busy poisoning children, who all seemed to suffer from 'gastric fever'.

Mary Ann Cotton was a district nurse who lived in West Auckland, County Durham. She lost four husbands and a dozen children in what seemed, at the time, to be a series of dreadful natural tragedies. Over a period of 20 years, some 21 people close to her died, and she was later suspected of a further 15 murders too. The motive appears to have been money, either from the insurance, or to allow her to marry again.

She first married in 1844, when she was only 22. She and her husband, William Mowbray, had five children, who all mysteriously died in infancy, one after the other, of 'gastric fever'. In 1865, William also died of 'gastric fever'. Mary then became a nurse, which gave her access to the supply of arsenic at the hospital where she worked.

While working there, she met George and married again, but her second husband died some 15 months later. He, like her first husband, had insured his life, and Mary collected the insurance money for a second time. She then became housekeeper for John Robinson, a widower who had five children. Three of these children died the following year of 'gastric fever'. Meanwhile, Mary became pregnant, and she and John were married before the child was born. This baby died of 'gastric fever' at the age of three months.

Robinson could be considered to be a lucky man, as he survived. Mary left him, taking his valuables with her, and went to look after her elderly mother. Her mother didn't live much longer though, dying of 'gastric fever' within a very short time indeed. Mary then went to Newcastle and married a rich widower, Frederick Cotton. This was husband number four. He soon died too, as did a child of the marriage, and three of the children from his previous marriage, as well as Mary's lodger and sometime lover, Joseph Natrass, in addition to the local excise officer, a Mr Quick-Manning, by whom Mary had become pregnant. At this point, a neighbour became suspicious and went to the police.

Mary's latest victim was a boy called Charles. The doctor, who had seen this child alive and well only the previous day, refused to sign the death certificate. A post-mortem was performed and, on analysis, the stomach contents were found to contain arsenic. Mary Cotton was charged with murder and taken to Durham Prison. The defence claimed that the dead child was poisoned accidentally, by arsenic contained in the green floral wallpaper in the home. Cotton's defence was that she had recently bought some arsenic, but that it was to kill bedbugs. The jury didn't believe either story and found her guilty. As she was pregnant (by the now deceased excise officer) at the time of the trial, they kept her in prison until the child was born, then took it from her and executed her five days later in March 1873.

Those who kill children may have a personality disorder. Munchausen's Syndrome by Proxy is a mental disorder in which the patient inflicts harm on others (often children) in order to attract medical attention. This is a rare mental disorder, first diagnosed in 1977. For some reason, a number of such cases have involved feeding the affected child excessive quantities of table salt.

One horrific example involved a mother and daughter in Scotland. In 2003, Susan Hamilton, who lived in Edinburgh, was found guilty of poisoning her daughter with salt. For medical reasons her daughter had to be fed by way of a tube, and it was claimed that Susan had injected strong salt solutions into this tube on many occasions. Repeated admissions to hospital were needed, and high levels of sodium in the daughter's blood were found. The doctors ran tests to eliminate a number of possible illnesses which could cause this, and then realised that the girl was being deliberately poisoned by her mother. Finally the poor child had a stroke, which left her permanently brain damaged. Susan Hamilton was diagnosed with Munchausen's Syndrome by Proxy. She had deliberately harmed her daughter, time and time again, simply to attract the medical attention she craved.

Twelve years earlier in 1991, Beverley Allitt was employed on the children's ward of a hospital in Lincolnshire, UK. They were short staffed and so employed her as a nurse, even though she had failed her exams and so was not qualified. She proceeded to murder a number of children, usually by injecting potassium chloride, which stopped the heart, but she also tried injecting insulin too. It was not until 1993 that she was tried and found guilty. She was sentenced to no less than 13 life sentences for her crimes, having been diagnosed as suffering from Munchausen's Syndrome by Proxy.

The Ultimate Poison Legacy: Catastrophes, Accidents, Murder and Suicide_

Chapter 14

Catastrophes, Calamities and Accidents

ACCIDENTAL POISONING ON A grand scale has happened as a result of many industrial accidents and numerous natural disasters. Accidents can happen for a wide variety of reasons, including miscalculations, design failures or even – it has to be said – the sheer stupidity of some people. A few of these accidents are recounted here.

The worst industrial disaster in history

This accolade must go to a chemical plant and an explosion in 1984, which is still causing problems today. The disaster occurred at the Union Carbide insecticide manufacturing plant, at Bhopal in Central Northern India. It happened on the night of the 3rd December 1984, following a massive leak of toxic gases.

No accurate record of the total numbers of people killed or injured could be made at the time, such was the scale of the disaster. However, it was estimated that somewhere between 16,000 and 30,000 people were killed, and a further half a million or so injured.

The chemical plant involved was used to make an organophosphate insecticide called carbaryl, for which the Union Carbide brand name was Sevin. The American company Union Carbide was very safety conscious, and when this manufacturing plant first opened at Bhopal in May 1980, there were many safety systems in place, all designed to prevent any leakage.

However, on the fateful night some four and a half years later, not one of these safety systems was still operational. A pressure gauge was broken, the storage tanks were at ambient temperature instead of being cooled down to a temperature of 0°C and a number of other safety systems were out of action due to ongoing maintenance.

As part of the production process, the plant produced the poisonous gas, phosgene, as well as a poisonous liquid called methyl isocyanate, a form of cyanide. This liquid, called MIC for short, was supposed to be maintained at a constant temperature of 0°C (the freezing point of water) to keep it stable. On the night in question, all the safety systems on the MIC tanks were inoperative, and the liquid was at 20°C, the usual nightly temperature in that part of India during the winter. At 20°C, the MIC gave off a number of gases, including hydrocyanic acid, otherwise known as the intensely poisonous Prussic Acid. Of the three storage tanks, one should have been completely empty, for use as a back-up in case of emergency, but on this night it was partly filled with MIC. The problems started when water and chemical debris also got into this back-up tank. This water and debris was the result of cleaning the inside of the pipe work at the plant.

Mixing the MIC, at the ambient temperature, with this chemical debris and water resulted in a number of chemical reactions involving the production of massive amounts of heat. This heating caused a massive pressure build-up in the tank, which then exploded out of its concrete casing, rupturing the pipe work in the process. The explosion released the MIC vapour, and, because it was twice as heavy as air, it formed a gas cloud at ground level. Other lighter gases, such as the poisonous phosgene gas and hydrogen cyanide gas, formed in layers above the MIC, together with yet another gas, monomethylamine, which smelt like ammonia.

Unfortunately the prevailing wind that night blew this toxic collection of gases over the nearby slums, and then over the city of Bhopal itself, killing whoever was in its path – men, women, children and their animals. Chemical breakdown of the methyl isocyanate released hydrogen cyanide, and this gas, when inhaled, instantly destroyed the cellular ability to transport oxygen, of any person in contact with it. This was how the hydrogen cyanide gas killed the majority of the victims that fateful night.

The corpses, when collected later, even smelt of bitter almonds, the characteristic smell of hydrogen cyanide gas. In 1997, 13 years later, it was estimated that at least 150,000 people were still alive and chronically affected by the Bhopal tragedy. Between ten and 15 of these patients still die each month, even now, nearly 25 years later.

These chronically ill survivors suffer from breathing difficulties, persistent coughs, anorexia, fevers, weakness, depression, cataracts, cancer and chronic gynaecological problems. The children born to these survivors suffer from retarded growth at best, but many are stillborn. Union Carbide have never said what they thought was in the toxic cloud – they probably never knew exactly, and no one will ever know for sure now– but this lack of precise information has made treatment of the survivors particularly difficult for the doctors who care for them.

Union Carbide agreed to pay $470 million in damages, but many survivors received little or nothing, and most soon found that any money they were given was quickly consumed by medical bills. Today, the factory is derelict, but toxins from it continue to leak into the water supply and into the soil. Babies are still being born deformed and children still fail to thrive.

Union Carbide is now part of the Dow Chemical Company, of napalm and Agent Orange fame. Dow is the world's richest chemical corporation but refuses to clean up the derelict factory, claiming that it is not responsible for it. The book *Five Past Midnight in Bhopal* by Dominique Lapierre and Javier Moro explains this whole sorry story in great detail. And its sales help to fund the charity for the Bhopal survivors.

The largest nuclear accident in history so far

Unfortunately, Chernobyl can easily claim this title. The effects of this environmental disaster were not only felt in this city near Kiev, in north-central Ukraine, but eventually reverberated throughout the world.

In April 1986, a series of explosions at the nuclear power station resulted in a serious release of radioactivity into the atmosphere. Because of prevailing winds, it spread quickly to the neighbouring republics of the former Soviet Union and to a number of countries in Europe. Radioactive caesium drifted over Western Europe, including the British Isles. Heavy rain then washed it down to earth, where it was absorbed by plants, through their roots. These plants were then eaten by livestock, including sheep and cattle.

The city of Chernobyl, with a population of about a quarter of a million people, was evacuated. Many died during the explosions or shortly afterwards, while others suffered from exposure to the radioactivity, and died from cancers they developed much later. Even in 1994, half a million sheep in Britain were still classified as affected and could not be sold for slaughter until their meat had been checked for contamination caused by the radioactive caesium, which was released at Chernobyl eight years previously.

Chernobyl happened while it was still part of the Soviet Union, only seven years after the Three Mile Island incident, an accident that occurred at a nuclear power station in the USA. This American incident resulted in damage to the uranium in the reactor core and led not only to a reassessment of safety standards, but also provoked a strong opposition to any future expansion of the nuclear industry in the West. However, Chernobyl, still part of the old communist regime of the USSR at that time, was governed from afar where they preferred to do things the 'Russian' way, and so there was no reassessment of their safety standards in those intervening seven years.

The Camelford disaster (detailed in Chapter 11), which poisoned 20,000 people in Cornwall through contaminated water, two years after Chernobyl, is another prime example of a man-made misfortune. And a final example: the rapid growth of industry in China in recent years, with factory owners greedy for profit, has resulted in mass poisonings of many families living close by. Factories and smelters have spewed out pollutants on a massive scale, without carrying out even the most basic environmental monitoring. Thousands of children and their parents have been found to be suffering from lead poisoning as a consequence. At least two regions, Shaanxi and Hunan, have such problems, yet the government's promises to move villages to safer areas or to control pollution have not been kept. Some of the children's lead levels are more than ten times the level considered safe in China and public anger is rising fast. Elsewhere in China similar problems near the smelters of other metals, including zinc, cadmium and indium, are causing further public anger and protest.

An African natural disaster

Another example of mass poisoning and mass fatalities on a slightly smaller scale comes from Africa. But this time it was no accident, for this was a natural disaster. On 21st August, 1986, more than 1,700 people died of carbon dioxide poisoning. They all lived within a 15-mile radius of Lake Nyos in Cameroon. Two years earlier, 37 people died in a similar way, at nearby Lake Monoun.

The cause of the problems was the underlying volcanic activity that resulted in the release of a large quantity of carbon dioxide gas from beneath the lakes. This gas is colourless, tasteless and odourless – it is a silent killer that asphyxiates its victims. This is likely to happen again if the lakes are disturbed by further volcanic activity in the future. According to scientists who have studied the problem, the safest solution is to permanently relocate people away from the lakes.

Carbon monoxide is another colourless, tasteless and odourless gas, which is poisonous and highly flammable. Carbon monoxide burns with a characteristic blue flame, combining with the oxygen in the air to produce carbon dioxide. But poorly maintained heating systems result in the incomplete combustion of gas and the production of carbon monoxide instead of carbon dioxide.

As little as one part in a thousand (0.1 per cent) of carbon monoxide in the atmosphere can be fatal. Both inadequate ventilation and incomplete combustion can result in the production or build-up of carbon monoxide. This type of problem is often seen with old and defective heating appliances, which have not been serviced regularly by a suitably qualified heating engineer.

The haemoglobin in our red blood cells binds to the oxygen in the air when we breathe in and carries it around the body for us. Carbon monoxide, when inhaled, takes over and binds to the haemoglobin in the blood in preference to oxygen, resulting in death by asphyxiation and tissues coloured a bright cherry red. This happens because carbon monoxide has a far greater affinity for haemoglobin than oxygen does. As a result it forms carboxyhaemoglobin, in preference to oxyhaemoglobin, fatally depriving the body of the essential oxygen that we all need to sustain life.

Malign menthol and poisonous peppermint

Many products are peppermint flavoured, but today this is frequently an artificial flavour rather than the naturally occurring peppermint oil, which was widely used in the past. Peppermint oil contains about 50 per cent menthol, so poisoning involving peppermint is generally regarded as being due to its menthol content.

Two patients who became addicted to peppermints about a century ago both suffered from atrial fibrillation, a disorder of the heart's normal rhythm. Once the peppermint sucking stopped however, the normal cardiac rhythm was restored.[1] An American patient suffered exacerbations of his asthma, with wheezing and difficulty in breathing, which were found to be related to his use of toothpastes containing peppermint, or wintergreen, as flavourings.[2]

In Australia, two patients had recurrent attacks of muscle pain when they consumed large quantities of confectionery flavoured with peppermint oil.[3] A dentist also reported a case where a patient suffered from an acute allergic reaction, affecting the mouth, neck and throat, which was eventually traced to the peppermint oil in toothpaste.[4] And, a few years ago, a baby died after being given peppermint water that was incorrectly prepared, at ten times the strength it should have been.

A 31-year-old woman was found to be hypersensitive to menthol; the symptoms produced were urticaria (hives), flushing and headaches. This woman reacted to a number of products containing peppermint, including a cream, peppermint-flavoured toothpastes and sweets and mentholated cigarettes, and she was so sensitive that she even reacted to mint jelly.[5]

A 13-year-old boy developed ataxia with unsteady gait and uncoordinated movements, confusion, euphoria, rolling eyes and double vision following the inhalation of 5ml of Olbas Oil, instead of the recommended few drops. It was considered probable that the menthol in the preparation was responsible for his symptoms, as the amount of menthol inhaled in this case was about 200mg. The fatal dose in humans is estimated to be about 2g.[6]

Amazing alkaloids

Colchicine is a potent alkaloid, derived from the meadow saffron corm, used in the treatment of gout. The precise details of this are explained in Chapter 17. One tablet is taken every two-three hours, until relief of pain is achieved, or until vomiting and/or diarrhoea occur (because vomiting and diarrhoea are the first signs of toxicity).

Despite using the recommended dosage, some patients have died. In 1966, a patient with obstructive jaundice developed fatal problems with their bone marrow and its production of blood cells.[7] Doctors considered that the liver problems enhanced the toxicity of the drug in this case. Another man who took a 500 microgram tablet daily for three years as a treatment for gout developed fatty stools due to reduced absorption of fat from the diet, as well as a lymphoma, or malignant tumour of the lymph nodes, in part of the intestines called the jejunum, as a result.[8]

In 1947, a 23-year-old woman took a 50mg dose of the alkaloid atropine by mistake. This was 50 times the therapeutic dose but fortunately, because of the prompt treatment given, she recovered.[9] A year earlier, another patient in America, who had taken 1g of atropine by mouth, was successfully treated with a potassium permanganate stomach wash-out, followed by intravenous dextrose and phenobarbitone sedation.[10] In the 1930s, a man who had been taking regular daily doses of belladonna tincture, which contains atropine, for many months with considerable benefit began to lose ground. After a time he suddenly showed great improvement again, which could not be accounted for until he admitted that he had started to take double doses of his medicine.[11]

Aconite is a plant that is very poisonous because of the alkaloids it contains. In the past it was used both internally and externally, but it is now regarded as too dangerous to use and other much safer agents are now available.

In 1911, a patient swallowed enough aconite liniment to kill six people. The treatment given to counteract this overdose included Ipecacuanha Wine as the first antidote, because it acts as a cardiac depressant. Mustard emetic was also given to save the enfeebled heart. The head was kept low and the feet raised, a mustard plaster applied to the heart and hot flannels to the extremities and the abdomen. Strychnine and digitalis were given by injection and brandy was administered as an enema. Artificial respiration was continued unceasingly for six hours, after which this extremely lucky patient recovered.[12]

In 1910, a patient accidentally took a tablespoon of a liniment that contained not only aconite, but also belladonna, chloroform, red pepper and wintergreen oil. The patient fortunately vomited and was then given an injection of strychnine plus an enema of hot, strong coffee, and as a result he made a remarkable recovery, considering that the amount of aconite taken was 35 times the maximum oral dose of the official tincture.[13]

Another case of acute poisoning was reported in India in 1935, where a patient accidentally swallowed about 25ml of A.B.C. Liniment, which contained Aconite, Belladonna and Chloroform. After all the usual treatment measures failed, the poisoning was successfully treated using intravenous saline. All the symptoms abated and recovery was complete within 72 hours.[14] Some 30 years later, a 77-year-old man who drank 60ml of camphorated oil by accident developed vomiting and convulsions. He was treated with haemodialysis using eight litres of soya oil for four and a half hours, during which time 6.56g of camphor was removed. Amazingly, he recovered.[15]

Solvents and pickles

Carbon tetrachloride was used as a solvent and dry cleaning fluid for many years. Cases of fatal kidney damage have occurred due to long-term exposure to, and inhalation of, the vapour. In 1946, three patients each took a mixture containing carbon tetrachloride, followed by a dose of Epsom Salts. An hour or so later, they were all vomiting and had diarrhoea, headache, general pains and, in two cases, cramps in the arms and legs. They were treated with a high carbohydrate diet, calcium lactate taken by mouth and calcium gluconate and glucose given intravenously, which led to their eventual recovery.[16]

In 1949, *The Lancet* carried a report of a 53-year-old man who died after drinking about 50ml of carbon tetrachloride while under the influence of alcohol. Death was delayed for 24 days, and the post-mortem showed that the kidneys suffered the most damage.[17]

Diluted formaldehyde is the liquid used to 'pickle' tissues to preserve them for future examination. In 1912, the *British Medical Journal* reported a case of formaldehyde poisoning. The patient drank about 75ml of a four per cent solution of formaldehyde and lost consciousness within three minutes. The stomach was washed out but the patient died. At the post-mortem examination it was found that the oesophagus and stomach had both been extensively corroded by the formaldehyde.[18]

Fatally cheap 'alcohol'

Chemicals frequently have more than one name. One such chemical is ethylene glycol, which also goes by the name ethylene alcohol. This substance is commonly used in antifreeze solutions. Unfortunately, it has also been used as a cheap alternative to the similar sounding ethyl alcohol, which is the 'alcohol' that we drink in beers, wines and spirits. There have been many cases of accidental poisoning, commonly among students seeking a cheap alternative to the real thing.

A number of such cases have already been recounted in earlier chapters, including the chapters on treatment and food. The problem with drinking ethylene glycol is that it is oxidised to the poisonous oxalic acid by the enzyme alcohol dehydrogenase in the liver, whereas ethyl alcohol is converted to the much safer acetic acid. While the ethylene glycol causes central nervous system depression, the oxalic acid produced causes extensive kidney damage, which follows on later. Similar cases of poisoning have happened time and time again, and each time people are killed trying to get drunk on the cheap.

Oxalic acid and its salts are all extremely poisonous. In 1968, the *Pharmaceutical Journal* reported the case of a 16-year-old girl who was accidentally given 1.2g of sodium oxalate intravenously. She died five minutes later, but her heart function was restored by cardiac massage and life support was maintained for four days, without any sign of her regaining consciousness. Post-mortem examination showed extensive damage to her kidneys. Ganglion cells throughout her central nervous system were also found to be similarly damaged.[19] Once given that fatal injection, she never had a chance.

Oh my, Spanish Fly

A hundred years ago, Spanish Fly, officially called cantharides, was in use as a medical treatment, but even then it was used sparingly. Many people have heard of the reputed claims of Spanish Fly as an aphrodisiac, but very few have ever personally tried it. However, in 1921, *The Lancet* carried a report about an inquisitive medical student who, out of misplaced curiosity, decided to give it a try. Unfortunately he used too large a dose and ended up with blood and albumin (a protein) in his urine as a result, which indicated kidney damage, but, happily, he eventually made a full recovery.[20]

Until the 1950s, Spanish Fly was mainly used medicinally on plasters, which were applied to raise blisters on the skin. The size of a blistering plaster was usually about 1 inch (2.5cm) square, and only rarely were larger ones used; such was the power of cantharides. After the 1950s, more modern treatments, such as ointments and rubs, replaced the crushed, dried beetles of Spanish Fly. Today self-heating pain relief pads are available, which will supply up to eight hours of warmth to soothe away the pain.

Having heard of its aphrodisiac reputation, people have still continued to try Spanish Fly on their own. In 1967, a 42-year-old man took a teaspoonful of a preparation which contained about 20mg of cantharidin, the active ingredient in Spanish Fly. He developed symptoms of kidney damage but fortunately responded to treatment, and his badly ulcerated mouth was treated by giving him hydrocortisone pellets to suck.[21]

In 1978, an 18-year-old woman swallowed about 2ml of a preparation containing Spanish Fly. She developed the symptoms of heart damage, as well as direct damage to her mouth, throat and pharynx. She was fortunate that her symptoms all responded to treatment, although in the past there have been fatalities due to cantharides poisoning.[22]

Even contact with just one Spanish Fly blister beetle or 1mg of cantharidin can produce distressing symptoms; for example, even brief skin contact can lead to the formation of blisters.

In the past, many substances have been regarded as aphrodisiacs and sold as such, but for most of them these effects are largely in the mind of the user, and have rarely been proven scientifically. Some are downright dangerous: indeed, one of the victims described in Chapter 2, James Maybrick, took small doses of arsenic, which he considered to be an aphrodisiac, with fatal results. Strychnine in small doses was also regarded as an aphrodisiac by the Victorians, and old habits die hard: it was only in 1989 that the American FDA put strychnine on their list of banned aphrodisiacs.

More myths dispelled

Word of mouth is a powerful thing. Myths may be based on no true fact at all, but they spread and wreak havoc when people believe them to be true.

During the 1950s in Puerto Rico, there were a number of accidental deaths caused by people eating rat poison containing phosphorus. They chose to eat it in the misguided belief that it had brain-enhancing effects or that it was an aphrodisiac – both untrue urban myths.

Two other chemicals involved in myths are dinitrocresol and dinitrophenol, which were used in the past as insecticides and to kill mites and ticks. They are both equally dangerous to use and are no longer in use because of adverse effects to agricultural workers. However, in the past, they were thought to speed up metabolism, and this possible quick fix for burning fat could not be ignored by some.

In the 1930s, dinitrocresol was used medicinally to speed up a patient's metabolism in the treatment of obesity. In 1934, a young dancer sought to reduce her weight by way of a course of treatment which included taking a single capsule of dinitrocresol daily. Her death from an overdose occurred within a few days. At the post-mortem the drug was found in both her stomach and her intestine. In her eagerness to lose weight, she had taken 17 of the 50mg capsules in a period of only three days, instead of one daily.[23]

Dinitrophenol is still available today, sold by disreputable traders heedless of its dangers. In 2003, the Food Standards Agency (FSA) issued urgent advice, to the bodybuilding community in particular, about the dangers of consuming products containing 2,4-dinitrophenol (DNP). DNP has even been available on the Internet, sold as 'fat-burner' capsules.

This chemical is known to have both very serious short-term and long-term adverse effects. As few as three or four of the capsules, if taken as a single dose, could be fatal. Smaller amounts, even less than one capsule per day, if taken long term can have serious side effects, such as blindness, due to the formation of cataracts in the eyes.

Sometimes word of mouth, like Chinese Whispers, can cause

dangerous confusion. In 1936, a fatal case of poisoning occurred in Malaya when Japanese star anise was taken in mistake for a different species, Chinese star anise. Chinese star anise, also called Aniseed Stars, is used in the East as a remedy for colic and rheumatism, and in China for seasoning dishes, especially sweets, with aniseed. The distinction between the toxic and the harmless species is not very marked, but the taste of the toxic species is pungent and bitter and its odour resembles that of oil of cajuput or cardamom. The poisonous Japanese star anise is smaller and less regular in appearance than the non-toxic version, and contains a poisonous principle called sikimin. In China, Japan and the Philippines, the poisonous variety, being much cheaper, is often sold and used as a substitute for the non-toxic variety, resulting in many cases of poisoning.[24]

Some skin problems

In America in 1936, a fisherman was found to have developed blood poisoning, with jaundice, of unknown origin. He began suffering from generalised itchiness because of the jaundice, and to treat it he was given a number of injections of ergotamine tartrate, one of the ergot alkaloids. This unfortunately resulted in him developing gangrene of the feet, which eventually necessitated amputation of his legs. A total of 19ml of the ergotamine had been injected within one week. This was a massive overdose.[25]

Today his itchiness would have been easily treated using an antihistamine, but, at that time, antihistamines had yet to be synthesised. Piriton, one of the earliest antihistamines, was first marketed in the United Kingdom in 1954, and is still widely used today.

Pyrogallol is not used internally, as it is known to be toxic to the liver. However, it was used externally for many years in the past to treat a number of dermatological complaints, despite the known possibility of poisoning if sufficient absorption of pyrogallol through the skin occurred. A fatal case of pyrogallol poisoning occurred in America in 1925. The patient was badly affected by the chronic skin disease psoriasis, which was affecting his whole body. He was given an ointment containing pyrogallol to treat it. Within five minutes of covering about two-thirds of his body with the ointment, he collapsed and died due to pyrogallol poisoning caused by skin absorption. It was estimated that the patient had probably absorbed a total of about 10g of pyrogallol.[26]

Accidents with antiseptics

Phenol used to be known as carbolic acid, and was widely used as an antiseptic for many years. It has been a long time now since it was superseded by far more effective, less toxic agents.

In 1922, a large glass bottle called a Winchester, containing about two litres of carbolic acid solution, was dropped. It smashed, releasing its contents all over the floor. A boy began to mop it up with a cloth and could not help but inhale the fumes of the carbolic acid. This quickly resulted in symptoms of severe poisoning. The boy's life was only saved by prompt treatment; his breathing improved once he was given oxygen, and then two pints (about a litre) of normal saline solution with added sodium bicarbonate were given intravenously. He made a full recovery.[27]

Mercurochrome was another toxic antiseptic and disinfectant widely used in the past. There have been many reports in the medical literature of deaths attributed to its use. In 1979, a 59-year-old woman, who had previously had surgery for an oesophageal stricture, had a two per cent aqueous mercurochrome solution applied to her surgical wounds and bedsores. By Day 22, the mercury level in her blood was very high and she died the following day of therapy-resistant shock. Aplastic anaemia was confirmed at the autopsy and tentatively ascribed to the mercurochrome treatment she had received.[28] Mercurochrome will be mentioned again in Chapter 17 on Potent Potions.

Eating arsenic and other unfortunate errors

In the early years of the twentieth century, the *British Medical Journal* reported a strange death caused by arsenic poisoning. It was known that the body could tolerate a certain amount of arsenic if taken regularly – if it was started at a low dose and gradually increased – but it was not realised that once such tolerance was achieved, in order to stop safely, a slow decrease in dose was also required. A man working in an arsenic factory was in the habit of eating 20 grains (about 1.3g) of coarse powdered arsenic each day. Wishing to give it up, the man simply stopped eating it, rather than cutting his consumption down gradually, and as a result he quickly developed stomach pains and diarrhoea, before he collapsed and died.[29]

Zinc sulphate has, for very many years, been used in astringent lotions for the skin. In 1947, a woman requiring a laxative mistakenly swallowed about 25g of zinc sulphate instead of Epsom Salts (magnesium sulphate). Vomiting and purging occurred, and were then followed by acute collapse and restlessness. Treatment with soothing liquids, morphine for pain and nikethamide, a stimulant for the central nervous system, produced a marked improvement within 12 hours. Progress was maintained until the sixth day, when the patient became semi-comatose, with symptoms of diabetes. Insulin therapy was tried but the unfortunate lady died.[30]

In 1967, a married couple both developed chronically sore skin on their wedding ring fingers, but nowhere else. Investigation revealed that the gold used to make their wedding rings contained radioactive impurities with half-lives of over 20 years.[31] If they continued to wear the rings, the soreness would persist, and they would be exposed to a continuous, though slowly reducing, dose of radioactivity. As a result of this case, it was recommended that all gold for either jewellery or dental use should be routinely screened for radioactive impurities.

Allergy to gold is rare, but it does happen. One woman developed persistent, small spotty sores on both her earlobes, which only appeared after she had her ears pierced and started wearing gold earrings. She subsequently developed hypersensitivity to gold.[32] And an unfortunate jeweller developed necrotising inflammation of the blood vessels in his fingers, induced by prolonged exposure to gold during his work.[33]

Insect powders containing fluoride became popular in the mid-twentieth century because they had the advantage over those containing arsenic in that they were cheaper, quicker acting and effective against a wide range of insects. There have been a number of fatalities resulting from people taking insect powder that contained either sodium fluoride or sodium silicofluoride by mistake, instead of their indigestion powder. Even half a teaspoonful taken in error for sodium bicarbonate has caused death within ten hours.[34]

In 1936, a patient intending to take flowers of sulphur, an old-fashioned laxative, took a dose of insect powder containing sodium fluoride. Nausea, vomiting, diarrhoea, pains in the arms and legs, problems swallowing and ocular paralysis with double vision were the symptoms exhibited. Oxygen and stimulants were used to treat the symptoms and later nikethamide, strychnine, atropine and radiant heat were all needed to cure the patient.[35]

In *The Lancet* in 1949, there was an interesting case of accidental lead poisoning. The patient had applied dressings soaked in Strong Lead Acetate Solution BP to areas of his body for some 16 weeks, which had resulted in a generalised shedding of the skin with mild lead intoxication. This was considered to be very rare as cutaneous absorption of non-volatile lead compounds was thought to be minimal until this case was reported.[36]

Some medical mistakes

Accidents can happen anywhere, and every year drug errors even occur in hospitals for a multitude of reasons: drugs are administered by the wrong route or not prepared correctly before administration; the dose may be wrong; even similar looking packaging and similar sounding names can cause errors.

In January 2001, a young man who had received some 18 months of treatment for leukaemia and whose cancer was now considered to be in remission was given an injection of vincristine, a potent anti-cancer drug, to complete his course of treatment, but unfortunately it was injected by the wrong route. The drug should have been injected intravenously, but it was injected intrathecally instead – directly into the spinal fluid, even though the syringe was clearly labelled that it should never be injected by that route. Although the error was realised within minutes of the injection being given, it was too late. There was nothing that could be done to save this patient's life. The injection led to an agonising creeping paralysis, with multiple organ failure, until the young man eventually died a month later. Sadly this traumatic incident was not unique; there had been at least 13 similar cases since 1985, all which had left patients either dead or permanently paralysed.

Drugs are frequently added to bags of intravenous fluids, such as those containing half or one litre of normal saline – that is, an isotonic solution of sodium chloride (ordinary salt) in water. This method is chosen so that any added drugs will be delivered slowly to the patient, and in dilute solution, over a period of several hours. However, if the bag of saline is not manipulated after the addition to ensure that the added drug is thoroughly mixed throughout the saline, the patient may receive the added drug in too concentrated a form. One nurse was seen to inject a concentrated solution of the antibiotic vancomycin into a patient's infusion bag of saline and not mix it in. Fortunately someone else noticed and fixed the problem, but if this error had gone unnoticed the patient would have received this antibiotic far too rapidly, which could have caused shock and even cardiac arrest.

In 1997, a man died when the anticonvulsant syrup that he should have been given to swallow was injected into him instead. There was even a case where a nurse tried to administer Calpol, the well-known children's paracetamol medicine, intravenously. Fortunately this was spotted in the intravenous-drip tubing before it entered the child's bloodstream.

Medical mistakes are not a new problem. There have been many incidents, in the past as well as in more recent years, of fatal errors, and no doubt many mistakes have caused death without anyone ever realising. But sometimes, the mistake is well known: in the churchyard at Hillswick on Shetland is a gravestone carrying the following epitaph:

Donald Robertson.
Born 14th January 1785, died 4th June 1848.
He was a peaceful quiet man,
and to all appearances a sincere Christian.
His death was much regretted,
which was caused by the stupidity of Laurence Tulloch,
who sold him nitre instead of epsom salts
by which he was killed in the space of five hours
after taking a dose of it.

Nitre was the old name for potassium nitrate, which was used in very small doses as a diuretic over 80 years ago, while Epsom Salts are magnesium sulphate, and are used in much larger doses as a laxative. The poor man probably died of heart failure.

A massive arsenic overdose

Sometimes doctors get carried away with a treatment. In 1961, a 28-year-old woman was the victim of an acute and fatal poisoning due to arsenic. Her doctor was treating her for a resistant vaginal infection caused by an organism called trichomonas, a sexually transmitted disease. Other than this infection, she was perfectly healthy when they took her to the operating theatre to administer a general anaesthetic, prior to packing her vagina with no less than 18 acetarsol vaginal tablets.

This arsenic-containing drug had been used for many years by the hospital without any serious problems. The usual starting dose was between one and four vaginal tablets, which were inserted twice a

day for the first few days and then at lengthening intervals for a period of up to two months. Why 18 were inserted to start this unlucky woman's treatment, we do not know.
By the next day, the poor lady was very ill, so the medical team gave her some chlorpromazine, a major tranquilliser which is also an anti-nauseant. On the third day, they inserted a further 12 of the vaginal tablets, making 30 in total, containing some seven grams of arsenic. The last 12 were inserted without the benefit of an anaesthetic, as the doctors were of the opinion that her illness might have been due to the earlier anaesthetic.
On the fourth day, she had a fit, her pulse was rapid and she appeared to be confused. The nurses washed out her vagina and found that all the vaginal tablets had been absorbed. She then had several more fits before falling into a coma. Towards the end, somebody finally realised that she was suffering from arsenic poisoning and administered an antidote, but by then it was far, far too late and she died.[37]

Awful handwriting pays the price

Many medical mistakes simply come down to bad penmanship or careless haste. Errors commonly occur because of the position of the decimal point, and milligrams are frequently confused with micrograms. And sometimes, the prescriber's handwriting is all but illegible.

In 1988, a patient visited his doctor to get his regular prescription for inhalers and tablets, and because he had a chest infection his doctor prescribed an antibiotic as well. Unfortunately this doctor's writing was so appalling that the pharmacist misread the prescription and dispensed Daonil, a drug taken by diabetics, instead of Amoxil, an antibiotic. The patient suffered irreversible brain damage as a result and was later awarded £139,000 in damages, with the pharmacist to pay 75 per cent and the doctor 25 per cent. This is just one example of many where drug brand names are similar or can be confused due to poor handwriting on the part of the prescriber.

When doctors make a prescribing error, the pharmacist will normally spot the error and then contact the prescriber to sort the problem out. Rarely do such errors go unnoticed and cause harm. However, sometimes errors do slip through. In 1982, a woman suffering from migraines ended up requiring excessive surgery for gangrene in both feet – all because of a prescribing error that was not spotted.
The doctor had written a prescription for Migril tablets to treat the

woman's migraine. These tablets contain ergotamine, one of the ergot alkaloids. Ergotamine is so potent that there is both a daily and a weekly limit to the dose that can be taken. Unfortunately the doctor had incorrectly prescribed a dose that was to be taken several times every day, a dose much higher than the limit.
The pharmacist dispensed the prescription complete with the incorrect dosage instructions, just as the doctor had written, with disastrous results. The lady far exceeded the manufacturer's recommended maximum dose, and by the time the overdose was discovered, several days later, she had developed gangrene. The owner of the pharmacy was deemed to be 45 per cent at fault and the doctor 55 per cent when the case came to court, and £100,000 damages were awarded to the lady.

In 1995, a handwritten prescription for the angina drug Isordil was misread as Plendil, which is a blood pressure drug. Plendil has a maximum daily dose of ten mg while the dose for Isordil can be up to 120mg when used for angina. The Plendil was dispensed complete with the dosage instructions for Isordil. The resulting overdose resulted in the patient's death following a massive heart attack. This was a tragedy for the patient's family, and was followed by some very expensive lawsuits.

In July 2000, the heart drug Amrinone was renamed Inamrinone in the USA. This change was made because the name Amrinone was being confused with Amiodarone, also a heart drug, but an anti-arrhythmic, used for totally different heart problems. There were a number of cases of serious illness, and at least three deaths, due to the confusion caused by these similar names.

Similar errors have occurred with one of the newer non-steroidal anti-inflammatory drugs called Celebrex. Originally the manufacturers had planned to call it Celebra, but this was vetoed due to the similarity with an anti-depressant called Celexa in the USA (fortunately called Cipramil in the UK). So the manufacturers decided on Celebrex instead, and the new drug was duly launched. However, it still caused errors, due to confusion with another quite different drug, an anti-convulsant called Cerebryx, which is used in Canada and the USA. This confusion was somewhat surprising, as this anti-convulsant drug is only available as an injection and Celebrex was a capsule to be swallowed.

Some years ago a woman was taking Plavix, an anti-platelet drug used to prevent future heart attacks. Unfortunately on her admission to hospital, when a doctor asked her what medication she was taking, she got the name wrong and said not Plavix, but Plaxil. There is no drug of this name so the doctor assumed that she meant the similar sounding Paxil, an antidepressant marketed under that name in Canada and the USA. In Britain, this drug has the brand name Seroxat. There followed severe disorientation for this patient, who was given the wrong medication for several days before the mistake was sorted out.

Another instance of error due to similar product names occurred when no less than five women were given injections of a steroid preparation normally used for rheumatic disease and other inflammatory conditions. This was the anti-inflammatory injection Depo-medrone, which was injected into these women in error for another drug, Depo-provera, which is a three-monthly contraceptive injection. Both preparations were long-acting (also called depot) injections, which were administered intramuscularly; both were made by the same company; they were packaged in the same company livery, in the same size of package; even the ampoules enclosed were of identical size. **Before administering a drug to a patient, the label must be read – on the box and on the glass ampoule inside – and checked by someone else. Checking afterwards is too late.** Some would say that this was an accident waiting to happen.

A doctor of my acquaintance was instructed, very early in his career, to write all his prescriptions in capital letters as his handwriting was so appalling. But his capitals deteriorated rapidly too and within a short space of time they also became virtually illegible. Fortunately, today, most prescriptions are printed out by computer, and pharmacists no longer have to try to second-guess what the doctor has written, apart from the few prescriptions that may be handwritten on emergency home visits.

Chapter 15

Mortal Combat, Conflict and Murder

POISONING DOES NOT ALWAYS happen by accident, and sometimes people clearly have murder in mind.

Poison especially comes into play in warfare and strategic assassinations. Weapons of biological and chemical warfare have been used to kill for thousands of years. Poison arrows and other projectiles, plagues of various types, together with a number of chemical weapons, including Greek Fire, naphtha and even gunpowder were used in the wars of the Ancient World. Greek Fire was a mixture of naphtha and quicklime which is set on fire on contact with water. It was used with great success against Arab ships when they were besieging Constantinople in 674-6 AD.

Unfortunately, the desire to destroy and conquer has extended beyond these battles of the past to modern warfare, and more: politics and even religion have led to assassinations and murder. In the personal realm, money and marriage have also been the cause of many, many murders.

Smoke bombs and burning bullets

Chemical weapons were first used as a means of mass poisoning in modern times in the First World War (1914-18), and have been used in many later conflicts during the twentieth century.

In World War I, burning white phosphorus produced dense white clouds of gas, which were used to hide troop movements from the enemy. Phosphorus was also used in tracer bullets and in Mauser bullets, which shot down the Zeppelin airships. These airships were filled with hydrogen, a highly inflammable gas that burns readily in air. The phosphorus-containing bullets were set alight as they sped through the air, reacting with the oxygen in the atmosphere and bursting into flames. The flaming bullets destroyed the airships by not only setting them on fire, but also causing them to explode because of the hydrogen they contained.

In WWII (1939-45), phosphorus was again used to make smoke bombs and shells, as well as Molotov cocktail bombs and phosphorus bombs, which could be dropped from aeroplanes flying over enemy territory to set fire to anything they might land upon. Phosphorus was also used in more recent times, in the late 1960s and early 1970s, during the Vietnam War. Then, the Geneva Treaty of 1980 stated that phosphorus was unsuitable for use as a weapon of warfare against civilians. However, such niceties were disregarded by the Israeli army in their bombardment of the Palestinian populace in the Gaza Strip in January 2009.

Phosphorus particles from flares and rockets have to be removed from wounds as soon as possible; otherwise they continue to burn through the flesh, causing far more extensive damage. The most effective treatment in the past involved dilute copper sulphate solution being first applied to the burning phosphorus. This was then washed away with saline solution, and sodium perborate was applied to try to prevent copper poisoning, although this could cause boron poisoning instead, which could easily occur due to absorption by the burnt areas of raw flesh.[1]

Blistering poison gases

In WWI, the Germans used chlorine gas, which was blown over the British trenches, causing some 5,000 deaths among the troops and wounding at least a further 15,000 men. Later in the war, phosgene, mustard gas and lewisite were also used. Phosgene, also called carbonyl chloride, was a poison gas chemically related to chlorine, but which had far more drastic poisonous effects.

Mustard gas was a chemical warfare agent, dichloroethyl sulphide, first used during WWI by the German army at Ypres, Belgium, in 1917. This colourless, oily liquid produced a vapour which caused blistering of the skin, swelling around the eyes and, if inhaled, stripping of the mucous membrane from the bronchial tree in the lungs. The effects of mustard gas took time to manifest themselves and led to a lingering death, some four or five weeks later, which frequently occurred in severe cases of gassing. Today, mustards are successfully used as medicines: they are used with very great care, in conjunction with a variety of other drugs, in the treatment of cancer, for which see Chapter 17.

Lewisite, an arsenic-containing substance, was another chemical weapon first used in WWI because of its poisonous nature. It was named after the American chemist, Winford Lewis, who developed its potential after its inventor, Julius Nieuwland, realised how poisonous it was in 1904 and refused to work on it any further. This chemical, which apparently smelt of geraniums, caused massive blisters on exposed skin and damaged the lungs if it was inhaled.

Fortunately, an antidote was quickly developed in Britain, and was simply called British Anti-Lewisite (BAL). When given by injection, BAL can remove the arsenic atoms of the Lewisite from the proteins, including enzymes, to which it attaches within the body. This antidote is an important drug still used today to treat patients who have been poisoned not only by arsenic, but also by mercury and other heavy metals. It is now called dimercaprol. Further details of it have already been given in the chapter on treatment of toxins.

A nerve gas is any chemical warfare agent that attacks the nervous system. The first nerve gas, called tabun, was made by the German scientist, Gerhard Schracher in 1936. Tabun was even more deadly than mustard gas because its molecule contained a phosphorus-cyanide chemical bond. During World War II, further nerve gases, including sarin, which contained a phosphorus-fluorine bond, and soman were developed. The Nazis used these agents as part of their policy to exterminate the Jews across Europe during the period from 1939 to 1945. The Holocaust resulted in the deaths of some six to seven million European Jews in concentration camps, such as Auschwitz, where they were gassed with these poisonous chemicals, or eventually died from disease or starvation.

And after the war

After the war, in the 1950s, British researchers developed even more such gases codenamed VX, VE, VG and VM, which all contained phosphorus-sulphur bonds. For several decades in the mid-twentieth century, the Ministry of Defence used the facilities at the Porton Down chemical research establishment to test nerve gases and other chemical and biological agents. Thousands of British servicemen participated in these tests and some even died as a result. Many of these human guinea pigs were under the impression that the tests being carried out were for the common cold.

All these poisonous nerve gases are organophosphorous compounds, which work by inhibiting the body's ability to break down the neurotransmitter acetylcholine. This ultimately results in signal failure, as the nerve commands, including those to the lungs, are no longer being transmitted, causing respiratory failure. Nerve gases are particularly nasty because they do not even have to be inhaled – they are absorbed directly through the skin, and some are so powerful that they can be fatal within minutes.

In 1969, there was a leakage of VX vapour from a plant in Newport, Indiana, USA, where American nerve agents were manufactured. The vapour drifted across the appropriately named Skull Valley and killed all 60,000 sheep that were grazing there. In the Iran/Iraq war (1980-88), the Iraqis, led by Saddam Hussein, used mustard gas and nerve gas to devastating effect on the Kurdish town of Halabja, killing some 5,000 civilians. These gases were also used in the offensive against the Kurds in 1987-88, when 100,000 Kurds – mainly civilians – were killed.[2]

More recent attacks

In April 1995, there was a sarin attack on the Japanese underground in Tokyo, killing 12 people and injuring 5,000 others. This was carried out by the Aum Shinrikyo doomsday sect, a sort of religious cult. This same group had carried out an earlier attack in Matsumoto in 1994, killing seven and injuring 200. Cult members even used VX to dispose of a former member in Osaka in December 1994. They stopped him in the street, squirted VX on the back of his neck and ran off. He tried to chase them but collapsed in a coma and eventually died ten days later.

Even today, the regime of Kim Jong Il in North Korea uses poison gases in its prison camps. Dissidents and their families are imprisoned, systematically tortured and experimented on in gas chambers, where the scientists test chemical weapons on the prisoners while also executing some of them.

The 700,000 troops who were sent to fight in the Gulf War in 1991 were not only at risk of nerve gases and other chemical warfare agents coming from the enemy; they were also at risk from the supposed protection provided by their own side – they were exposed to organophosphate poisoning due to everything being drenched in insecticides in an effort to prevent malaria. The American and the British troops were all injected with a cocktail of vaccines and given tablets to take as well, but not told exactly what they were being given.

Tens of thousands of the troops, American and British, developed Gulf War Syndrome as a result. This collection of mainly neurological symptoms, including chronic fatigue, dizziness, amnesia, digestive upsets and muscle wasting, has caused many of the troops to become so disabled that they have had to be retired from the services. Many of them are now so disabled that they cannot work at all.[3] However, it is only very recently that the British Ministry of Defence has even admitted that Gulf War Syndrome exists, let alone paid out compensation for it.

Tear gases are another group of gaseous substances. They specifically affect the mucous membranes of the eyes, causing irritation and copious watering. Designed as a means of riot control, these agents are used by police forces in many countries. Usually tear gases are organic halogen compounds. Among the most widely used are Mace and CS gas, with the more recently introduced pepper spray as a safer alternative. Their effects are designed to be quickly disabling to the recipients, but unlike the chemical warfare agents, they leave no lasting damage once the lacrimatory – tear inducing – effect has worn off.

Murder most politic

Political assassinations using poison are almost as old as politics themselves. The Egyptians, Greeks and Romans all murdered in the name of politics, and some examples have been given in earlier chapters. Centuries ago, poisons such as arsenic were used, especially in Europe, in so-called 'succession powders'. These were used to remove kings, dukes and even popes from power, in order to allow a preferred successor to reign.

Generations of the Borgias turned the Vatican and the Pontificate into a long-running scandal. It is believed that they used a white powder, La Cantarella, which probably contained arsenic, lead and phosphorus, to poison their enemies. In the fifteenth century, Cesare Borgia's father became Pope Alexander VI, and Cesare became a Cardinal at the rather young age of 18.

This Pope had a cabinet in his Vatican apartments, with a stiff lock, and he would ask his victims to unlock it for him. There was a sharp point in the handle of the key, which would prick the person's hand and so poison the soon-to-be-dead victim.

Eventually the Pope and his son Cesare were both poisoned with some toxic wine. There are a number of variations among historical sources as to how this happened, but the outcome remains the same. The father died, but his son survived the poisoning and continued killing his enemies. Cesare had a large ring, which had twin lion heads with sharp teeth on it. His victims were greeted with a hearty handshake, and the poisoned teeth did the rest.

In the early eighteenth century, a Sicilian woman named Toffana was arrested in Naples, tortured and then executed by strangulation. This was in retribution, as it was said that she had murdered 600 men with a poisonous liquid. This liquid came to be known as Aqua Toffana and was claimed to be a transparent, tasteless liquid of which five or six drops would be fatal.

The drink's composition remains unknown, although it probably contained arsenic, as either sodium or potassium arsenite, and possibly a little belladonna. It produced death slowly, without pain, inflammation, convulsions or fever. Instead, it caused a gradual decay of strength, disgust for life, lack of appetite but constant thirst, leading eventually to consumption and death. Some used this liquid to get rid of their romantic or political enemies.

Some well-known victims

Generally the names of the victims of crimes tend to be forgotten, while those of the criminals go down in the annals of history. The political murder of Georgi Markov using ricin has already been presented in Chapter 3. Here are a few more stories about some well-known victims, what – and who – poisoned them and why.

In 2008, analysis of the remains of China's second-last Emperor found lethal levels of arsenic in the hair and bones. The Guangxu Emperor died in 1908 at the age of only 38. The analysis suggests that he may have been poisoned to stop him introducing reformist plans. It may never be known whether it was the dying Empress Dowager Cixi, her faithful eunuch courtier or her commanding army general who was responsible, but the Guangxu Emperor was certainly poisoned with arsenic and died in the Water Terrace Pavilion in the Imperial Palace complex.

Cixi died the following day in the nearby Graceful Bird Pavilion, also within the Imperial Palace complex. The Empress Dowager had imprisoned Guangxu for over a decade as a punishment for a show of independence, as he had ambitions to reform and revive the faltering Qing dynasty, which ended three years after his death. The modern analysis was able to rule out chronic low-dose poisoning, by comparing his arsenic levels with those of his wife, which were over 250 times lower.

In the early twentieth century, poison was still considered to be a useful means of disposing of people who could be regarded as political enemies. Rasputin was a peasant monk who arrived in St Petersburg at a time when mystical religion was fashionable in Russia. He obtained an introduction to the royal household and quickly gained the confidence of the Tsar, Nicholas II and the Empress Alexandra during World War I.

It appeared that, through hypnosis, he was able to control the bleeding of the haemophiliac heir to the throne. However, he caused a public scandal due to his sexual and alcoholic excesses, as well as using his influence politically. He was murdered in 1916, by poisoning, probably with cyanide. His murderers were a group of aristocrats, led by a distant relative of the Tsar.

In late August 1978, Pope John Paul I was elected following the death of Pope Paul VI. Only 33 days later he was dead, supposedly of a heart attack, despite a medical examination three weeks previously that had shown him to be in good health and with no family history of heart trouble. Many suspected that he was poisoned, with all the evidence carefully removed in a Vatican cover-up. It is believed that he was handed a poisoned drink in a champagne glass by the now deceased Cardinal Villot, who at that time was Secretariat of State and who orchestrated the cover-up.

Some medical murders

A number of doctors in the past have used their knowledge of the poisonous properties of various medicines and other preparations to dispose of unwanted girlfriends, wives and other relatives. A number of them have already been mentioned earlier in this book, and here is another big one.

Doctors and nurses have been known in the past to end the lives of terminally ill patients by injecting potassium chloride, which results in rapid death by heart attack. Indeed, this is the 'non-toxic lethal injection' used quite legally in some states in America as a form of capital punishment. The prisoner is injected with a barbiturate to induce sleep, with a muscle relaxant and then with a concentrated solution of potassium chloride to stop the heart.

Dr J. Milton-Bowers likely poisoned both his second and third wives similarly, with phosphorus; his first wife may well have gone the same way, but we shall never know. As a doctor he had easy access to phosphorus as it was a common, though useless, ingredient of medicines at that time.

The doctor was born in 1843 and married his first wife, Francis Hammet, when he was only 21 years old. The marriage didn't last too long as she died in 1865. Wife number two was an actress called Theresa Shirk, whom he met in New York. She died in 1881.

Within barely six months of her death, he had married his third wife, Cecilia Benhayon-Levy. This lady's parents disapproved of the marriage because they had heard rumours that the doctor had poisoned his second wife, Theresa. They were right to be suspicious,

as their daughter, the third Mrs Bowers, died in agony in 1885. Her body was grossly swollen at the time of her death and her family demanded a post-mortem to determine the cause. This showed that she had died from phosphorus poisoning.
The doctor already had a fourth wife lined up but was arrested, tried and found guilty of murder before he could marry her. Two years later, Cecilia's brother committed suicide by taking cyanide and left a suicide note confessing to the murder of his sister, the third Mrs Bowers. Investigations revealed that this brother had rented an apartment to the husband of the doctor's housekeeper. The police suspected that this man might have poisoned the brother and forged the suicide note, acting on behalf of the doctor.
However, at his trial, the housekeeper's husband was acquitted. This meant that the brother's suicide and the confession that he had murdered his sister were the truth. So the doctor's first trial was then set aside and a retrial was dismissed. Once released from prison, Dr Milton-Bowers sought out the young lady he had been seeing before the trial and they married. She, at long last, became wife number four, and the couple lived happily ever after, until the doctor died in 1905.

The case of the pinpoint pupils

Many doctors use morphine, laudanum or heroin to kill, including the notorious Dr Shipman, whose case is described at the end of Chapter 20. Today the use of morphine and other opium-derived drugs is controlled by the Dangerous Drugs Act (1965) and the Misuse of Drugs Act (1971), but this was not so in the past.

Dr Robert Buchanan qualified in Edinburgh, and in 1886 took his wife to New York, where he set up in general practice. His practice prospered by day, but by night he frequented clubs and brothels. It was in one of these brothels that he met Anna Sutherland, a brothel madame, who was also one of his patients. At that time she was unmarried. In 1890, Dr Buchanan divorced his first wife because he had found out that she had had an affair.
Soon after this, he persuaded Anna to make a will leaving her money to her husband, should she be married at the time of her death or, if she died unmarried, to her friend Buchanan. Having ensured that either way he would inherit her wealth, he then married her a few weeks later, even though she was twice his age. The doctor continued to enhance his professional reputation but feared he would lose patients if they discovered the embarrassing nature of his wife's

business.
In 1892, he told his new wife that he proposed to return to Edinburgh for further study, and that he intended to go alone. His wife was not pleased at being left behind and threatened to cut him out of her will. Nonetheless, he bought only one single ticket, rather than two return tickets. Four days before he was due to set sail, he told some of their friends that Anna was seriously ill. Another doctor was called to examine her, but she lapsed into a coma and died within a few hours. The cause of death was given as cerebral haemorrhage and Buchanan inherited $50,000, with little sign of grief.
The following month, a business partner of the deceased brothel keeper, who also happened to be a disappointed suitor, visited the coroner. He had suspicions about the cause of death and told the coroner that he suspected murder by the doctor, in order to inherit the money. The coroner was not very interested in his suspicions, until a New York newspaper heard about the story and began asking questions.
Only the previous year, in 1891, the paper had followed up on a murder case, in which pinpoint pupils were a feature in the diagnosis of morphine poisoning. In the case of Dr Buchanan, the newspaper had discovered that he had remarried his first wife within but a mere month of the death of his second wife, and so printed a news story about it. The newspaper made a connection between the two stories, and this led to an exhumation, which was followed by a post-mortem. This examination showed that the cause of death was not cerebral haemorrhage at all, but was indeed morphine poisoning.
But the pinpoint pupils, which were a known feature of such poisoning, were absent. With or without this one sign, the doctor was put on trial in March 1893. During the trial, someone remembered that at the time of the 1891 case, Buchanan had been consulted and had said that pinpoint pupils could be disguised by using belladonna eye drops, which have the opposite effect on the pupils. Which is precisely what he had done. Buchanan was found guilty of murder and died in the electric chair in 1895.

Pinpoint pupils again

Dr Robert Clements married his first wife in 1912, when he was only 22. She died in 1920 of sleeping sickness, and the following year he remarried, but he was widowed again in 1925 when his second wife died of endocarditis. In 1928, he married for the third time; this marriage lasted somewhat longer, as his third wife didn't die until

1939, supposedly of cancer. And, in due course, he married yet again.

The doctor was a Fellow of the Royal College of Surgeons, and this case illustrates well how some people can abuse their position of trust to their own ends. One evening in May 1947, Clements called a doctor to examine his current wife (number four) at their home in Southport, Lancashire. Two doctors came but the lady was unconscious by the time they arrived. After examining her, they then arranged to transfer her to a nearby nursing home, where she died the following morning.

Her husband had told the doctors who attended her that she had myeloid leukaemia, and so that was the diagnosis entered on to the death certificate. A third doctor then performed a post-mortem and confirmed the diagnosis as myeloid leukaemia. The other two doctors were somewhat suspicious of this as a cause of death, as they had noticed that her pupils were pinpoints. So, suspecting an overdose of morphine, they contacted the coroner.

Investigations began and it became clear that Mrs Clements' health had deteriorated over a considerable period of time, with the symptoms suggesting that she was a morphine addict. However, it was then discovered that Clements had prescribed large doses of morphine for one of his patients, but this patient had never received them. A second post-mortem was then ordered and the pathologist was surprised to find that some of the body's organs were missing. They were conveniently destroyed after the first post-mortem. Despite this, further examination showed that death had indeed been due to morphine poisoning.

The police then called on Clements but found him dead. He had committed suicide – using morphine, of course. Then the doctor who had performed the first post-mortem also committed suicide, but he had used cyanide. He left a note saying he could not live with the terrible mistake he had made, confirming the cause of death as myeloid leukaemia when it was really murder by morphine.

Clements was in due course found to have murdered his wife. She was heir to a fortune that he would have inherited. Further investigation showed that all his marriages were for money. In each case, the doctor signed the death certificate on each of his four wives. Some suspicions were raised after the death of wife number three, but she had been cremated by the time anyone official thought to investigate. The authorities decided not to exhume the bodies of his first two wives.

Agatha Christie's book

Thallium has been mentioned a number of times in earlier chapters, and has been used as a poison both in fiction by Agatha Christie and in real life by political regimes to remove opponents, as was the case in the 1980s and 90s by the Iraqis.

Christie worked in the dispensary at University College Hospital in London during World War II and became well aware of chemicals and their poisonous effects. In 1961, her book *The Pale Horse* was published, in which the plot involved an accurate description of thallium poisoning. In 1977, some years after the author's death, a young Arab girl was flown to the UK and admitted to hospital. After several days of further deterioration, she began to lose her hair. The doctors were mystified by her condition, but one of the nurses caring for her happened to be reading *The Pale Horse*, and was struck by the similarity of the symptoms. Subsequent testing of her urine revealed thallium, and following treatment for poisoning she was able to return home three weeks later.

In 1971, a man called Graham Young used thallium sulphate, purchased from a chemist's shop, to poison his workmates. Several were taken ill and two of them died. The full story can be found in Chapter 2.

Arsenic, that most popular poison

Amy Hutchison had a violent husband, whom she had married 'on the rebound', having been jilted by her first love. Some years later when she met up with this previous lover, she decided to get rid of her brute of a husband. So, she poisoned him with arsenic and then ran off with her lover. In 1750, this British murderess was caught, tried and then executed, first by strangulation and then by being burnt, for killing her violent husband.

In the nineteenth and early twentieth century, arsenic was readily available as a weedkiller as well as on flypapers, from which it could be easily extracted by simply soaking them in water. Many a poisoner used it in repeated small doses so that the resulting illness was prolonged and mistaken for some sort of wasting disease, as shown by the Maybrick and Seddon cases discussed earlier in Chapter 2.

Fortunately, in 1836, James Marsh developed a chemical test that could detect very small amounts of arsenic. It took some years for this test to be accepted by the authorities, and so it was not until much later in the nineteenth century that the Marsh test became more widely used. This test meant that, at last, it was possible to look for the presence of arsenic in suspected cases of poisoning, and the poisoner could be brought to justice.

But arsenic has still been a powerful tool in the past, as the following five stories show.

A holiday romance

Madelaine Smith was the eldest daughter of James Smith, a successful Glasgow architect. She was an educated girl who filled her days with artistic pursuits. In the spring of 1855, she began a secret affair with a packing clerk from Jersey, Pierre Emile L'Angelier, whom she met while on holiday at the family's country home. Her maid acted as the go-between during the holiday as the couple exchanged love letters.

Once back home in Glasgow, contact was rather more difficult and her father, not knowing about the affair, was on the lookout for a suitable husband for her. The young people eventually quarrelled and Madelaine requested the return of her letters. The suitor threatened to send them to her father instead. Madelaine begged him not to and renewed the affair in an attempt to appease him.

Two years after the affair started, in March 1857, Pierre suddenly became ill and died. A post-mortem showed his body contained a lot of arsenic. Once the love letters were found and the affair discovered, Madelaine was arrested and charged with murder. She made a statement admitting that she had bought arsenic, but claimed that she only used it as a cosmetic.

The prosecution claimed that she had grown tired of her lover and had administered the arsenic to him in a chocolate drink. The defence made the dead man out to be a seducer and blackmailer, who was known to take arsenic and who also had a history of gastric complaints. The jury brought in the Scottish verdict of 'not proven', which can be interpreted as 'she almost certainly did it but–'. Madelaine went on to live to the ripe old age of 93, marrying twice, the first time to an artist in London and then later again in America.

The fortune-hunting doctor who got away

Thomas Smethurst was born in Lincolnshire in 1805. He had two brothers and a sister, and his father is believed to have been a herbalist. Thomas was a self-styled doctor, having obtained his degree from a university in Bavaria. In 1827, he married one of his patients, a lady some 20 years older than him. She brought a modest fortune to the marriage, which Thomas used almost 20 years later to buy a hydropathic establishment at Moor Park in Surrey.

Such spas were very fashionable at the time and, after only six years, this successful and profitable venture allowed him, in 1852, to retire and become a gentleman of leisure. By 1858, the couple were living in a boarding house in Bayswater, London, and Mrs Smethurst by now was a semi-invalid in her seventies. In the autumn of that year, a new lodger arrived at the boarding house: a spinster of some means, Isabella Bankes. The doctor and Miss Bankes were instantly attracted to each other, and soon their friendship resulted in gossip amongst the other guests. Indeed, the gossip led the landlady to suggest to Bankes that she should move elsewhere, which she quickly did. The doctor left his wife and went with her.

In December 1858, Dr Smethurst bigamously married Bankes at Battersea Parish Church, and they took rooms and lived happily together until the following March. Isabella now became ill, the symptoms suggesting dysentery. She was so ill that Smethurst called in some other medical colleagues. However, the medicines prescribed failed to help, and so a move to healthier quarters was suggested. They moved quickly to new rooms in a house on Richmond Hill.

The two doctors attending Isabella now thought that poison of some sort might be responsible for her illness. But Thomas did not seem like a poisoner. He was most solicitous in his care of Isabella, never permitting anyone to see her unless he too was present. He wrote to her sister, who visited but did not stay long. He called in another prominent doctor and, at the end of April, he summoned a solicitor, who was shown a draft of a will. This draft, wholly in Thomas Smethurst's writing, left all Isabella's property to her 'friend' Thomas.

A final will was then drawn up by the solicitor and signed quite willingly by Isabella. This disturbed the two local doctors, who now became suspicious enough to test the bodily evacuations of their patient. They contacted the police as a result of their findings and Thomas was arrested. Following a search, more than 20 medicines were removed from their lodgings, but despite everything, the poor lady died early in May 1859.

Smethurst was arrested, but the local magistrate was not impressed by the evidence against him and set him free. He was not free for long though, as the local coroner reviewed the evidence and issued a warrant for his re-arrest. An early forensic test for arsenic proved positive and Smethurst was sent for trial at the Old Bailey. It was revealed in court that Isabella had been two months pregnant at the time of her death. The doctors were of the opinion that arsenic and possibly antimony, bismuth or potassium chlorate may have been taken by the deceased. It was clear from his summing up that the judge thought Thomas guilty, and so did the jury, and so a sentence of death was passed.

However, the public now decided this was wrong and there was a great outcry. Further investigation led to Smethurst's reprieve two days before the date set for his execution. On his release, he was promptly re-arrested for bigamy, for which he was sentenced to a year's hard labour. After his release he went to live in a house off the Vauxhall Bridge Road, London. Although he was quite comfortably off, he then sued for Isabella's legacy, and much to the distress of her family, he won.

The solicitor's tyrannical wife

Major Herbert Armstrong was a small, mild-mannered ex-soldier and provincial solicitor who practised law in Hay-on-Wye. His marriage was not a happy one; his wife was somewhat of a tyrant and when she died, early in 1921, everyone judged the Major's loss to be a happy release. She died after an attack of abdominal cramps, which the doctor had diagnosed as neuritis. A few months later, the Major had a dispute with another local solicitor. The rival was invited to tea and ate a buttered scone and a slice of currant loaf and became ill with severe stomach pains shortly after returning home.

The doctor was summoned and diagnosed a bilious attack and appropriate medication, but the sickness continued and the doctor decided to investigate. On analysis he found arsenic and then realised that the symptoms of arsenic poisoning were very similar to those of neuritis, for which he had treated Mrs Armstrong before she died. The doctor went to the police with his suspicions. The wife's body was exhumed and tests found a quantity of arsenic in the stomach. The Major was arrested and duly tried, pleading 'not guilty', but the jury convicted him and he was sentenced to death. He was executed by hanging in 1922.

The gypsy lover

Charlotte Bryant was an ugly-looking young woman, illiterate, slightly retarded and she drank too much. She was married to a very poor farm worker called Frederick who, people said, was not the father of all her five children. They lived in the village of Coombe in Dorset, where Charlotte was much in demand by the men of the village. Her prostitution was a way of earning a little extra money to help make ends meet. Charlotte dreamt of meeting her ideal man one day.

In December 1933, she met a gypsy called Leonard Parsons, and fell head-over-heels in love with him. She even invited him to dinner on Christmas Day, without even consulting her husband. Leonard and Frederick fortunately got on together so well that within days Leonard was invited to move into the spare room, as he needed a place to stay. Needless to say, it did not take long before Charlotte and Leonard became lovers. They quickly began jumping into bed as soon as Frederick left for work each day. When Frederick eventually found out, he told the gypsy to leave immediately, which Leonard did.

Three days later, Charlotte received a telegram from her lover, asking her to meet him, but as Charlotte could neither read nor write she showed the message to her husband, asking him to read it. Frederick accompanied Charlotte to the meeting place and told Leonard to stay away from his wife, but, after further discussion, the two men made friends again and shook hands on it. Just two days later, Leonard returned to the Bryants' spare room, and also to his place in Charlotte's bed.

Charlotte was madly in love and told Leonard how wonderful it would be for them to spend the rest of their lives together. At the thought of this, poor Leonard's ardour began to cool. Charlotte became aware that he might leave and decided she must act quickly to ensure that he would stay. She went to Yeovil the next morning and bought a tin of arsenical weedkiller from a chemist's shop, signing the poison register, as was legally required, with a cross.

She hurried home and laced her husband's lunch with the weedkiller. Frederick was seized with violent stomach pains that afternoon and so Charlotte sent for the doctor. By the time the doctor arrived, it was evening and Frederick was now complaining of not only stomach pains but also of cramps in his legs; the doctor, however, diagnosed a bout of gastro-enteritis. Frederick recovered and returned to work but then went down with another bout of the same symptoms.

The gypsy lodger then announced that he had decided to leave, and after he had departed Charlotte became quite desperate to get him back. Her husband's third bout of gastro-enteritis happened the next day and, this time, it was so severe that the doctor put him on a special diet. A few days later, just before Christmas 1935, Charlotte prepared her husband's lunch, again heavily laced with arsenic, leaving it out for him.

She set off to visit Weston-Super-Mare, as she had heard that her beloved gypsy was camped near there. She could not find him and by the time she got home again her husband was in a far worse state. The following day, he was taken to the hospital at Sherborne, where he died a few hours later. A post-mortem revealed the arsenic in his stomach and the police duly arrived at the Bryants' cottage.

They sent Charlotte and the children to the local workhouse while they searched the house and garden. Although they found traces of arsenic in dust and dirt swept up in the cottage, there was not enough to charge Charlotte, as arsenic was commonly used on farms in those days. They did find an empty green-coloured tin, which had contained weedkiller, in the garden, but had no way of proving that Charlotte had bought it.

A friend of Charlotte's, however, was her undoing. She told the police that Charlotte had shown her a green tin, saying that she must get rid of it. Several days later, this friend found the same tin in the ashes of the boiler, and had thrown it into the backyard. She also told the police how Charlotte had told her that she hated her husband. Charlotte was also reputed to have said 'if they can't find anything, they can't hang me' shortly after Frederick's death.

Charlotte was now arrested and sent for trial at Dorchester Assizes. It transpired that the ashes of the boiler had also been analysed, and an unnaturally high level of arsenic was found in them. The jury found her guilty and her appeal was eventually rejected. Her hair turned quite white in the few months while she waited for her execution, by hanging, in July 1936.

Another murderous woman

Audrey Marie was born in Alabama in 1933. In 1945, her family moved from Blue Mountain to Anniston, and Marie went to junior high and then on to the high school there. She met Frank Hilley when she was only 12, and in due course they courted and married in May 1951. Their first child, Michael, was born in November 1952 and Carol, their daughter, in 1960. Marie began spending more than they could afford, and, by the end of 1974, she had even rented a post office box and diverted all bills to that address to hide her overspending from her husband.

It was at about this time that Frank found her in bed with her employer. During this year, Frank had been sick rather a lot, with the doctor prescribing various medicines to treat him. None of them seemed to help much, so eventually the doctor carried out some tests, which showed that Frank had liver failure. New medications didn't help, his condition worsened and he became jaundiced and died. The cause of death was given as infectious hepatitis. The death certificate stated that the death was due to natural causes so Marie had no trouble collecting the life insurance of about $31,000. Marie then went on a spending spree.

At about this time, there were a series of mysterious fires in her house, as well as in neighbouring ones. There were also harassing phone calls and many other complaints to the police. Unknown to the family, Marie had begun buying insurance, not only for herself, but also to cover her children too. Her daughter, Carol, became ill and was eventually confined to hospital. Carol's brother, Michael, began to suspect that his father's illness was not what it seemed and was now suspicious of his sister's illness. By now, cheques were bouncing and the bank was filing charges.

Eventually Michael told the wider family about his suspicions, and they managed to convince a doctor, who took them seriously and examined his patient very carefully. He found tell-tale white lines on Carol's fingernails, which are characteristic of arsenic poisoning and indicated that the poisoning had been going on for a period of months. Michael now contacted the coroner about his father's mysterious illness and death. The body was exhumed and found to have arsenic present in the tissues. Marie's mother had died from cancer some years previously, and this body was also exhumed. Although she had definitely died of cancer, her tissues also contained high levels of arsenic.

On searching the house, a bottle of liquid, which proved to contain arsenic, was found. Another similar bottle was even found in Marie's handbag at the time of her arrest. She was charged with murder, but was allowed bail and disappeared about a week later, in November 1979. Other members of the family were also tested and they too were found to have high arsenic levels. Marie had poisoned at least four other members of the family. She went on the run, assuming an alias, bleaching her hair and losing a lot of weight. In January 1983, she was arrested in New Hampshire and returned to Anniston for her trial.

The jury found her guilty of the murder of her husband and the attempted murder of her daughter. She was sentenced to life for the murder and 20 years for the poisoning. Throughout, Marie protested her innocence. She was sent to a women's prison and, after a time, she was reclassified as a minimum security prisoner which meant that she was allowed passes and even leave from the prison. She absconded, of course, in February 1987, and it was assumed that she had left the state, but only a week later she was found suffering from hypothermia near her birthplace, on Blue Mountain, and there she died.

Phosphorus-flavoured food

In Germany in 1838, a woman prepared some phosphorus-laced soup for her husband and left it out for him to eat on his return from work. In the meantime she set off to visit relatives. But it was night-time before he arrived home, and so he could not help but notice that the soup glowed in the dark – and it smelt rather strange too. Suspecting that the soup had been tampered with, he put some in a bottle and took it to a public analyst the next day. The soup was shown to contain phosphorus. His wife was arrested, tried for attempted murder and sent to prison.

Today, there is a brand of rat and mouse bait available called Rodine. The active ingredient of this product is a coumarin anticoagulant, which will kill rodents after they eat it for several days. This well-respected brand name was in use for more than a century, although the active ingredient has changed with the times. In the mid nineteenth century, the Rodine brand of rat poison contained phosphorus. At that time, phosphorus was actually introduced as a safety measure to replace the arsenic-based rat poisons which were then available. In 1843, the government was so alarmed at the use of arsenic-based products, not only by murderers, but also by the suicidal, that they passed a law stating that only phosphorus could be used to kill vermin. It did not take long for people to discover that the new rat poisons were every bit as effective as the old ones for killing people as well as rats. And once lucifer matches were banned in 1910, rat poison became the only readily available source of phosphorus.

Phosphorus for use as a poison had a few problems: indeed, a person killed with phosphorus was easy to spot, as they glowed in the dark. It also smelled of garlic and the taste was very difficult, if not impossible, to disguise.

The poisoned cake

Mary Ann Ansell was working as a maid servant in a house in a fashionable part of London in 1899. She was engaged to be married, but could not marry until she and her fiancé had saved up seven shillings and sixpence to pay for the marriage licence. Mary Ann devised a plan to raise the money. She took out an insurance policy on the life of her sister, Caroline, who was insane and confined to an asylum in Hertfordshire.

For a premium of only three pence per week, Ansell was able to insure her sister's life for £11. She had said that her sister 'worked' at a hospital in the country when she took out the policy. She then purchased a tin of rat poison and proceeded to bake a cake, including the rat poison as an extra ingredient to the recipe. She sent the cake to the asylum, requesting that it be shared by the inmates of Ward 7, which included her sister. In fact, Caroline ate most of the cake, and although all of those who ate a piece of the cake became ill, she was by far the worst affected.
At that time, there were several cases of typhoid fever in the asylum, so the doctor was rather busy and did not manage to examine Caroline for several days. When he did, he immediately sent her to the asylum infirmary, but she died soon afterwards. The insurance company did not pay out the £11 to Ansell, as a post-mortem revealed that the death was due to phosphorus poisoning.
The police were called and investigations showed that it was Ansell who had sent the poisoned cake, and that she had earlier purchased the rat poison. She claimed that she had bought it to kill vermin, but her employer denied that there was any need for her to purchase rat poison. Ansell was arrested and tried, found guilty and hanged in July 1899. It would seem that her fiancé had a lucky escape.

She killed for the money

Louisa Highway was born in Wigan, Lancashire in 1907. She was one of eight children, and her father was a miner. She left school at 13 and went to work in the local cotton mill. At about this time she also joined the Salvation Army. When she was 20, she got a job as a kitchen maid at a nearby hospital, and five years after this she married for the first time. She had nine children over the years, although five of them died in infancy, and the rest were taken into care as she preferred to spend her spare time drinking in the local pub to caring for her children.

During World War II, she was sent to prison for stealing ration books. And then, in 1949, her husband died in suspicious circumstances, but investigations found nothing, and his death was put down to kidney failure. Louisa now married a man she had already been seeing for some time – the recently widowed and somewhat elderly 78-year-old, Richard Weston. He sold his house and put the proceeds into his bank account. He moved in with Louisa and married her in early 1950. He was dead only ten weeks after the marriage and his money, some £2,400, had already been spent by Louisa by the time he died.

Louisa now married for a third time to a gentleman called Alfred Merrifield, in August 1950. It was no surprise to find that his money was soon gone too. Louisa Merrifield now got herself a position as a live-in housekeeper to Sarah Ricketts, a 79-year-old widow who lived in a bungalow called Homestead on the North Shore at Blackpool. Alfred lived there too, doing odd jobs about the place.

One day, his wife sent him by train to Manchester to buy a tin of Rodine rat poison. The chemist's assistant later remembered him well, because he was wearing a hearing aid.

As time went on, Mrs Ricketts was obviously very pleased with the arrangement and began talking of rewriting her will in favour of the Merrifields, rather than leaving everything to the Salvation Army as she had planned. Rather stupidly, Louisa began telling some of her friends that the old lady had died, while to others she said that she soon would be dead.

One evening, Louisa phoned the doctor saying that Ricketts was ill. The doctor came to visit somewhat reluctantly that night, and finding that his patient only had mild bronchitis, remonstrated with the housekeeper. Louisa protested that she was frightened that something might happen to the elderly lady during the night. She called the doctor out several times that week, but he never found anything seriously wrong with his patient.

The following weekend, she phoned the doctor again, saying that Ricketts was now seriously ill. This time the doctor's partner, who was making all the home visits that weekend, decided to visit other patients first, and so eventually arrived to visit Ricketts at about midday. He was shocked to find the patient near to death, and examination showed that there was nothing he could do to save her. Only 33 days had passed since the Merrifields had arrived at Homestead, and it was only 13 days since the will had been rewritten.

A post-mortem found a mixture of rum and rat poison in the stomach. It was clear that Mrs Ricketts had been poisoned with phosphorus. An intensive search of the house and grounds followed, as the police looked for the tin of rat poison. They couldn't find it, and they knew Ricketts could not have disposed of it herself as she could hardly walk and rarely left her bungalow. Soon, other incriminating evidence came to light. The Merrifields were quickly arrested once the news emerged that they were the sole beneficiaries of the will and friends and neighbours remembered what Louisa had said a few days earlier.

The jury found Louisa guilty, and she was hanged in Manchester at Strangeways Prison in September 1953. The jury could not agree a verdict against her husband and eventually he was discharged. Alfred Merrifield received one-sixth of the estate, with the remainder being divided equally between Mrs Ricketts' two daughters. The Blackpool Waxworks Museum later paid the sum of £500 to Merrifield for agreeing to allow effigies of himself and his wife to be displayed in their Chamber of Horrors.

Poisoned With HP Sauce

Mary Wilson may have got rid of her first husband in the same way as she did with the second and third, although we'll never know that for certain. We do know that for the latter two deaths, she mixed a phosphorus-based beetle poison in with HP sauce to disguise the taste of her poison of choice. She also mixed some with a patent medicine called Chlorodyne, which contained a lot of chloroform, to mask the garlicky smell.

Ernest George Wilson, her third husband, died in November 1957. Suspicions were aroused because of the similarity with the death of her second husband. Two weeks later, the third husband's body was exhumed and a post-mortem was performed, which showed fatty degeneration of the liver, a sure sign of poisoning. Various tissues were removed for analysis. At the same time, the body of Oliver James Leonard, her second husband, was also exhumed. He had died the previous year supposedly of heart disease, but his liver also showed fatty degeneration, so various other tissues were removed for analysis too.

Extensive tests showed that both men had died of phosphorus poisoning; indeed, the tissues of both men glowed in the dark. In March 1958, Mary Wilson was tried in Leeds. Her defence team asked if Damiana pills, a popular nerve tonic at that time, which the men might have taken, could explain the presence of phosphorus in the bodies. The answer was no, as the pills also contained strychnine and none of this had been found. Mary was found guilty and sentenced to death, but this sentence was later commuted to life imprisonment.

Euthanasia or for personal gain?

Dr John Bodkin Adams was a physician and alleged poisoner. He was born in Northern Ireland and trained as a doctor. He became a general practitioner in Eastbourne, and may have murdered several of his patients. He was tried in 1957 after one of his patients, Edith Alice Morrell, died in suspicious circumstances. Like many elderly patients grateful for the ministrations of their doctors, she had made him a beneficiary of her will. She died of an overdose of morphine, but the jury found the doctor not guilty, not least because of the poor case advanced by the prosecution. However, after the case, the doctor was struck off the Medical Register for some years. Many considered that he had practised a form of euthanasia, but others believe he murdered at least nine old ladies for personal gain.

Some poisoners do it just for kicks. Male nurse, Benjamin Geen, of Banbury, was diagnosed as a psychopath with a narcissistic personality trait. He murdered two patients and took 15 others to the brink of death because he loved the excitement and thrill of trying to revive them after he had injected them with drugs that resulted in respiratory failure and so required emergency resuscitation. Following his trial in 2006 he was sentenced to life imprisonment.

In January 2008, a Czech nurse was on trial, accused of the murder of seven patients and the attempted murder of another ten in the summer of that same year. He administered huge doses of heparin, a drug used to prevent and break down blood clots, and his patients suffered massive haemorrhages. He too was motivated by the thirst for action.

A final failed conspiracy

In March 1917, at the Old Bailey in London, Alice Wheeldon, her two daughters, and Alfred George Mason, the husband of one of the daughters, were found guilty of a murder conspiracy. It was claimed that they had intended to poison the Prime Minister, David Lloyd George. Their trial and conviction were based largely on the evidence of two 'agents provocateur' and even today raise questions of a set-up.

Mason, a chemist and druggist who lectured on pharmacy at a technical school in Southampton, had supplied to the Wheeldons phials containing a preparation of curare and a solution of strychnine. The intention was definitely murder, but the defence claimed that the poisons were intended to kill dogs.

The unmarried daughter, Hettie Wheeldon, was acquitted but the others were all found guilty and sentenced to imprisonment. Wheeldon was sentenced to ten years, Mason to seven years and his wife Winnie to five years. In later years the extreme left-wing leanings of the Wheeldon family and their communist activism came to light, so maybe they did intend to assassinate the Prime Minister on the golf course, with a poison dart, after all.

Chapter 16

Perilous Personal Poisoning

DELIBERATE, SELF-INFLICTED POISONING using an overdose is the choice for more than one in four people intending suicide. Intentional killing of oneself – suicide – is far more common to men, and only one quarter of all suicide fatalities are woman. Unsuccessful attempts, regardless of sex, may be a cry for help, which may later lead to a fatal attempt.

Intentionally killing yourself is regarded as praiseworthy in some oriental societies. In Japan, hari-kiri is a ritual disembowelling, done either to escape humiliation or even to demonstrate loyalty. In India, although abolished 175 years ago, some Hindu widows in the remote regions of northern India still practise sati, in which they are voluntarily burnt to death on their husbands' funeral pyres.

Legal niceties

Christianity, Judaism and Islam all forbid suicide, and for many centuries in Christian countries it was regarded as a crime, and those who committed suicide were refused burial in consecrated ground. It was only in 1961 that the criminal penalties for attempting suicide were abolished in England and Wales.

Today in the UK, it is an anomaly in the law that it is not a crime for a person to end their own life, but it is a crime to assist someone in doing so. This 'crime' is punishable by 14 years imprisonment. However, the Director of Public Prosecutions currently considers that it would be against the public interest to prosecute a relative who might have 'assisted' a suicide. This topic is currently the subject of much debate.

The Voluntary Euthanasia Society, recently renamed Dignity in Dying UK, has campaigned for many years for the right to assist suicide in the case of the terminally ill. But, at present, such cases must travel to other parts of Europe such as Switzerland and the Netherlands where euthanasia is allowed. There are currently over 800 people in the UK who have joined Dignitas, the Swiss medical charity that assists such patients who wish to end their life at a time of their own choosing. So far, more than 115 Dignitas members from the UK have ended their lives in this way.

Organisations such as the Samaritans train volunteers to listen sympathetically to the distress of the suicidal.

Termination of a tiny life

Frequently in the past, self poisoning was used as a means to procure an abortion. A small number of herbal medicines came to be used for contraceptive purposes or as abortifacients (substances that induce abortion). Such medicaments were in use from ancient times throughout the Middle Ages and up to the first half of the twentieth century. These herbs included savin, tansy, myrrh, cow parsley, juniper, wormwood, pennyroyal, white poplar bark, ivy and rue. These herbs had many other uses too, but would not have been given to those women trying to get pregnant (in the case of those herbs with contraceptive properties), or to the already pregnant (in the case of those herbs that act to terminate a pregnancy). Even back in the mists of time, herbalists appreciated the importance of side effects and contraindications in their practice.

In the nineteenth century, women who wanted an abortion would even eat match heads, which in those days used to contain phosphorus, although this act was often fatal to the mother as well as the expected child. It was not until the 1930s that information and access to effective methods of birth control began to be available to the working classes. Until then, abortion was a major form of birth control, with infanticide not far behind.

Even in the 1950s and 1960s, medical journals were continually reporting cases of attempted abortion. Doctors and nurses call any early loss of an embryo or foetus from the womb an abortion. This terminates the pregnancy. Non-medical people tend to make a distinction between abortion and miscarriage, considering the former as being deliberately induced and the latter as an accidental occurrence. In modern times, with the introduction of newer, innovative methods of conception, such as In Vitro Fertilization (IVF), the injection of potassium chloride has also been used in cases of multiple pregnancy, where medical staff considered that selective reduction was necessary.[1] Further details are given in Appendix V.

Abortion – performing or procuring one – was a criminal offence until the passing of the Abortion Act in 1967. It had been made a statutory offence in 1803 and further legislation led to the 1861 Offences Against the Persons Act, but the Abortion Act changed all that overnight. But even before the new act, most working women regarded it as neither immoral nor illegal to attempt abortion as long as it was before the 'quickening' – the time when the baby's first movements are felt by the mother, usually at about the 16th week of pregnancy. Even today, abortion is still illegal in some countries, such as the Republic of Ireland and Iran.

In 1913, *The Lancet* carried a report of colocynth poisoning. More commonly known as bitter apple, colocynth is a drastic cathartic, causing violent vomiting and diarrhoea. A teaspoonful and a half of it has been known to be fatal. It was often used in the early twentieth century to try to induce an abortion. One woman, who took about five grams in powder form, suffered vomiting and diarrhoea for several hours, although she had little abdominal pain. She was treated with arrowroot, morphine and atropine, and after two days she was given the antiseptic, chlorbutol, in olive oil every six hours, together with a milky diet, and she ultimately recovered.[2]

Potassium permanganate was also used in the past for its supposed abortifacient action, by inserting tablets, crystals or douches containing it into the vagina. Unfortunately, this could result in corrosive burns, severe vaginal haemorrhage and perforation of the vaginal wall, sometimes leading to peritonitis. The fatal dose of about ten grams could result in death, which was long delayed, sometimes for up to one month after such poisoning.[3]

Lead plaster, sometimes known as diachylon plaster, is made by the combination of olive oil and lead oxide heated together in the presence of water. Such plasters have been in use since the time of the ancient civilisations of Greece and Egypt, where women were known to have eaten them in an attempt to procure an abortion.
Even a hundred years ago, lead plasters were still being used as abortifacients, as were diachylon pills. In the 1920s, there were reports of women who took a teaspoonful or two of lead oxide for the same purpose, resulting in serious and prolonged ill-health.[4] Considering that the poisonous nature of lead has been known for centuries, it seems amazing that a diachylon bandage, Lestreflex, was still prescribable on the NHS until about 20 years ago.
In 1938, *The Lancet* reported the case of a woman who recovered from severe lead poisoning after ingesting a quantity in excess of 100g of lead acetate over a period of about a month, in an attempt to kill her unborn child and abort the pregnancy. She was lucky as she was given treatment with a high calcium diet and calcium lactate tablets, which, together with injections of parathyroid extract, led to her recovery.[5]

Ergoapiol was a medicine, available as capsules, which contained ergot extract, aloin, oil of savin and parsley extract, used in the first half of the twentieth century for the treatment of menstrual disorders. Regrettably, some women misused it in an attempt to induce an abortion, sometimes with fatal consequences. One woman took 17 capsules of ergoapiol in an attempt to induce an abortion, and died.[6]

An American journal reported on 37 women suffering from toxic polyneuritis following the use of apiol, an alcoholic extract of common parsley, as an abortifacient. This misuse of apiol led to its inclusion on an official list by the British authorities in 1939, a list of substances likely to be used to procure an abortion, still a crime at that time.

The antihistamine, dimenhydrinate, used to be taken for travel sickness, nausea and vomiting, and, because drowsiness is a major side effect, also as an aid to sleep. In 1962, a 24-year-old woman died of an overdose which she had taken in an attempt to procure an abortion. She took 7.5g (150 tablets), which caused vomiting, vertigo and convulsions, and this then led to coma, cyanosis and death from respiratory failure just an hour and a half later. Post-mortem showed that massive damage had occurred to both her lungs and kidneys.[7]

Quinine, derived from Cinchona bark, has been used to treat malaria for several hundred years. In 1935, the *British Medical Journal* reported that quinine was often used in Greece for the purposes of suicide. While in some cases, doses of ten grams might not lead to serious intoxication, doses of 20g or more often proved fatal. Even smaller doses of less than ten grams could be fatal though, as shown by the woman who took only six grams of quinine sulphate to procure an abortion and died as a result.[8]

It is hardly surprising that she tried quinine because, at that time in the 1930s, it was very popular among doctors, who gave small daily doses (300mg) to expectant mothers in the last few weeks of pregnancy, in the belief that it shortened labour and led to an improved childbirth generally.[9] Needless to say, this useless and ineffective 'fashion' went out of the window over 50 years ago.

A breath of fresh poison

In the past, when the household gas supply was manufactured from coal at the gas works, the gas produced was poisonous. Many people were killed accidentally by gas leaks and explosions. But many used gas to commit suicide. Even though the natural gas we use today may not be poisonous, it still manages to cause deaths and injuries from leaks and explosions.

In 1924, dressmaker Florence Martha Miller, who was living in

London, committed suicide by putting her head in the gas oven. The cause of death was coal gas poisoning. The French surrealist painter, René Crevel, committed suicide in the same way in 1905. In 1963, the poet Sylvia Plath also killed herself in this way, having previously made an attempt on her life some years earlier with an overdose of sleeping pills.

Major James Dunning was a 59-year-old retired American Army officer living in London. He had financial interests in the City and lived in fine style in Chelsea. He also owned a farm in Sussex. He was very proud of his 1913 Rolls-Royce, which he liked to repair himself, if he could. Any major repairs were undertaken by a specialist garage.

In February 1931, he sent his car to the specialist garage and several days later went to Birmingham by train for a business appointment. He sent his wife a telegram saying that he would collect the car on his way home to London. He did collect the car, but did not return to his wife, who was waiting at their home in Chelsea. Instead, he drove to the farm in Sussex and told the maid there that the car had some mechanical trouble which he proposed to work on after his evening meal.

The maid and the cook slept above the garage. They both heard the car engine running when they went to bed and thought that it continued for perhaps another 30 minutes. Early the next morning, the Major was found dead underneath his car. Tests showed that he had died of carbon monoxide poisoning, as there was a high level in his blood. The coroner gave a verdict of death by misadventure. It then transpired that the Major's life was heavily insured for accidental death, but the terms of the policy excluded suicide. The insurance company refused to pay out, as they felt he could have killed himself, and the case eventually reached the court.

It appeared that the Major was having some income tax difficulties, which had put his financial position in doubt. The insurance company thought this to be a sufficient cause for him to commit suicide, hence their refusal to pay out. However, the medical experts supported the original verdict of accidental death. Calculations showed that the Rolls-Royce could easily produce the required lethal level of carbon monoxide in the garage, with the doors and windows all closed. The court found in favour of the widow and the insurance company then had to pay up the £10,000 due, which was a small fortune in 1931.

Accident, overdose or suicide – who knows?

Thomas Chatterton was born in Bristol in 1752. He was apprenticed to an attorney but was released from his apprenticeship in 1770 and went to London to seek his fortune. He was a poet and spent his time writing essays, satires and even a burlesque opera. Sadly he killed himself with arsenic that same year, at the age of only 18. Presumably he had found the streets of London were not paved with gold.

Chloroquine, like quinine, is used both to treat and to prevent malaria. In the 1950s, a young man committed suicide by taking a massive overdose of chloroquine. The accepted toxic limit was usually considered to be 20mg per kg of body weight, but this young man took the equivalent of about 100mg per kg and died about two and a half hours later of respiratory failure.[10] Another man also died following overdose; he was taking three or four chloroquine tablets every day to prevent malaria, instead of just two tablets each week.[11]

In the earlier chapter about household products, cases of poisoning by the use of formic acid for descaling kettles were mentioned. Many of these poisonings were undoubtedly attempted suicides. A report from only 20 years ago concerned some 53 such cases, among which there were 15 fatalities.[12]

In 1936, the *British Medical Journal* reported an amazing case of strychnine poisoning and its treatment, at a time when treatment lacked all the monitoring devices and intensive care facilities that are available today. A man had taken strychnine for suicidal purposes, but he was treated promptly. He was given zinc sulphate within 20 minutes of taking the poison, anaesthetised using chloroform and given a morphine injection. He became cyanosed and was then given amyl nitrite to relax the blood vessels and so lower his blood pressure. His stomach was washed out with dilute potassium permanganate solution.

Later, more morphine and atropine were injected, while both chloral hydrate and potassium bromide were administered rectally. All these medicaments were continued as well as oxygen until the spasms diminished, the cyanosis lessened and the respiration improved. Amazingly the patient eventually awoke, drank some water, vomited, then slept again, and was discharged from the hospital, having apparently recovered, the following day.[13]

People have even taken rat poison containing thallium deliberately, in order to poison themselves. In 1946, a 26-year-old French woman ate about 40g of rat poison, which contained just over two grams of thallium. Two days later, she had gastro-intestinal symptoms, and by the fourth day she had developed peripheral neuritis in her legs, and then a severe stomatitis (a sore inflamed mouth) on the fifth day. She eventually died on the ninth day due to pulmonary congestion.[14]

Certainly cyanide

Everyone seems to know that cyanide is poisonous, having watched films and read books about spies, James Bond and the like. Cyanide poisoning can even occur accidentally with products such as Cymag, an agricultural product, which releases hydrogen cyanide gas when used to kill rabbits and rodents. However, many people have deliberately taken cyanide for personal reasons. Cyanide capsules were also used by secret agents during World War I and other wars since.

Heinrich Himmler, a German Nazi leader and Commander of the SS, committed suicide in 1945 by taking poison, rather than risk trial and execution at the end of WWII following his capture by the British near Bremen. He committed suicide at Luneberg by swallowing a dose of cyanide which he had concealed in his mouth. As the initiator of the gas ovens, with responsibility for the concentration camps and the butchery of seven million people in what we know as the Holocaust, he would undoubtedly have been executed after a trial at Nuremberg after the war. So, he took the easy way out.

Another deliberate, but non-military, use was in the case reported in *The Lancet* in 1978. The patient had taken 413mg of potassium cyanide and only survived because of the aggressive treatment and supportive measures given. These included intracardiac injection of adrenalin, 100 per cent oxygen, correction of his body's acid/base balance and blood volume replacement.[15]

Another man, 21 years old, was admitted to hospital unconscious. As the doctors had no idea of what had caused his unconscious state they could only treat him symptomatically. The man developed pulmonary oedema and acidosis and was treated with oxygen, the diuretic frusemide and various intravenous fluids. It was nine hours after admission before it was discovered that the patient had deliberately taken three 200mg capsules of potassium cyanide. No specific treatment was given apart from responding to his symptoms, and he made an uneventful recovery.[16]

In the 1970s, the Reverend Jim Jones was the leader of The People's Temple, a religious sect founded in California. The faith-healing preacher had gathered a flock of social misfits of various kinds, including the mentally ill, drug addicts and ex-convicts. In 1978, when a group of relatives wished to investigate the sect's activities at their base in Guyana, South America, he persuaded no less than 913 members of his sect to commit mass suicide, by taking cyanide. He then shot himself.

More recently a young man used cyanide to end his life only a week before he was due to start a degree course at the University of St Andrews in 2003. He left a note in which he described the pain and anguish he had suffered since the age of 13 because of the particularly aggressive type of acne that he had suffered from, which had not been amenable to any form of treatment he had tried.

And in 2006, a toxicology lecturer suffering from depression, who must have known about the effects of cyanide, swallowed the poison in front of his partner. No post-mortem was carried out because of the potential risk to the hospital staff at the very same hospital, Arrowe Park, Birkenhead, where another suicide by poisoning occurred, mentioned in Chapter 9.

Sleeping finality

In the past, when barbiturates were used mainly as sleeping pills, many people took them in an attempt to commit suicide. Some were successful, although not as many as might be supposed. In 1934, *The Lancet* carried statistical tables detailing all fatalities associated with barbiturates that were recorded up to the end of 1932. Of 5,147 recorded suicides in 1931, only 13 of them were attributed to barbiturates, and there were none at all in 1932.[17]

In 1936, an American medical journal reported the case of a small man, weighing only 54kg or about 8½ stone, who survived a total dose of 18g of sodium barbital, which he had taken with suicidal intent. He was in a deep coma for six days, with a very high temperature, a rapid pulse and increased respiration rate. This was in contrast to the severe depression of these measurements in most similar cases of overdose, although why his were raised was never explained. However, supportive treatment was successful and he survived.[18]

In 1970, the *British Medical Journal* reported the case of a 25-year-old woman who took no less than eight grams of phenobarbitone. She did not regain consciousness until four days later. Her blood-phenobarbitone level fell only slowly, and restlessness, rapid-eye-movement sleep and insomnia increased over the next 19 days following the overdose.[19]

The introduction of the benzodiazepines as an alternative to the barbiturates in the late 1950s and early 60s was considered a major breakthrough at the time, as even large overdoses were survived. Patients who took as much as two grams of diazepam made a rapid recovery. As diazepam is available as tablets of 2mg, 5mg and 10mg strengths, 2g equals 200 tablets of the highest 10mg strength. Most intending suicides would, no doubt, expect to be successful after taking such a large quantity of tablets.

Downing your poison

Poisons can come in all shapes and sizes, and some suicides have occurred as a result of liquids instead of tablets.

A 42-year-old man suffering from depression drank 200ml of the solvent acetone in 1966. He went into a coma for 12 hours but then recovered. He was found to have slightly raised blood sugar during the next four months, and there was sugar in his urine, indicating that he had become slightly diabetic as a result of drinking the acetone.[20]

A 38-year-old man drank between one and two pints (approximately 500ml to 1L) of Nitromors paint stripper, which contained methylene chloride, and other ingredients including detergent. The man's condition was initially very serious[21], but with supportive treatment, his recovery was rapid, and kidney damage was prevented. Although the man developed ulceration of his intestines and diverticulitis, amazingly he went on to make a full recovery.

The American poet, Vachell Lindsay, suffered from depression and committed suicide in 1931 by drinking a quantity of Lysol. This product was similar to Jeyes Fluid, containing a mixture of tar derivatives, and is also used as a disinfectant.

One man who attempted suicide about 30 years ago drank 150ml of camphor liniment. He suffered peripheral circulatory shock and severe dehydration due to vomiting. He also had three attacks of severe and prolonged grand mal epileptic fits. His treatment began with a stomach wash-out and intravenous diazepam to control the fits. Intensive supportive treatment followed, as a result of which he duly recovered. This was believed to be one of the highest doses of camphor to be followed by survival.[22]

Lethal law

In June 2008, an abusive father killed himself with poison while on trial, after learning that he would go to prison for a brutal assault on his baby daughter. He mixed a solution of crushed sleeping tablets in a Coca-Cola bottle and gulped it down as he stood in the dock. The Sri-Lankan-born father, 40-year-old Anandakumar Ratnasabapathy, had repeatedly threatened to commit suicide if he was found guilty of the assault on his daughter.

The three-month-old girl was left quadriplegic, epileptic and partially blind after he deliberately blocked her nose and mouth. He was taken down to the cells and a doctor was called, but he died there later. This was believed to be the first suicide committed in a British courtroom.

However, this was not so, as in 1795 the Rev. William Jackson, a former curate of St Mary le Strand and editor of *The Morning Post* (1784-86), took arsenic while in the dock of the King's Bench court in Dublin, which was then part of the British Isles. He died before he could be carried out of the court.

The Poison Penalties: Uses, Abuses and Consequences

Chapter 17

Potent Potions of the Past

MANY POISONS HAVE BEEN used in the past as medicines, and some are still used today because the benefits of treatment outweigh the risks of their use. Some of these poisons were used a great deal in the past, even though we now know that they were totally ineffective and did nothing at all.

For example, in the nineteenth century, phosphorus was widely used in medicine. Doctors had discovered that the brain contained a lot of phosphorus and so thought that more would be a good idea. It was given in small doses for a wide variety of conditions, from epilepsy to cataracts, and even pneumonia and cholera. Tonics and toothache treatments containing it were sold, and phosphorus was still being prescribed for bone conditions in the 1920s. By the 1940s, however, it was rarely used, and eventually doctors realised that phosphorus was absolutely useless as a medicine, having no therapeutic value at all.

Would you reuse your laxative?

In the Middle Ages, pills made from the metal antimony were sold, believe it or not, as reusable laxatives. This metal has a toxic effect on the bowel, tending to cause diarrhoea. So, if you were constipated, you swallowed the antimony pill, which was about the size of a pea, and then waited for the desired effect, which duly returned this perpetual pill for reuse at a later date, providing you remembered to look for it.

Antimony was also used to make special goblets, called emetic cups, into which a little wine was poured and left overnight after an evening of feasting and drinking. This wine, laced with antimony, was drunk the following morning, as a cure for the hangover and the overindulgence from the feast of the night before. The irritant effect of the antimony on the gut induced vomiting, and so emptied the stomach.

Paracelsus vs. Galen

Antimony became very popular in the late Middle Ages when the famous physician Paracelsus advocated its use. Paracelsus was the

Latin title adopted by a gentleman with an impressive name: Philipus Aureolus Theophrastus Bombastus von Hohenheim (1493-1541). He could certainly sign his prescriptions faster using the name Paracelsus. He adopted this name because he considered himself to be 'above Celsus', who was a great Roman medical philosopher and forerunner of Galen. Until this time the works of Galen were the accepted version of medical authority and treatment.

Paracelsus challenged the Galenic authority and, while teaching at Basle University, he publically burnt Galen's works and offered instead his alternative theory based on the use of metals and minerals rather than just herbal remedies. The description 'bombastic' came into use from his name, because he was so full of himself. However, it was Paracelsus who realised the importance of the dose of medication required, in relation to both the beneficial therapeutic effects and adverse toxic effects. He knew that a small dose of 5mg of antimony potassium tartrate would stimulate the metabolism and induce sweating, while a larger dose of 50mg would cause copious vomiting and damage the liver. Tartar emetic (antimony potassium tartrate) was consequently commonly used in the sixteenth century and thereafter to treat all kinds of ailments.

And so it was that mercury, lead, antimony, arsenic and their salts came to be used extensively as medicines for the next several centuries.

Dr William Palmer was a genial young man, a physician by profession, but a gambler too. He became known as the Rugeley Poisoner. He loved the good life, but he lacked the money to support his lifestyle, even though he owned several racehorses. His life appeared to be littered with a number of most unfortunate, though lucrative, fatalities.

First his mother-in law died, leaving him some property, and then a racing friend to whom he owed £800 died, thus wiping out the debt. Then the unfortunate doctor's wife died, not so very long after he had insured her life for £13,000. He had only just made the first payment on the policy. It was not until some years later that these deaths came to be regarded as suspicious. His undoing was the death of his friend and fellow racehorse owner, John Parsons Cook. Mr Cook was taken ill and died in agony after a visit from Palmer.

They had attended Shrewsbury races together in November 1855, where Palmer, who was already in deep financial trouble, lost heavily while his friend Cook was on a lucky streak. They went to celebrate his winnings in the bar but John suddenly collapsed. The

doctor showed some concern for his friend and then volunteered to pick up his winnings for him. He used this money himself, to pay off his most pressing debts, and then arranged for his friend to be taken to the Talbot Arms Hotel, which just happened to be directly across the road from Palmer's home in Rugeley, Staffordshire.
He continued to treat his friend for several days until he died on 21st November 1855. Cook's stepfather, however, became suspicious and a post-mortem was performed by a leading toxicologist, Professor Dr Andrew Taylor. This doctor made his name by writing the first textbook on medical jurisprudence. He found traces of antimony in the dead man's stomach. Palmer was duly found guilty of murdering his friend Cook, and was hanged outside Stafford Jail on the 14th June 1856, at the age of only 32. It is thought that he may have murdered as many as 14 people in his short medical career.

Patent medicines

In medieval times, the spicers (the retail sellers of spices) also began compounding and selling medicines. By the start of the fourteenth century, they had joined forces with the pepperers (the wholesale sellers of spices) to form the apothecaries. The first records of an apothecary shop date from 1345, and in the eighteenth century the chemists and druggists appeared. They sold medicines, such as their own 'nostrums', as well as patent medicines, in addition to dispensing prescriptions.

In the seventeenth century, Nehemiah Grew, a botanist and physician, patented magnesium sulphate. This was the very first patent medicine. He had it extracted from the spa at Epsom, which was in fashion at the time. He called his medicine Epsom Salts and sold it as a cure-all, for this was the heyday of 'quack medicines'. It has been used ever since in skin preparations, or added to bathwater, but even today is still used by some older people as a treatment for constipation. Magnesium hydroxide, also known as Milk of Magnesia, is also used for indigestion and constipation, but overuse of this apparently innocuous digestive aid can lead to toxicity, resulting in lethargy, muscle weakness, confusion and even, eventually, to death.

Patent medicines, with royal letters patent, granted to an individual, by the sovereign, the sole rights to make a unique product. These were deemed respectable, unlike the many useless 'quack medicines' sold widely since the 1600s. James's Powders, comprising one part antimony oxide to two parts calcium phosphate, were patented by Dr Robert James in 1747. The patent cost him £150, a great deal of money in those days. The investment in this patent was obviously worthwhile as James's Fever Powders were still on sale in chemists' shops in the 1920s. Today the Patent Office confers such rights, but patents now generally only last for a term of 20 years.

Antimony for infections and parasites

Antimony preparations, together with those of arsenic and mercury, were used for hundreds of years as a treatment for syphilis. About 80 years ago, it was also found that antimony salts were effective medicines for a number of parasitic infections common in the tropics, such as schistosomiasis, leishmaniasis and trypanomiasis.

The chosen medication was given by injection, in carefully calculated doses, so that it killed the parasite, but not the patient. Fortunately, since then, most antimony-containing compounds have been superseded by far less toxic and far more easily administered drugs, such as praziquantel, in the treatment of some of these parasitic infections in the tropics which, even today, are still so common.

In November 2008, the television presenter and adventurer, Ben Fogle, contracted leishmaniasis while filming in Peru. The parasitic protozoan left him with an inch-deep hole in his arm, for which a course of daily injections of an antimony-containing drug was given for several weeks. The side effects of this type of drug are serious and make people feel very ill indeed; however, it is the only way to kill off this flesh-eating parasite.

Arsenic solutions

The effects of arsenic have been known since prehistoric times, and it has been used in medicine ever since for all kinds of ailments, including such widely diverse complaints as rheumatism, diabetes, malaria and consumption (the old-fashioned name for tuberculosis). Arsenic sulphide was used by Hippocrates, Dioscorides, Galen and Paracelsus and, during the heyday of quack medicine in the eighteenth century, arsenic became very popular indeed.

In 1780, a Dr Fowler concocted his 'Solution', which was quite simply a mixture of arsenic trioxide and potassium hydroxide dissolved in water, to which lavender water was added. The lavender water was included for its smell, which was intended to act as a warning to prevent accidental overdose. The dose, of only a few drops, was added to water or wine and taken as a general tonic, a popular cure-all, and was even used as an aphrodisiac. Doctors prescribed it to aid convalescence. Even Charles Dickens is claimed to have used it. Fowler's Solution was still in use in the mid-twentieth century, although by then it was used only for skin complaints, and today it is no longer used.

In small doses, arsenic seems to boost the metabolism and increase the formation of red blood cells; however, prolonged exposure can cause dermatitis, and breathing arsenical fumes can cause lung cancer. The stimulating effect of arsenic was used in the past by unscrupulous trainers who doped their horses in the hopes of winning a race. Such doping is easily spotted, as a simple urine test will detect the presence of arsenic.

A derivative of arsenic acid, roxarsone, was used to fatten pigs and poultry, again utilising the stimulating effect that arsenic has on the metabolism. This food additive was removed from the animals' food about a week before they were due to be slaughtered, to allow sufficient time for the arsenic to be excreted before the meat reached the butcher's block. This tonic effect can be tried out by you, dear reader, simply by drinking Vichy mineral water, which contains about two parts per million of arsenic, derived from the particular rock strata Vichy water passes through.

In 1909, Paul Ehrlich found, after many years' work, that an arsenical compound – the 606th he had tried! – could cure syphilis. He named this substance, Salvarsan, and it quickly became the standard treatment for syphilis until penicillin arrived in the 1950s. Salvarsan, also known as arsphenamine, was still being used to treat dysentery and sleeping sickness until eventually, in the 1970s, much better, more effective and less toxic agents were found for these diseases.

Arsenic is still in use today, but it is now only used in properly tested and approved medicines for treating a particular form of acute leukaemia. In this situation, when it is given in combination with other drugs, the ability of arsenic to stimulate the production of normal red blood cells is put to good use.[1]

Deadly cyanide revisited

Although sodium nitroprusside contains cyanide, it is used to treat hypertensive crises and to reduce blood pressure during anaesthesia, in certain special types of surgery. It is also used to treat ergotamine poisoning and has even been tried as a treatment for Raynaud's disease.

However, it is little used today, except in the case of an emergency, due to the risk of cyanide poisoning with its use. This same substance was used for many years by diabetics to test their urine for the presence of ketones that result from the partial oxidation of fats, which can occur if lack of insulin prevents glucose being used as a fuel by the body. Nowadays, most diabetics monitor their blood sugar level instead, as this gives a far more accurate real-time result and allows them to adjust their diet or insulin dose accordingly.

In 1982, a 12-year-old girl, four women and two men in the Chicago area of the USA all collapsed suddenly and died after taking Tylenol capsules. Tylenol is an American brand name for the painkiller paracetamol, or acetaminophen, as Americans call it. These poisonings were due to product tampering. The perpetrator was never caught, despite a large reward of $100,000 being offered by Johnson & Johnson, the manufacturers of Tylenol.
Investigation revealed that someone had apparently shoplifted packs of Tylenol from a number of small stores, emptied some of the capsules and then refilled the bottles with potassium cyanide, before replacing them on the shelves of different stores. The Tylenol bottles examined were found to contain different quantities of poisoned capsules – some as few as five, some up to as many as 30 capsules per bottle.
A wave of copycat tampering followed, with Lipton Cup-A-Soup in 1986; Tylenol again in 1986; Excedrin, a combination painkiller, also in 1986; Sudafed, a decongestant in 1991 and Goody's Headache Powders in 1992; and there were yet more deaths. Before the Tylenol murders, tamper-evident packaging was little used, but since then the technology of this aspect of packaging has evolved and is now commonplace, for medicines and for many other products.

Marvellous or malicious mercury

Mercury compounds have been used in medicine for thousands of years, although today we now recognise the poisonous properties of mercury and its compounds. Calomel (mercurous chloride) was widely used from the late fourteenth century for several centuries as both a treatment and cure for syphilis. This treatment was effective in killing the causative micro-organism but the cure was risky, causing intense salivation and other toxic symptoms to the patient.

Calomel has long been used as both a laxative and a diuretic, as well as in teething powders, which were still on sale in the first half of the twentieth century; this latter use is described in detail in Chapter 13, The Noxious Nursery. The even more poisonous mercuric chloride, known as corrosive sublimate, was also widely used in the past, as a disinfectant. Mercury is still used today in traditional Chinese medicine, in cinnabar sedative pills and antidotal pills.

Metallic mercury was once widely used for skin complaints, made into ointments by being mixed with fats. Such salves first appeared in the thirteenth century. These applications also resulted in excessive salivation, a symptom of poisoning, because the mercury was absorbed through the skin and into the bloodstream, and so was distributed throughout the body.

Years ago, a nine-month-old baby girl with a bad case of nappy rash was prescribed ammoniated mercury ointment by her doctor. Although it worked well in clearing up the nappy rash, it gave her mercury poisoning too, which then also needed treating. Ointments containing mercury were phased out several decades ago, in favour of the far more effective antibiotics and antifungals we use today

Yellow mercury oxide ointment was sold as 'Golden Eye Ointment' for many years, until the early 1980s when it was discontinued in the United Kingdom because of its mercury content. However, some years later the brand name 'Golden Eye' was re-introduced, although the ointment and drops now sold under this name in Britain no longer contain mercury.

Most dentists still use mercury in amalgam fillings, this use being invented by an American dentist in 1895. Today, the alloy powder that is mixed with the mercury contains silver, tin, copper and even a trace of zinc. These fillings are considered safe for the patient; the main hazard is actually to the dentist and his workers who are in daily contact with the amalgam and the mercury vapour produced.[2]

In the nineteenth century, mercurochrome was widely used as a general household antibacterial and antiseptic agent. Mercurochrome was a weak disinfectant, bright yellow in colour, which contained both mercury and bromine – its full chemical name was mercurodibromofluorescein. It was used medicinally, in solutions for bladder irrigation, as well as for bathing wounds, but deaths occurred due to skin absorption and shock. There were also cases of contact dermatitis and epidermal cell toxicity, as a result of which the use of mercurochrome was discontinued.

Thiomersal, or thimerosal in the USA, is a mercury-containing compound that was used as a preservative in biological products, including vaccines. This was a matter of controversy in recent years, for some parents believed that this source of mercury may have been the cause of autism in young children. Autism tends to manifests itself at about the same age as when the child receives the vaccination. The work of Dr Andrew Wakefield, which sparked this controversy in 1998, has now been discredited, but it is undeniable that there was a marked increase in the number of cases of autism diagnosed in the last decade or so.

Many causes for this increase have been suggested, but no single cause has yet been proved. Earlier and more accurate diagnosis is probably the best explanation. Dr Wakefield's legacy came home to roost in 2008, when measles broke out in a number of places across the UK and some fatalities occurred. This only happened because so many parents decided not to let their babies have the MMR series of inoculations, a decade ago, which would have protected them.

Although research into this area continues, mercury as a vaccine preservative is now being phased out. The medicinal uses of mercury and its compounds are such that, today, the therapeutic benefit is far outweighed by the hazards. Consequently, clinical use has now been largely abandoned, except for those few mercury compounds still used in homoeopathic medicine.[3]

Lead galore

Despite its toxicity, doctors have been using lead compounds as medicines for thousands of years. Medicines containing lead were used in the past for everything from gonorrhoea to neuralgia, from coughs to hysteria, although now we know better.

In the Roman Empire a physician to the emperor Tiberius is said to have invented diachylon plasters, which contain lead oxide and olive oil, for skin complaints. Unfortunately even then, women made use of the poisonous lead for their own reasons – they ate the plasters, as a means of procuring an abortion to terminate a pregnancy, as was mentioned in the previous chapter.

In the eighteenth century, in the heyday of quackery, lead acetate was used for skin complaints, for piles and even for cancer. Victorian doctors used it together with opium for diarrhoea: the lead paralysed the gut while the opium killed the associated pain.

The medicinal use of lead as an astringent skin lotion in the past has long since stopped; the lead acetate solution has been superseded by far more effective, though less toxic, preparations. However, lead acetate is still in use today in cosmetic lotions used to 'restore' colour to greying hair. Preparations such as 'Restoria' and 'Grecian 2000', which also contain sulphur, are sold for this purpose, the lotion meant to be applied to the greying hair to darken it. The solution works because the lead and the sulphur combine to produce black lead sulphide on the hair. Hair is dead tissue, so this treatment is not harmful; only a minute amount of lead would be absorbed through the scalp, and then only if any of the lotion came into contact with it.

Better with bismuth

Bismuth is a heavy metal that is only present in the human body in a few parts per billion. It is not involved in our metabolism in any way, but its compounds have found a specific use in medicine. In the 1780s, doctors began to prescribe bismuth mixtures for gastric complaints and peptic ulcers. Compounds of bismuth are still in use today in products such as Moorlands Indigestion Tablets and Pepto-Bismol, both for gastric complaints, as well as in some preparations for haemorrhoids.

Some bismuth-containing compounds are also active against *Heliobacter pylori*, the organism that can colonise the gut and cause gastric or duodenal ulceration. Although bismuth's mode of action is still not fully understood, it appears to act on the mucus that protects the lining of the stomach, as well as de-activating the protein-digesting enzyme pepsin. While bismuth preparations are considered to be very safe, excessive use can cause liver damage and a characteristic blue line on the gums, which can persist for years. Injections of bismuth-containing compounds were used in the past to treat syphilis, and one containing both bismuth and arsenic was used as an oral treatment for amoebic infections; however, newer, safer and more effective drugs are now available. Bismuth compounds are also still used in homoeopathic preparations.

Happy halogens

Fluorine is a very potent antibacterial. It is the most reactive of the group of elements called the halogens; it is far more effective as a treatment than chlorine, bromine or iodine. Various drugs contain fluorine atoms in their molecules. With some, the action is so potent that a single dose is all that is required for treatment. Diflucan One, fluconazole, an anti-fungal agent, will treat vaginal thrush with a single oral dose of one 150mg capsule.

This same drug is also used for treating the life-threatening fungal infections suffered by transplant patients.[4] This is not a new idea: flucloxacillin, an antibiotic of the penicillin family, also contains fluorine, and has been available for over 30 years. The antibiotics known as fluoroquinolones, such as ciprofloxacin, are used for infections that are difficult to treat with other conventional antibiotics. They also contain fluorine.[5]

Bromine is another useful halogen. In the nineteenth century, mixtures of bromide salts were prescribed by doctors for the treatment of nervous complaints and epilepsy. They were the Victorian equivalent of tranquillisers, used to depress both mental activity and the sex drive. The reference during wartime to 'bromide in the tea' of the troops was a suggestion that they were dosed to ease tension and take their minds off their womenfolk back home. It is interesting to note that some spa waters owed their popularity in the eighteenth and nineteenth centuries to their high bromine content.

Bromides were widely used until the mid-twentieth century as sedatives for the mentally ill, but precise doses were needed for optimum effect. These are no longer used since benzodiazepines, such as Valium, better known now as diazepam, became available over 40 years ago.

Iodine was used for over 150 years as an antiseptic and disinfectant and is still much used today, particularly by operating theatre staff, as povidone-iodine for its germ-killing properties during surgery. Colourless iodine, marketed because it would not stain skin or clothing, does not have the same disinfecting properties as povidone-iodine, which is amber in colour, like the ordinary iodine solution. A great deal of iodine, as the radioactive isotope Iodine-131, was released into the atmosphere in the Chernobyl disaster, mentioned in the treatment chapter and Chapter 14. Taking potassium iodide tablets as soon as possible is the recognised treatment to protect the thyroid gland from damage by the radioactivity. Regrettably, such tablets were not readily to hand when Chernobyl blew up.

Quack cures with radium

Radium was regarded as a wonder drug a century ago, and quack cures using it were marketed within a few years of its discovery. Some of them remained on the market for almost 30 years. Radium Salve was a very weakly radioactive ointment marketed in Britain about 100 years ago.

In America, the 'Cosmos Bag' was a supposed radiation cure that was to be applied to arthritic joints. 'Revigorator' was a flask lined with radium, in which you left water overnight to drink the following morning. Another product was 'Raithor', a weak solution of radium salts, claimed to prevent disease. A steel magnate from Pittsburgh, Eben Beyers, drank a bottle of Raithor every day for four years, by which time he had severe radiation sickness and cancer of the jaw. He died in 1931, and the subsequent publicity about his death marked the end of such quackery.

Lithium for gout?

In the nineteenth century, taking the waters was still fashionable. In 1843, Dr Ure advanced a theory that lithium could be used to treat gout. He believed that taking certain lithium-rich spa waters would aid recovery. Because gout is exceedingly painful – being caused by the formation of uric acid crystals between the joints, particularly of the bones of the feet – Dr Ure probably made a tidy fortune from his spa water.

It was not until 1912 that Dr Pfeiffer showed that lithium actually slowed down the elimination of uric acid from the body, so making the condition worse rather than alleviating the problem. It was also shown that the concentration of lithium in the spa waters was far too low to have any therapeutic effect anyway – another quack cure discounted.

Lithium, however, does have its uses in modern medicine. It has been used for the last 50 years in the treatment of manic depression. Its use for this was discovered by accident by Dr John Cade, an Australian doctor, in 1949. He was trying to discover if manic depressive patients were producing an excess of some body chemical which might be causing their symptoms. He noticed that the experimental animals he was using had become very calm when injected with a lithium-containing solution. He then tried it on a patient who was so ill that he had spent five years in a secure unit.

This patient improved so much within a few days of starting this treatment that he was transferred to a normal ward and was eventually able to return home and even take up his old job again. Other doctors tried the treatment with equally impressive results, and this is now the standard treatment for manic depression (now called bipolar disorder).[6]

Great care must be taken to get the correct dose of lithium for each patient. Regular blood tests are needed to monitor patients' lithium levels, as the therapeutic level is not far below the toxic level. Patients taking lithium need to be aware that many other types of drug can interact with lithium. If the lithium level in the blood is allowed to rise too high, it can lead to toxicity, the symptoms of which are confusion and slurred speech. If left untreated, such toxicity can be fatal.

In the first half of 2009, a clinical trial started in which 220 sufferers of motor neurone disease (MND) will be given a daily lithium carbonate tablet for 18 months. There is evidence that this treatment might help sufferers of MND. At present there is no effective treatment for this progressive illness, but such a trial may show evidence of a slowing of the course of MND. This slowing may occur by promoting new nerve growth or by protecting existing nerves.

Normally, MND patients are only expected to live for about 15 months after diagnosis. Current treatment with riluzole can extend life by only about two months. Professor Stephen Hawking, now aged 66, discovered he had the disease over 40 years ago, and his survival is an exception to the normal rapidly progressive nature of this disease.

Danger! Treatment risks

Nitrogen mustards were originally developed as chemical warfare agents, but a number of them are now used as anti-neoplastic agents in the treatment of cancer. Great care must be taken by all the healthcare workers involved, such as doctors and nurses, when treating patients with these agents. Even the urine from patients receiving such treatment must be handled with protective clothing, including gloves, for 48 hours after treatment; such is the toxicity of these drugs.

Silver nitrate, which has antibacterial and antiviral properties, has been used since medieval times, when it was known by the impressive name of 'lunar caustic'. It has long been used as a remedy applied to warts to remove them, and works by killing the virus causing the wart. Caustic pencils are still on sale today for this purpose.

The antimicrobial property of silver is also used in the treatment of burns, with compresses soaked in silver nitrate solution and cream containing silver sulphadiazine. Unfortunately, with large areas of raw tissue exposed following extensive burning, some cases of argyria – due to silver poisoning – have occurred because of the direct absorption of the silver by the raw tissue, which would not occur through normal skin.

Dressings and catheters containing silver are used today, both in hospitals by nurses treating patients and in the community, where the

antimicrobial activity can be utilised against antibiotic-resistant strains of infective organisms. It is now possible to buy first aid plasters containing silver as an antibacterial from any local pharmacy.
Thallium was used in medicines in the past – as a hair remover for people about to undergo treatment for ringworm of the scalp. Hair loss is a symptom of thallium poisoning, so this 'side effect' was utilised, with thallium coming into general use as a depilatory for over 50 years. A number of cosmetic preparations containing seven per cent thallium acetate in a depilatory cream were available to purchase in the 1930s. However, following many fatalities from both systemic and topical treatment it was no longer used for such purposes. A number of cases have been described in earlier chapters.

Big bad Bs: brimstone, barium and boron

Sulphur was used in medicine since prehistoric times and has been prescribed by doctors for over 3,000 years. The old name for sulphur was brimstone, and a popular remedy for both digestive upsets and regularity in the past was 'brimstone and treacle'. Its mode of action was as a laxative, the sulphur being converted to irritant sulphides in the gut. The modern equivalent, sulphur lozenges and tablets, were still on sale in the 1980s but have since disappeared from pharmacies. Sulphur was also used in creams, pastes and ointments to treat acne and scabies, a parasitic infection, as well as in lotions made with various organic solvents such as ether, alcohol and chloroform. These lotions were abused by 'sniffers'.
Barium and many of its salts are very poisonous if absorbed by the body. In the early days of X-ray investigations of the gut, deaths occurred due to the unfortunate administration of some unsuitable barium salts. Barium sulphate was used for many years as a contrast medium in X-ray investigations of the stomach and intestines. This is perfectly safe because this particular barium salt is insoluble and so it cannot react with the stomach acid. It remains in the gut, passing along until it is excreted in the faeces.

Other barium salts, however, even if insoluble in water, may be dissolved in the stomach acid and then absorbed, causing poisoning and ultimately death. This happened in a number of fatal cases in the past, where barium thiosulphate was given in error for barium sulphate during an X-ray investigation, and barium carbonate, as a medicine for gastric problems, in error for bismuth carbonate.[7, 8]
Boron-containing compounds were once widely used as medicines. Borax – sodium borate – and boric acid were both used as

disinfectants, because they had weak bacteriostatic and fungicidal activity. They were also used in mixtures for the treatment of epilepsy. However, boron compounds are toxic; a dose of 5g of boric acid is enough to make a person ill and 20g would put their life in danger. Details of a number of boron poisonings are given in the Household Horrors and Noxious Nursery chapters.

Migraine, gout and dropsy

Ergot is a fungal infection that grows on the cereal crop rye. In the past, people who unwittingly consumed bread made from flour of infected rye grain suffered a condition called 'St. Anthony's Fire'. The symptoms that they developed (described in Chapter 3) were due to the ergot alkaloids, produced by the fungus, infecting the grain. This fungus produces a large number of different ergot alkaloids, a dozen of which are chemically related to LSD.

Some of these alkaloids have been used medicinally, including ergotamine, which was used in medicine to treat migraine, and ergometrine which was used in childbirth. Today newer, more effective agents are increasingly used.

Ergotamine causes constriction of blood vessels and so relieves the spasm and subsequent overdilatation of certain arteries in the brain, which cause the throbbing recurrent headaches, usually one-sided, that we call migraine. The dose that can be taken is restricted, as overdose can lead to such drastic vasoconstriction that gangrene of the extremities, particularly of the toes, can occur.

Ergometrine acts specifically on the uterus, stimulating contractions during childbirth to assist labour, and also helping to control bleeding afterwards. Less than a milligram of this potent alkaloid is needed, and is usually given by injection.[9]

Colchicine is a potent alkaloid derived from the meadow saffron corm, which is still used today in the treatment of gout. Precise instructions are required to be followed to prevent toxic side effects. The dose to treat an acute attack of gout is usually between three and six milligrams. No more than ten mg in total should be taken during each attack. A further course should not be taken within three days; such is the danger of toxicity. The tablets each contain 500 micrograms (half a milligram), and the dosage is to take one tablet every two to three hours until relief of pain is achieved, or until vomiting and/or diarrhoea occurs, these being the first signs of toxicity.

In the late 1960s, two women took, respectively, 26mg and 39mg of colchicine in suicide attempts and somehow survived. They suffered nausea, vomiting, colic and diarrhoea, followed by cardiovascular, neurological and dermatological signs of toxicity. Throughout, neither patient lost consciousness, and after many months of illness, they both eventually recovered.[10]

About a hundred years ago, 'Laville's Gout Cure', which contained both colchicine and quinine, could be bought from any pharmacy. The *British Medical Journal* reported a case in 1915 in which a patient took so much of this patent remedy over a period of 19 days that he became paralysed. This paralysis persisted for nearly a month, and even six months later the patient had still not fully recovered.[11]

Digoxin is another potent medicine derived from a natural source, found in the leaves of the foxglove plant. The dose of digoxin, like colchicine, is measured in millionths of a gram. Digoxin intoxication was common years ago, before variations in bio-availability were properly understood in the pharmaceutical industry: Digitalis leaves collected in the late afternoon contain far higher levels of digoxin than leaves collected early in the morning, due to chemical changes happening in daylight.

Today digoxin tablets are standardised and available in three strengths: 62.5, 125 and 250 microgram. Patients starting to take digoxin always start on a low dose. The drug is introduced very carefully, with the patient's digoxin level monitored at regular intervals, to ensure that the dose is optimal for that specific patient, as everybody is different.

Murder by chloroform

In the 19th century, chloroform was used as an anaesthetic. Today, small quantities are used as flavouring is some medicines because in very dilute aqueous solution it tastes far more pleasant than the often very bitter active ingredient, so making the medicine far more palatable.

Adelaide Bartlett was tried at the Old Bailey in 1886 on a charge of murdering her husband using chloroform. When she was 19, Adelaide had married Edwin Bartlett, a prosperous grocer who was ten years her senior. In 1885, Adelaide and Edwin became friendly with a young Wesleyan minister, the Reverend George Dyson.

Edwin made a will, leaving everything to his wife and making the minister his executor.
Within a few weeks of making his will, Edwin became ill with a gastric complaint. He became an invalid and eventually died on New Year's Day, 1886. On examination of his body, a large quantity of liquid chloroform was found in his stomach and Adelaide was charged with his murder. Try as they might, however, the prosecution were unable to explain how Adelaide had administered the chloroform, and she was acquitted. Nothing is known about what happened to her after the trial.

Opium and onwards

In the first half of the nineteenth century, opium was refined into morphine, codeine and papaverine, all highly addictive painkillers. Morphine was the most important of these alkaloids, and was named after Morpheus, the Greek God of Dreams.

Laudanum was the commonest painkiller available at this time; everybody used it and many became addicted to it. Laudanum was simply an alcoholic solution of opium, containing a mixture of more than 20 alkaloids that were to be found in raw opium.

Today morphine and other opioids are widely used to relieve pain, particularly for the terminally ill, by injection or as patches, tablets, capsules and oral solutions. Pain management at its best is now highly sophisticated because the hospice movement has pioneered major improvements in pain control. However, continuing research for similar, yet more powerful, painkillers is needed as, even today, some 20 per cent of terminal patients do not have complete pain control – some pain is still intractable, despite the best efforts of doctors and nurses.

Research for new and improved medicines is a continual process. Naturally occurring substances, many of which are poisonous, are being utilised in the search for more effective medicines to replace existing sub-optimal ones, as well as for medicines to treat those conditions for which there is as yet no suitable treatment. Tetrodotoxin, derived from a number of fish, such as the puffer fish, is far more poisonous than cyanide, and is currently being investigated for possible future use in medicine. Derivatives of tetrodotoxin have been found to be 3,000 times more powerful than morphine in the control of extreme pain by stopping the pain signals from reaching the brain.[12]

Galantamine is an example of this more recent research. This substance was derived initially from snowdrop and daffodil bulbs, and is now in use in the treatment of Alzheimer's disease. Supplies of the bulbs are being imported from Bulgaria and China; however, there are plans to plant some 30 to 40 acres of daffodil bulbs in the Black Mountains area of Wales to provide a homegrown supply of galantamine.

Botulinum toxin is now used in medicine, with minute quantities being injected to alleviate tics and spasms that arise from muscular overactivity. These work by paralysing the affected muscle. Botulinum toxin is probably best known to the general public as Botox, a beauty treatment for the temporary removal of wrinkles. Whatever the use, the injections need to be repeated every two or three months to maintain the effects.

Adulterated consequences

Over the years, a number of incidents have come to light where medicines have been adulterated with dire consequences. Unbelievably, ethylene glycol, perhaps better known as antifreeze, was found in Paracetamol Elixir, a children's liquid painkiller on sale in Bangladesh in 1995. This caused an epidemic of poisoning in which some children died from kidney failure as a result.[13] Ethylene glycol is sweet and cheaper than sugar – but deadly. Several such cases were described in Chapter 13.

Arsenic has been found in adulterated herbal preparations. This is an ongoing problem, common in the past, in some parts of the world and still causing problems in the twenty-first century, particularly in India and China. In 1975, some 74 cases were reported in Singapore. These cases resulted from the use of herbal preparations which were intended to treat asthma and other illnesses.[14] Another report involved a man who had taken a proprietary asthma relief preparation, which contained arsenic, for 55 years – its continued use eventually resulted in a massive haemorrhage.[15]

Also in the 1970s, a 59-year-old man died following the application to his face of an arsenical paste, which was sold to him as a quack cancer cure.[16] While using this product, he developed arsenical peripheral neuropathy, with an acute respiratory involvement and chronic bronchitis. During the 1970s there were also a number of reports about arsenic trioxide being used in India as a supposedly aphrodisiac addition to opium.[17]

And, unbelievable as it may seem, there is even an Ayurvedic medicine containing arsenic, which is commonly taken during pregnancy by women in some parts of India, in the hope of producing a male child.[18]

In the latter part of the twentieth century, many instances of medicines containing heavy metals and other poisons came to light. Lead and mercury have been found in Indian ethnic medicines. Bal Jivan Chamco Baby Tonic[19] was found to contain lead, as were many Chinese herbal medicines from Hong Kong.[20, 21, 22] Ointments supplied to treat skin complaints, containing Chinese herbal remedies, have also been found to be adulterated with steroids.

A report in the Archives of Disease in Childhood in 2004 told of creams, purchased by parents, for use on children who attended a specialist skin clinic in Birmingham, UK. The parents had purchased a variety of Asian herbal creams because they were worried about the side effects of the prescribed steroids that their GPs or consultant dermatologists wished them to use, as prolonged use of steroid creams or ointments can cause skin thinning and growth retardation. The parents were very pleased with the results of using the supposed natural alternatives. However, analysis of the creams found that 20 out of 24 creams contained potent or very potent steroids, which were even stronger than those prescribed by the doctors. No wonder the results were so impressive.

Other Chinese herbal remedies were also found to be adulterated with western drugs so that their efficacy was greatly enhanced. Sometimes these remedies were toxic because a cheaper but poisonous herbal ingredient had been substituted in a mixture. Similar reports of toxicity due to the presence of arsenic and mercury as adulterants, in a variety of Indian ethnic remedies, were made in 1993.[23]

Even in the 21st century, such practices continue. As many as one fifth of all Ayurvedic medicines imported from India have been found to contain arsenic, lead and other heavy metals. Sometimes these are accidental contaminants, but in other cases they are deliberately included, in the belief that they have health benefits, such as promoting vitality, rejuvenating energy levels and controlling blood sugar. Metals may be present due to the practice of *rasa shastra* by some practitioners, where herbs are combined with metals, minerals and even ground-up gemstones. Although Ayurvedic medicine has a long tradition of use in India – hundreds of years – there is little scientific evidence of its efficacy.[24]

In the past, many substances we now recognise as poisons were used as medicines. As their toxic nature became known and more effective, safer drugs have been discovered and developed, and these poisons have thankfully fallen from use.

Chapter 18

Malevolent Medicines

APART FROM THOSE MEDICINES that cause problems such as dependence, there are also medicines that have been withdrawn from use or have had special precautions applied to their continued use because of a variety of other problems. Some have been found to damage the unborn child, some cause muscle damage, including heart damage, as the heart is a muscle, others may damage the liver or kidneys, or cause blindness or deafness, while yet others can cause death by affecting the blood or the brain.

Our regulatory authorities continually monitor reports of adverse drug reactions and act quickly when a potential problem becomes known. All new medicines are evaluated by the Committee on Safety of Medicines (CSM), a part of the Medicines and Healthcare products Regulatory Agency (MHRA). In the United Kingdom, this current system was set up following a drug disaster of immense proportions that occurred over 50 years ago.

As safe as houses

In 1956, a new sleeping pill was launched on to an unsuspecting world. This drug was claimed to be 'outstandingly safe' even in overdose. Doctors were pleased that at last they had something safer than the barbiturates to prescribe for their patients. The barbiturates were available for about a hundred years and had two major faults: they were highly addictive and they were very dangerous in overdose, particularly to children.

The new sleeping pill was developed by a German drug company called Chemie Grunenthal, and was available in several countries in Europe, as well as in many other countries worldwide, including Australia. In both Australia and the United Kingdom, the new drug was marketed by the Distiller's Company, under licence from Chemie Grunenthal, using the brand name Distavan. It was marketed as a great advance by the medical sales representatives to every doctor, and these doctors duly prescribed it to soothe the stomach and calm the mind. Because it worked so well, doctors also happily prescribed it for pregnant women suffering from morning sickness as well as those needing a sleeping pill. The name of this new drug was thalidomide.

In November 1959, an Australian obstetrician and gynaecologist sent a letter to *The Lancet* about a number of babies with birth defects, stating that the only thing that the mothers of these children had in common was that they had all taken thalidomide during their pregnancy. This was the first warning of what was to follow. Thalidomide was eventually to become known as the most potent teratogen (a substance that induces developmental abnormalities in a foetus) ever synthesised by man.

Thalidomide was withdrawn from the market in December 1961, but in the five years it was available it caused immense damage. Eventually it became known that some 8,000 children in 46 countries, including 400 in the United Kingdom, were born affected in one way or other by this drug.

At the time of its withdrawal elsewhere, the drug was still awaiting approval by the Federal Drugs Agency in the United States. The FDA were worried about peripheral neuritis, a disabling and irreversible side effect of thalidomide affecting feeling in the fingers and toes, that had been known about for some years. So apart from a few free sample tablets, which had been given to a few doctors and passed on to their patients, thalidomide was not yet available to the general public in America. Needless to say, once the birth defects came to light, approval to market thalidomide in the USA was denied.

For every damaged child born alive, there were about twice as many infants born dead to mothers who had used thalidomide, the drug having caused so much damage to the babies' internal organs that they had no hope of sustaining independent life outside the womb.

Some of the affected children who did survive only had mild effects, a missing or shortened finger or two, but many were horribly deformed. Some had no arms or legs, with perhaps a hand or finger protruding from their shoulders, or a foot or toes protruding from their hips. Some had flipper-like limbs, instead of arms and legs, a condition called phocomelia. Babies were born with ears missing and heard through an opening on the side of their head; some had no internal hearing organs in their head, as well as no ears. Some were born without eyes, some with a cleft lip or palate, some unable to raise their eyes or head to look at things. Some were born with deformed genitals or none at all. Some babies were born without an anus, and so could not pass bowel movements. There were many with multiple handicaps, with damaged internal organs and damaged brains.

Careful investigation of the damage done to each child showed that the damage was dependent on the stage of the child's development in the womb, when the mother took the drug and the number of doses taken[1], which explained why such a wide range of problems were seen in the surviving babies. Between days 20 to 35 of the pregnancy, the drug affected the production of growth factors involved in the formation of blood vessels. However, beyond day 35 of the pregnancy, the mother was not affected by thalidomide and neither were those babies who had been exposed to the drug after day 35.

Animal testing performed during the drug's development before its launch had not shown these effects as the testing was only done using rats, which are rodents. If monkeys, much more similar to humans, had been used in the testing as well, the damage would have been seen. It was only after the thalidomide tragedy that the regulation of new drugs became a priority.

In the UK, the Dunlop Committee was initially set up and later was replaced by the Committee on Safety of Medicines (CSM) in 1964, which was part of the Medicine Control Agency (MCA), the UK regulatory authority at that time. The USA already had the Federal Drugs Agency (FDA) in place to oversee all new and revised drugs before they were marketed. In April 2003, the Medicines and Healthcare products Regulatory Agency (MRHA) replaced the Medicines Control Agency and the Medicines Devices Agency. It is the MRHA that now licenses all new medicines and medical devices. The CSM continues as before, but now within the MRHA, advising on the licensing of new medicines, and on their safety once they are marketed in the UK.

Manufacturers are now required to supply a mass of information about any new medicine, the conditions it can be used to treat, the results of the clinical trials and data about its safety and any side effects or contra-indications that have come to light during the clinical trials. Only when the authorities are fully satisfied will the new medicine be issued with marketing authorisation, previously known as a Product Licence. Once a new medicine has been given a marketing authorisation number, it is allowed to be used by doctors, who can then prescribe it to their patients. All new drugs are intensively monitored for several years for side effects, contra-indications and drug interactions.

Thalidomide today

It may seem surprising, but thalidomide is still in use even today – but now only with very careful safeguards to protect against damage in the future. In America, it is now marketed under the brand name of Thalomid. In the United Kingdom, thalidomide is only available on a named-patient basis under specialist supervision, for use by men and post-menopausal women only. Strict conditions are imposed on its use; it may not be used at all to treat women of child-bearing potential, and male patients must use a barrier form of contraception, as the drug may be in the seminal fluid, or may have affected the sperm. Male patients may not donate sperm and patients of both sexes may not donate blood while taking thalidomide.

Thalidomide has been found to have both an anti-inflammatory and immunomodulatory affect, preventing the immune response flare-ups so common in auto-immune disorders. Thus it is now used in the management of the severe mouth and genital ulceration that can occur in Behçets disease or in association with HIV infection. Various serious skin disorders involving light-sensitive dermatoses, such as actinic prurigo and discoid lupus erythematosus, have also responded to treatment with thalidomide, where other treatments in the past have failed.[2]

The drug has also been found to be effective in graft-versus-host disease, a complication that can occur after bone-marrow transplantation. In this disease, donor lymphocytes attack the recipient's body, causing a chronic scleroderma-like condition, with thickening of the skin, resulting in waxy, ivory-coloured areas. This is steadily destructive, and has often, in the past, been fatal, despite treatment with massive doses of steroids and other immunosuppressants, whereas now treatment with thalidomide can deal with the condition very effectively.

Thalidomide has also been found to be a particularly useful treatment for certain cases of leprosy. Unfortunately, in various South American countries, a new generation of thalidomide children have been born as a result of this use. The drug was sourced from the USA and so the warnings about its effects on the foetus were in written English and not understood by the patients taking it to treat their leprosy.

A new role for thalidomide in the prevention of weight loss in some cancer patients is currently being studied to determine whether the results found relate to all cancers and whether the slowing of weight loss prolongs survival.[3]

However, the peripheral neuritis that worried the FDA so long ago remains a major side effect, which can unfortunately be severe and irreversible. The debate continues.

Bendectin and Debendox doubts

Morning sickness in pregnancy is common, usually lasting for an hour or two each morning for a few weeks in early pregnancy. Unfortunately for some women, it can be severe and debilitating and, despite the name, can last all day, and sometimes for the whole of the pregnancy. An anti-nausea drug called Bendectin in America, and Debendox elsewhere in the world, which originally contained three ingredients, doxylamine, dicyclomine and pyridoxine, was used for several decades starting in the 1950s as a treatment for morning sickness in pregnancy.

Following on from the thalidomide tragedy, concerns arose from anecdotal reports of malformations in babies whose mothers had taken Debendox during pregnancy. Investigations over several years showed that one of the ingredients in this product, dicyclomine, might possibly cause problems. In 1977, dicyclomine was removed from the product in the USA, and then later in the UK, but still the rumours and reports continued.

Because of threatened litigation, the makers decided in 1983 to withdraw the product from the market. However, during the 27 years that it was available, the product had been used in some 33 million pregnancies. Both the CSM in the United Kingdom and the FDA in the USA reviewed the literature relating to all the studies done over the years. They concluded that, despite the lack of any scientific evidence of an increase in birth defects, the risk of teratogenicity could not be definitely excluded because of the particular combination of drugs contained in Debendox.[4]

It was not until some years later that William McBride, a leading Australian gynaecologist who had earned much praise earlier in his career as one of the doctors who had revealed the dangers of thalidomide, admitted that he had faked research data in order to force Debendox off the market. He claimed that he did it 'in the interests of humanity'.

Other tragic teratogens

Following the thalidomide tragedy, the CSM was very careful when new drugs were given marketing approval. When, in the 1980s, a new product to treat acne was launched, the British authorities insisted that it should only be prescribed by consultant dermatologists, and with very careful safeguards in place, because of possible birth defects it might cause. This new product was an oral retinoid, a vitamin A-like substance, called Roaccutane.

Some years later, in 1988, the FDA announced that they had received reports that over 600 babies had been born either mentally retarded or with severely deformed features after their mothers had taken the same product, called Accutane in the USA. America had escaped the thalidomide tragedy in the early 1960s, but not this time. Belatedly, the FDA now insisted that the Accutane product packaging must carry a picture of a deformed baby, to scare off pregnant women.

The British were only too aware of the threat of birth defects following thalidomide and so the CSM had insisted that the pack should carry clear warnings of the possible horrific side effects of this new drug. Today, all female patients are obliged to sign consent forms confirming that they are not pregnant before they can be given the drug. To exclude any possibility of pregnancy, it is mandatory that a pregnancy test be performed before treatment with this drug can be started and that effective contraception must be practised for one month before, then during and for one month after the 16-week course of treatment. Patients taking longer acting oral retinoids need to avoid conception for even longer periods, in some cases for up to two or three years.[5]

Stilboestrol is a synthetic nonsteroidal oestrogen, which today is called diethylstilboestrol and used with careful safeguards in the palliative treatment of breast and prostate cancer. However, in the late 1940s and 50s, it was widely used, particularly in America, by pregnant women. It was not until the 1970s and 80s that it was realised that the children of these women were suffering side effects resulting from the medication taken by their mothers before they were even born. The female offspring of these mothers, who received stilboestrol while pregnant, developed cervical and vaginal abnormalities, including glandular growths and cancer. The male offspring similarly exposed *in utero* also developed genital abnormalities and abnormal spermatazoa.[6]

The tale of the Dalkon Shield

The Dalkon Shield was an intra-uterine contraceptive device (IUD) introduced in the USA in the early 1970s. In the first year alone some 16,000 of them were fitted. Many doctors realised that this was a very useful alternative form of contraception. While the spread of cheap and effective contraception in the form of the contraceptive pill had been a great advance, it also had its problems: doctors were most concerned about the risk of thrombosis, as prolonged use of the pill led to a statistically significant increase in this risk. This was because the type of contraceptive pills that were available at the time contained a much higher dose of oestrogen than is the case now, over 30 years later.

Women who could not take the contraceptive pill for various reasons turned to alternative methods of contraception, one of which was the intra-uterine device. The Dalkon Shield was a new and supposedly improved type of IUD. Similar devices on the market at the time had a number of problems. For a start, they were less effective than the contraceptive pill. Doctors found that the devices then available were difficult to insert, and sometimes insertion was impossible in those women who had not yet had children. The devices also tended to be expelled from the womb, sometimes becoming lost even without the woman realising, which then put her at risk of an unwanted pregnancy.

The Dalkon Shield was designed to overcome these problems. It was a 3cm-long piece of plastic, a sort of diamond shape, like a kite, with three short blunt fins angled down each side. These were intended to act like the barbs on fish-hooks to provide a safe grip and prevent the device from slipping out through the cervix at the entrance to the womb. At the bottom end of the device, there was an eight cm length of cord, with a knot 3.5cm from the end.

This cord was there to enable the device to be removed and was to become the source of the problem. The device gripped so well that it needed a considerable force (of about 5kg) to extract it. So a strong multifilament cord was needed and such a cord was eventually developed. This multifilament cord could absorb fluid (and any bacteria in the fluid), which would pass along it, in the manner of paraffin along a lamp wick, a process called 'wicking'.

The makers realised that this was a potential hazard, and so decided to enclose the cord in a black plastic sheath. The sheath was supposed to prevent the problem of wicking, but in fact it didn't at all; what the coating did do was make the cord very stiff, which caused significant production problems. The makers then started to receive complaints from women who were fitted with the device. They complained that their partners could feel the stiff cord in its plastic sheath protruding from the cervix during intercourse. These ladies were not happy, and neither were their partners.

Nonetheless, by 1972, the Dalkon Shield was the best-selling IUD in America. However, there were more problems to come. Only a year later, following reports of seven septic spontaneous miscarriages, including one where the mother also died, a number of ectopic pregnancies, and one child who was born deformed, the Dalkon Shield was no longer marketed in America. The complaints and lawsuits that followed led to the makers paying out $6 million to one woman who had a near fatal miscarriage which resulted in a hysterectomy.

Even more money was paid out for many other out-of-court settlements. The device had been marketed around the world, including Britain, and although production had stopped in the mid 1970s, doctors still had stock on their hands in England, Australia and America and continued to fit them until 1979. It was not until 1982, following a victory in the American courts by a class action started by a group of British women, that the shields finally fell into total disuse.

Opren the awful

In 1980 in Britain, a new anti-inflammatory painkiller, Opren (benoxaprofen), was launched by Dista. Dista was a subsidiary of the American pharmaceutical company Eli Lilly, who later in the decade would launch the antidepressant drug Prozac. The launch of Opren was accompanied by the usual flurry of advertising in the medical and pharmaceutical press. It was a very effective painkiller for those with arthritis and was claimed to be even more effective than the existing non-steroidal anti-inflammatory drugs (NSAIDs) such as Brufen (ibuprofen) and Froben (flurbiprofen).

At a conference the following spring, some consultants were already warning that some of their elderly patients were showing signs of liver and kidney problems. The problem was that the elderly patients taking Opren had a reduced capacity to metabolise the drug, and reduced kidney function to excrete it, simply because they were old. As a result, they were being poisoned, more and more, day by day, as the drug built up in their bodies.

Opren was available in only one strength, while other similar drugs were available in two or more strengths. As the months went by, some patients taking the 'one strength is suitable for all' Opren developed jaundice, a sign of liver failure, and several of them died. Post-mortem examination of their livers showed it to be a drug-induced jaundice, and the one common drug taken by all the patients was Opren. By the spring of 1982, doctors began noticing strange symptoms caused by a build-up of the drug in their patients. Bald men grew hair, ladies grew whiskers, and nails grew very fast and then started to come loose, before falling off. This happened to about

one in 20 patients.
One of the most common complaints, which affected at least half the patients, was of a burning sensation in the skin after only a few moments in direct sunlight, while longer exposure tended to result in blistering. Other skin disorders also occurred, such as pronounced reddening with lesions and the life-threatening Stevens-Johnson syndrome – in some patients great patches of their skin literally fell off. Many suffered gastro-intestinal disturbances, including peptic ulceration and bleeding. Some had blood disorders, as well as the jaundice, other liver and bile disorders, and even kidney failure as mentioned above, and they were all side effects of this one drug.
It was fortunate that these early reports of side effects came to light so soon after the launch of Opren, as this meant that other countries where the drug had not yet been launched were able to prevent its use. The FDA did grant a licence in the USA on 19th April 1982, where benoxaprofen was given the brand name Oraflex. But, within six months, the number of deaths across the world had risen to 61, and the drug was withdrawn globally in October that year.
In the 22 months that Opren was used in Britain it was estimated that, out of 4,000 patients who suffered side effects, some 100 patients died. Many patients continued to suffer long after stopping the drug, living in a twilight world behind closed curtains, to prevent the burning and blistering caused by exposure of their skin to sunlight, which continued long after they had stopped taking Opren.[7]

Bute and blood

Phenylbutazone is a potent non-steroidal anti-inflammatory painkiller that has been in use for decades, and is still widely used in veterinary practice, where it is known as 'Bute'. However, its use in humans is now severely restricted in the UK, being limited to the hospital treatment of ankylosing spondylitis (a form of arthritis mainly affecting the spine) specifically under the supervision of a specialist consultant.

This restriction to its use is due to occasional, but very serious, side effects that could be potentially fatal. Blood counts are needed both before and during treatment due to the risk of blood abnormalities, including aplastic anaemia, Stevens-Johnson syndrome, toxic epidermal necrolysis and pulmonary toxicity. The most serious of these are associated with bone marrow depression. In the 30 years between July 1963 and January 1993 the CSM received over 100 reports of blood disorders, of which over 40 were fatal, associated with the use of phenylbutazone.[8]

Lasers no less

Indomethacin is an example of a similar, but much safer, drug still in use today. It is a very useful NSAID, particularly valuable in treating the excruciating pain of gout. The principal and most frequent side effect found with this type of drug, which has always limited its use, is that it can cause gastro-intestinal problems, including ulceration.[9]

Over the last 30 years, the pharmaceutical industry has tried various strategies in an attempt to reduce this problem, such as formulating sustained release or modified release oral preparations or suppositories for rectal administration. In December 1982, a novel dosage form of indomethacin was launched: 'Osmosin' was a very special sort of sustained release tablet.

At the centre of one side of each pale blue tablet was a small hole, which the makers said was drilled by a laser. At that time, lasers were a new innovation to the general public, and had even been featured in the latest James Bond film. The idea behind the novel dosage form was that fluid from the gastro-intestinal tract, including the stomach, would be drawn into the tablet by osmosis (hence the brand name). Some of the drug within would dissolve into the fluid and would then be slowly expelled in solution through the laser-drilled hole, where it could be absorbed into the blood stream, without upsetting the stomach.

It sounded brilliant, but unfortunately in the same way as bread always seems to land butter side down, if the tablet landed hole side down, against the stomach or gut wall, the indomethacin would speed up the development of an ulcer in double-quick time. There were many cases of perforations of both the small and large bowel, and of haemorrhage too. Osmosin was withdrawn less than a year later in 1983 after some 650 reports of serious side effects and 20 deaths. However, this novel technology of osmosis was later successfully used with a number of other modified release medicines, using different types of drugs that did not cause gastric problems.

New but not improved

At the end of September 2004, another NSAID called Vioxx was taken off the market by its manufacturers, Merck. Vioxx (rofecoxib) was a cox-2 inhibitor, which acted more specifically on pain, while causing less upset to the stomach than older painkillers of this type. The older drugs worked by preventing two enzymes, cyclo-oxygenase 1 and 2 (cox-1 and cox-2, for short) from working, but while this resulted in stopping the pain, it could upset the stomach too. The newer cox-2 type of painkiller only stopped one enzyme

working (cox-2), and this stopped the pain, without upsetting the stomach.
Vioxx was launched in the UK in 1999 and at the time of its withdrawal was prescribed for at least 400,000 people in Britain. Although Merck had known for at least four years that this drug had problems, it kept quiet as the drug's global sales rose to $2.5 billion. When trial results published in 2000 showed an increased risk of heart attack by those taking rofecoxib, these were explained away by Merck, and the FDA took no action, even though it was sent further data about this problem.
This drug caused thousands of heart attacks and deaths worldwide before its withdrawal and no doubt will lead to many years of litigation. A few months after the withdrawal of Vioxx, doctors in the UK were told not to prescribe similar cox-2 inhibitor type drugs to patients with known coronary heart disease, following the realisation that all drugs of this type increase the risk of heart attack in such patients. It was estimated that coronary heart disease patients had a 34 per cent increased risk of heart attack when taking Vioxx.

Good old Aspirin

Aspirin was first marketed by the German drug company Bayer as long ago as 1899. Today we use it as a painkiller for headaches and rheumatics, and as an antipyretic to reduce fevers. People also take it in small daily doses to prevent heart attacks. So not only is it analgesic and antipyretic, but it is also anti-inflammatory and thins the blood as well.

In the mid 1980s, the British Committee on Safety of Medicines suddenly announced that Aspirin could no longer be given to children under 12 years of age.[10] Aspirin tablets in the UK at that time came in two strengths: the adult 300mg tablets and the children's 75mg tablets, known as 'Junior Aspirin'. Research in America had suggested that a very rare illness called Reye's syndrome, occasionally suffered by babies and children, was caused by Aspirin. This syndrome has already been mentioned in the chapters about foodstuffs and the noxious nursery, but in cases caused by other substances.

The illness usually occurred after the children had had a viral infection, such as influenza, chickenpox or gastro-enteritis, and the syndrome was devastating in its effects, attacking the liver by causing fatty degeneration, and affecting the brain too. Half the affected children died and those who survived were badly brain damaged. An Australian doctor, Ralph Reye, first reported this collection of symptoms as long ago as 1963 but it was not until 1982 that the surgeon-general in America ordered that Aspirin should not be given to children or teenagers. Within four years, the incidence of Reye's syndrome in America had halved.

The British Committee on Safety of Medicines decided to investigate further and delayed action until a full review of the literature and further testing had been carried out. And so it was not until June 1986 that the use of Aspirin in children aged 12 or under was eventually restricted in the UK. Since then, the age below which Aspirin may not be used was raised from 12 to 16.

Aspirin should also not be taken by breastfeeding mothers because it then passes to the infant in the breast milk. There is only one application for which Aspirin is allowed in children, and that is in juvenile arthritis (Still's Disease), and then it may only be prescribed for the child under the supervision of a specialist consultant.

The lower strength Aspirin (75mg) is no longer called 'Junior Aspirin' but is still very much in use, but now by older people as 'a tablet a day keeps heart attacks away'. This strength is also used in stroke prevention and for its blood thinning qualities to prevent clots forming after cardiac bypass surgery.

Phenacetin has analgesic and antipyretic actions similar to Aspirin. This drug was used in headache tablets together with Aspirin, caffeine or codeine. In use since the end of the nineteenth century, it was withdrawn in the UK over 20 years ago because of adverse blood and kidney problems, which were associated with its prolonged use and large doses.[11] Many years ago, Phensic tablets used to contain phenacetin, in addition to Aspirin and caffeine, but this product was reformulated following the withdrawal of phenacetin by the authorities, although the related brand name lives on.

Synergy at its worst

Bactrim, made by Roche, and Septrin, made by Wellcome, were two brands of the same medicine, a combination of a sulpha drug, sulphamethoxazole, and another antimicrobial chemical called trimethoprim. The two ingredients were said to act synergistically – that is they were claimed to be even more effective in combination than the simple additive effect of either taken alone. This combination was given the generic name of co-trimoxazole.

Twenty-five years ago, co-trimaxazole was widely prescribed for an assortment of infections including cystitis. However, it had some rare but very nasty side effects associated with its use. These included Stevens-Johnson syndrome (which had also happened with Opren and Bute), in which there was severe blistering of the skin and bleeding in the mucous membranes of the lips, eyes, mouth, nose and genitals. The skin could literally fall off in great patches. Other side effects could also occur, including various blood problems, particularly bone marrow depression and agranulocytosis, which especially affected the elderly. These devastating side effects were attributed to the sulpha drug component of the combination.

Since then it has been found that trimethoprim, the second component, is a very effective anti-bacterial agent in its own right, and today this is the treatment of choice for urinary tract infections, such as cystitis. It will come as no surprise that the use of co-trimoxazole today is now restricted to certain specific, very serious infections,[12] while trimethoprim is considered so safe that it may soon be available from pharmacies without the need for a prescription.

Grey babies and other problems

Chloramphenicol is an antibiotic still in regular use today, in the form of drops for ear and eye infections and a lotion applied to the skin to treat acne. These uses are all diluted, external solutions. However, apart from these, it is now reserved for use only in life-threatening infections, such as meningitis and typhoid fever, when it is given by infusion (a drip). Its use is limited because, although it is a potent wide-spectrum antibiotic, it was associated with serious haematological side effects when given systemically.

The blood disorders associated with its systemic use include both reversible and irreversible aplastic anaemia (sometimes resulting in leukaemia), peripheral and optical neuritis, erythema multiforme (also mentioned with Opren), and 'grey syndrome' which occurs in small babies, resulting in abdominal distension, circulatory collapse and pallid cyanosis, which make the baby look grey. Chloramphenicol came into use in the early 1950s and these appalling side effects associated with its use became clear within a few years, leading to severe limitations in its use.[13]

In France, also in the 1950s, a new drug called Stalinon, an organotin compound (mentioned in Chapter 2), was used by doctors to treat staphylococcal skin infections such as boils. Stalinon capsules caused massive poisoning, in which at least 100 patients were permanently affected, and 102 died, thought to be due to contamination of the medicine with a slightly different organotin substance. Further investigation later revealed that during the clinical trials – necessary to get government approval for the drug's use – a much lower dose was used and so the toxicity had not been apparent.[14]

Hexachlorophane is a disinfectant that was once widely used in soaps, creams and dusting powders, including products for nappy rash. However, during the 1970s, it was discovered that hexachlorophane could be absorbed, particularly through the skin of babies, in amounts sufficient to produce spongy lesions of the brain, which were sometimes fatal. It was also found to be teratogenic. Nurses who had used hand washes containing hexachlorophane while at work were later found to have given birth to deformed babies. Consequently, it may no longer be used at all in preparations designed to be used on babies.[15] Limitations were also placed on its use in cosmetics and for personal hygiene products.

Along came Valium

In the 1930s, a Polish chemist discovered a new group of chemicals called the benzodiazepines, but it was not until 1961 that the first one went on the market. It was launched by the pharmaceutical manufacturer, Hoffman La Roche. Chlordiazepoxide, brand name Librium, was the first of many similar drugs. Hoffman La Roche launched Valium, or diazepam, the following year. Doctors had been searching for a safer type of tranquilliser, to use as an alternative to the barbiturates. The barbiturates had been available since the previous century and were known to be very dangerous, as they caused kidney damage and even fatalities in overdose.

Ten years after the launch of Valium there had not been a single fatality due to the benzodiazepines, and they were regarded as totally safe – the perfect tranquilliser – but by 1980 there were reports of problems. Some patients found that when they tried to stop taking their benzodiazepines, they suffered the most appalling withdrawal symptoms. They suffered personality changes, sudden rage and extreme mood swings, as well as anxiety, depression and agoraphobia. The first official warning did not come until January 1988 when the Committee on Safety of Medicines issued a report on the dangers posed. This was somewhat late in the day as many journalists had been writing articles in newspapers and magazines about the problems with benzodiazepines for several years previously.

Another drug of this type was Halcion, which became the

world's most widely prescribed sleeping pill. It was a short-acting benzodiazepine and so did not give patients a hangover the following morning, which was a problem with the older, longer-acting ones, such as nitrazepam. Halcion received much bad publicity when it was implicated in a number of murders in America, committed by people while under its influence. The British Department of Health banned the use of Halcion in October 1991.

By 1990, many sufferers, who blamed Valium, Ativan and other benzodiazepines for their condition, attempted to sue the makers of the various drugs.[16] Eventually, there were 5,000 litigants in the group action which, fortunately for the manufacturers, failed. In 2002, Valium was discontinued in the UK, as usage of this more expensive branded version of diazepam had dropped to a very low level.

In the summer of 2003, the father of two French tennis prodigies was arrested in connection with the death of one of their opponents. It transpired that a number of rival tennis players had under-performed and complained of drowsiness while playing against them. Their father, Monsieur Fauviau, was arrested following a car crash that killed Alexandre Lagadere, a 25-year-old teacher. He died following a match against Maxime Fauviau. There was no alcohol in his blood, but a benzodiazepine called lorazepam was found to be present in his bloodstream, even though he had never been prescribed it or been seen to take it. Such is the ruthless and competitive world of sport these days that this father was prepared to drug his children's opponents to enable them to succeed.

After Valium

Following the realisation in the 1980s that the benzodiazepines could cause problems when taken long term, the CSM advised that they should only be used for short-term relief of anxiety, or insomnia, and then only when it was severe and disabling or causing unacceptable distress to the patient. They advised that the benzodiazepines should no longer be used at all for mild anxiety or insomnia.

As an alternative to the benzodiazepines, doctors began to use very low doses of thioridazine for anxiety and distress, including treating the elderly who were suffering from confusion and dementia. Thioridazine, called thorazine in America, was a major tranquilliser and antipsychotic, in use since the early 1960s in Britain. When the doctors began using it in place of the benzodiazepines, it was used in a far wider population of patients than previously, especially in the elderly. This led to an increase in certain side effects, particularly those affecting the heart.

The cardiac problems caused a particular type of irregular heartbeat, or arrhythmia. This was due to prolongation of the QT interval of the heartbeat, which will cause even more problems later in this chapter. This was so serious that it led the CSM to advise that thioridazine should be reserved specifically for use as a second-line agent and only prescribed by consultants in the treatment of adults with schizophrenia, with stringent safeguards in place, to prevent heart problems in those taking it.[17]

Before Prozac

Depression is a mental state characterised by extreme sadness. Sufferers may be agitated and restless or slow and retarded. Pessimism and despair, problems with sleep, appetite and concentration are all common symptoms. Treatment is with antidepressant drugs, cognitive behaviour therapy and/or psychotherapy. The main group of antidepressants, in use for the last 50 years, have been the tricyclics, mentioned in numerous murders throughout this book as well as in the chapter on treatment. However, these have been superseded in the last 20 years by the serotonin re-uptake inhibitors, which affect the levels of the neurotransmitter serotonin in the brain, and are far less sedative than the older tricyclics.

Zimelidine, brand name Zelmid, was an antidepressant introduced in the UK in 1982. It was a serotonin re-uptake inhibitor, a forerunner of the selective serotonin re-uptake inhibitors (SSRIs) such as Prozac and Seroxat that are used today. Zelmid was marketed as being very safe in overdose and as having fewer side effects than other antidepressants available at that time. However, it was withdrawn worldwide in September 1983 because of the risk of Guillain-Barré syndrome, which is more usually associated with viral infections.

This syndrome causes a rare form of damage to the peripheral nerves, which then become inflamed, leading to paralysis. The paralysis starts in the legs and spreads progressively upwards to the arms and on up to the face, to affect speech, swallowing and breathing. Total paralysis needs close monitoring in hospital, with intubation and ventilation, until the symptoms subside. Fortunately most people recover completely without any specific treatment.[18]

Another antidepressant, nomifensine, whose brand name was Merital, worked by preventing the re-uptake of the neurotransmitters dopamine and noradrenaline in the brain, but had relatively little effect on serotonin. Its mode of action in depression was not fully understood; however, it worked well, had relatively few side effects and appeared to be safe in overdose. Unfortunately there were soon many cases of blood disorders including acute haemolytic anaemia, in which red blood cells are destroyed. Some patients also developed renal failure. Merital was withdrawn worldwide in January 1986, as the risks from these horrific side effects now far outweighed the benefits of treatment.[19]

Slimming pills: too good to be true

Fenfluramine, as Ponderax, and dexfluramine, as Adifax, were marketed by the pharmaceutical industry for the treatment of obesity several decades ago. They were slimming pills, or in medical jargon, anorectics – drugs that reduce the appetite. They were chemically related to the stimulant amphetamine, but the usual dose of these anorectics tended to depress the central nervous system rather than stimulate it.

Immediately there were problems, including primary pulmonary hypertension, already known to be attributed to anorectic drugs and reported in the early 1980s. Both reversible and irreversible cases of this problem were reported, some of which were fatal. Drugs used to treat ordinary high blood pressure were ineffective in this type of hypertension. Investigations by the CSM in 1992 led to the advice that, as the condition appeared to be linked to the length of treatment, in future any course of treatment should not exceed three months. In 1997, the recommendations were revised for fenfluramine and dexfluramine to allow treament for up to 12 months, under certain conditions, but the use of another drug, phentermine, was still limited to three months.

Shortly after this revised recommendation, another report was published about an even more serious side effect. Twenty-four patients had developed damaged heart valves as a result of taking these slimming pills. By September 1997, the FDA in America had received a total of 144 reports, including the original 24, about valvular heart defects associated with fenfluramine or dexfluramine used alone or in combination with phentermine. None involved the use of phentermine alone.

Subsequently, all these drugs were withdrawn worldwide. As a consequence, the US authorities made recommendations regarding the screening of all patients who had received either fenfluramine or dexfluramine in order to detect any heart valve lesions and provide future care. Further studies suggested that prolonged exposure or higher than normal dosages appeared to increase the risk of heart valve damage. In a review of 53 cases of overdosage published in 1979, nine of these patients had died of cardiac and respiratory arrest.[20]

In the USA, slimming pills containing these drugs were marketed under the brand names Phen-fen and Redux by a company called American Home Products. This company also sold its products to many other countries throughout the world. It was discovered during the FDA investigations, and the court cases which followed, that the manufacturers had been aware of these appalling side effects all along, but had chosen to cover them up. If they had been honest and told the FDA about the side effects, the drugs would have been withdrawn a lot sooner, or maybe not even marketed at all. Since then, they have paid out in excess of $17 billion in damages to the women they damaged, but no single individual employee of American Home Products has ever been prosecuted.

In 2007, a weight loss drug, Acomplia (rimonabant), widely available in Europe, including Britain, was banned from use in the USA when the FDA rejected a licence to market the drug. This followed research which indicated that the drug doubled the risk of suicidal thoughts and behaviour. Other drugs known to increase suicidal thoughts and behaviour, particularly in young people, are the anti-viral Tamiflu (oseltamivir), and the SSRI antidepressant Seroxat (paroxetine). Such drugs now carry warnings to alert prescribers that close monitoring is required.

New drugs, old problems

The pharmaceutical industry worldwide is always on the lookout for likely new drugs, particularly of the mood-altering kind. Whether drugs for schizophrenia, depression or dementia, there is a massive market to be tapped, and you can strike it lucky, like with Prozac. However, all drugs have side effects, some of which are worse than others. Sometimes even with potentially fatal side effects, a drug may still be allowed to be used, but with very careful supervision and monitoring, as we shall now see.

Clozapine, brand name Clozaril, is an atypical antipsychotic used in the treatment of schizophrenia, but it is reserved for use only in those patients who are unresponsive to, or intolerant of, the conventional antipsychotic drugs. This restriction is because of the risk of a fatal blood disorder, agranulocytosis, developing in patients using this drug. All patients taking it must be registered with a monitoring service run by the drug's manufacturer.

The risk is so great that a white blood cell count and a differential blood count must be done before starting treatment, and then every week for the first 18 weeks, after which the counts can be reduced to every two weeks. Only after a year may they be reduced again, to every four weeks. Even after use of the drug has stopped the blood counts must continue for a further four weeks.[21]

Another atypical antipsychotic, remoxipride, with the brand name Roxiam, caused aplastic anaemia. By November 1993, the CSM in Britain had received eight reports associated with this drug, five of which were in the UK. The patients, one of whom died, had only been taking remoxipride for between three and eight months. The drug subsequently remained available from the manufacturers for the compassionate treatment of psychosis in individual patients intolerant of other antipsychotics, but only with very careful monitoring.

The CSM recommended that it should not be given to patients with a history of blood disorders and that prospective patients should have blood counts before treatment commenced and then every week for the first six months, and monthly thereafter. Patients and their carers were advised to seek immediate medical attention if any bruising, bleeding, sore throat or fever developed. At the first sign of any blood disorder the treatment with remoxipride had to be stopped. Roxiam is no longer marketed.[22]

Paraldehyde, a trimer of acetaldehyde, was used in the past as both a hypnotic and sedative, and for its antiepileptic effects. Unfortunately, it has a solvent action when in contact with plastics. And it tends to decompose on storage, particularly once opened. It is dangerous to use if it has a brownish colour or has the sharp vinegary smell of acetic acid, as these are both signs that it has begun to decompose. Deaths have occurred in the past due to the use of paraldehyde that had deteriorated during storage.

The administration of paraldehyde is problematic too, as both gastric and rectal administration may cause irritation, and there are even greater hazards when it is given by injection. Intramuscular injection is not only painful but also associated with tissue necrosis, nerve damage and sterile abscesses, while intravenous injection is extremely hazardous as it may cause many other problems such as haemorrhage, hypotension and circulatory collapse.

Needless to say, due to these hazards associated with its administration, its tendency to react with plastics – so that all-glass syringes need to be used – and the risks associated with its deterioration, it is not surprising that paraldehyde has largely been superseded by other more easily administered drugs.

Gas doing more harm than good

Halothane is an anaesthetic that would normally be administered together with oxygen, or with mixtures of nitrous oxide and oxygen. Unfortunately it sometimes damages the liver, an adverse drug effect first recognised many years ago. The liver is the body's detoxification system so, if it is damaged, the consequences can be very serious. The first obvious sign of a problem is the development of jaundice, where the patient starts to turn yellow.

The CSM received 84 reports of liver damage associated with the use of halothane in the UK between 1978 and 1985, and so issued guidelines on the precautions to be taken by anaesthetists before using halothane. In 1997, these guidelines were reiterated after a further 15 cases of acute liver failure were reported, all of which required transplantation. An anaesthetic whose use results in the need for a liver transplant is not to be recommended.[23]

Another gas needing precautions is oxygen. While oxygen is vital to our survival, too much can be disastrous. Premature babies used to be kept in a pure oxygen or high oxygen concentration atmosphere, within incubators, until it was realised that some became blind as a result. The blind singer-songwriter, Stevie Wonder, is one such victim. Nowadays, patients, including pre-term infants, receiving oxygen therapy do not breathe in pure oxygen, but instead breathe in air enriched with oxygen.

Too much sugar

Today we call it Type 2 diabetes, but this condition used to be called maturity onset diabetes because it tended to be diagnosed in middle-aged patients who had put on a lot of weight. This sort of diabetes can usually be treated with tablets rather than with insulin. Phenformin was such an oral hypoglycaemic drug – that is, it lowered blood sugar levels. It was in use in the UK from the mid 1960s until the late 1970s. It is still in use even today in Italy and Spain, but was discontinued in the United Kingdom due to an unacceptably high incidence of cases of lactic acidosis and coma, which were often fatal.

Phenformin was also implicated in reports concerning excessive cardiovascular mortality associated with its use.[24] Buformin was a similar drug with similar problems, and it was also discontinued in the UK in the 1970s. However, a third drug of this type, called metformin, is still widely used in Britain and around the world to treat Type 2 diabetes, with relatively few problems. Metformin is also used with great success to treat the gynaecological condition called polycystic ovary disease.

Another oral hypoglycaemic medicine that was only on the market for a short time before being discontinued in Britain was troglitazone, brand name Romozin. Its withdrawal was due to severe liver reactions in certain patients, some of which were fatal. The average time before the onset of liver damage was only three months. The CSM was aware of more than 130 cases worldwide, including six deaths, by December 1997, although only one case had occurred in the UK.[25] Drugs cost a great deal to develop, so in America the manufacturers and the FDA agreed a schedule of routine monitoring of liver function in November 1997 rather than withdrawing the product, but more and more cases occurred. The drug company recommended ever more intensive monitoring, but eventually they had to withdraw troglitazone worldwide in March 2000.

Stimulant dynamite

Phenolphthalein was a stimulant laxative in use from the early years of the twentieth century. It was available alone or as an ingredient of compound tablets which were composed of phenolphthalein, aloin, strychnine and belladonna. These tablets were described at the time as 'a useful combination'; a more accurate comment would have been 'dynamite' – they were very good at what they were meant to do.

Some people may remember phenolphthalein from their schooldays, where they may have used it as an indicator in chemistry lessons: it is colourless in acid solutions but bright pink in alkaline solution, so if someone with alkaline urine took a laxative containing phenolphthalein, their urine would be coloured pink. About ten years ago, phenolphthalein was reclassified as a 'prescription only medicine' following animal tests in which rats and mice developed tumours after being fed very high doses of the drug.

This was a precaution as there does not appear to be any evidence of carcinogenicity in humans. However, action was taken in a number of countries because of concerns about the long-term safety. In the UK, all the 'over the counter' constipation remedies containing phenolphthalein were withdrawn, and although some of them were reformulated (most now contain senna instead), the rest of them were discontinued.

Beta-blockers

The general public have been aware of beta-blockers for many years because of their use in treating high blood pressure, heart problems and even anxiety, but like most other drugs, they have had a rocky road to success. The very first beta-blocker, nethalide, was never even marketed, as prolonged animal testing showed it to be carcinogenic.

Another of the early beta-blockers to be marketed in the 1960s was called practolol, brand name Eraldin. Shortly after its launch, serious adverse effects severely restricted its use. A series of severe side effects involved the eyes, ears, skin and mucous membranes. These led to blindness, deafness and numerous systemic effects resulting in painful joints and growths. By the end of 1974, nearly 200 reports had been received about the eye effects, and many other reports were filed related to the other side effects. Subsequently, use of practolol was restricted to the treatment of heart attack in hospital.[26]

Another beta-blocker, propranolol, developed by the same manufacturer as practolol, was given the brand name Inderal, an anagram of Eraldin. Propranolol and many other beta-blockers have been used successfully for over 30 years without the dreadful problems that were seen with practolol.

Lipid lowering

Medication to lower cholesterol has been in the news a great deal over the last few years. The main group of drugs used to treat high cholesterol are called the 'statins'. These drugs act by inhibiting an enzyme in the liver. Put simply, this reduces the total cholesterol in the body, as well as the low density lipoprotein (LDL) cholesterol concentration and the very low density lipoprotein (VLDL) concentrations in the plasma. See Appendix IV for further explanation. These drugs also tend to reduce the triglycerides while increasing the high density lipoprotein (HDL) cholesterol concentrations (the 'good' cholesterol), and so reducing risk of heart attack.

But all the statins and fibrates (another type of lipid-lowering drug) have caused some muscle problems to a greater or lesser extent.

Rhabdomyolysis is the name given to a rare side effect that leads to kidney failure, resulting in the destruction of muscle tissue together with blood problems. This initially causes weakness, then temporary paralysis progressing to total paralysis and death. There was a marked increase in such cases with one particular statin, cerivastatin, particularly when it was taken in combination with another lipid-lowering drug, gemfibrozil. This combination was commonly used in the USA but not in the UK. Since its initial authorisation and launch in the United Kingdom, there had been five reports of rhabdomyolysis reported to the CSM in Britain as well as many reports from other countries. So, in August 2001, Bayer plc, manufacturer of cerivastatin, brand name Lipobay, suspended marketing of the drug, first in the UK, and then worldwide.[27]

In November 2004, the CSM advised that patients taking another drug of this type, simvastatin, should not drink grapefruit juice because grapefruit juice inactivated a liver enzyme important in metabolising this statin. Regular consumption of grapefruit juice could lead to high blood levels of simvastatin, resulting in life-threatening muscle toxicity. This advice about not drinking grapefruit juice has now been extended to all the 'statins'. Eating a grapefruit for breakfast does not have this effect.

At about the same time, the CSM also advised patients taking the anticoagulant warfarin that they should not drink cranberry juice. This followed the second of two deaths of patients taking warfarin and drinking cranberry juice.

In both cases the problem is thought to be caused by something in the grapefruit and the cranberry juice that interferes with the way the body's enzymes metabolise drugs.

Nothing but heartache

About ten years ago, astemizole, an antihistamine used to treat hay fever and sold under the brand name Hismanal, was discontinued in Britain. This was because of cardiac side effects similar to those of thioridazine, mentioned earlier, due to the prolongation of the QT interval. Initially, the reports of severe life-threatening cardiovascular effects were the result of cases of astemizole overdose. However, there then followed further reports, which showed that even a single tablet a day – the normal adult daily dose – was sufficient to cause problems in susceptible patients and that some patients with liver problems were also badly affected when taking it.

Terfenadine, better known as Triludan, was yet another antihistamine used to treat hay fever, but with a twice daily dosage. During the 1980s and 1990s, this was a best seller in Britain, both on prescription and, after a few years, over the counter. However, once it became available to purchase over the counter, it was then being taken by a much larger pool of patients than had been the case previously. All sorts of additional side effects, which had been largely unrecognised until then, emerged.

One of these side effects came from an interaction with grapefruit juice: if Triludan was taken together with grapefruit juice, the same problems that had caused astemizole to be discontinued in Britain appeared. Since then, a number of other drugs, including the cholesterol-lowering drug simvastatin described above, have also been found to interact negatively with grapefruit juice.

When taking terfenadine, patients already taking certain other drugs, or those with impaired liver function, found that their metabolism was impaired, even when taking normal doses, and this was found to be a risk factor in developing cardiac arrhythmias.[28] Terfenadine is converted in the body to the active form fexofenadine, which does not have this effect, so Triludan was withdrawn and a new product, Telfast, containing fexofenadine, was launched instead.

It is interesting to note that another antihistamine, Piriton, chlorpheniramine, first marketed in 1954, still has massive sales today even though it is one of the older types of antihistamine, causes drowsiness and has a three times daily dosage.

Tummy troubles

Tummy upsets and mild food poisoning can seriously mar holidays, particularly in far-away places, where travellers are not used to the local water or the local bacteria. In 1976, a lady who was unable to get hold of a bottle of good old-fashioned Kaolin & Morphine Mixture, which she normally took for such upsets, used instead a product called Quixalin. It cured the tummy upset but had many other effects.

First her feet began tingling, then walking became difficult, and she became effectively blind, all due to the adverse effects of Quixalin. Many other people suffered similar effects from taking the same medicine. Quixalin, which contained a drug called halquinol, was manufactured by E. R. Squibb & Sons. The product was withdrawn from the UK market following reports about the neuropathy and optic atrophy associated with its active ingredient halquinol.[29]

Another similar product containing a related substance, clioquinol, had been on the market since the 1930s, when it was originally used to treat dysentery and amoebiasis. In the 1960s and 70s, Enterovioform, the brand name used for clioquinol, was regarded as a modern alternative to the old-fashioned remedy Kaolin & Morphine

Mixture, which had been used for decades for diarrhoea and holiday tummy. Nowadays we use loperamide and oral rehydration mixtures to treat diarrhoea, but in the 1960s, loperamide had yet to be invented and clioquinol was the drug of choice.

Clioquinol was very popular in Japan in the 1960s, and in that country there developed an epidemic of SMON (sub-acute myelo-optic neuropathy), as the illness came to be called. SMON was associated with taking high, or even normal, oral doses of clioquinol, but for prolonged periods of time. The symptoms included great abdominal discomfort and diarrhoea, pins and needles in the legs, sometimes progressing to paraplegia (total paralysis). There was also loss of visual acuity, which sometimes led to irreversible blindness, sensory disturbances and a strange but characteristic green pigmentation to the tongue, urine and faeces.

There were more than 30,000 cases of blindness and/or paralysis in Japan alone, and thousands of deaths worldwide, all caused by this one drug. Cerebral disturbances including amnesia and confusion had also been reported as adverse drug reactions to clioquinol.

Although a few similar cases were reported from several other countries, it was felt that the epidemic in Japan may have been due to a genetic susceptibility in the Japanese. Nevertheless, the Ministry of Health and Welfare in Japan prohibited the production and sale of clioquinol in September 1970, and subsequently oral preparations of the drug were banned in the United Kingdom and in most other countries because of the severe neurotoxicity associated with taking the drug by mouth.

Clioquinol is a chemical that contains an atom of iodine in each molecule and so can, rarely, cause iodism in sensitive patients. The local application of creams or ointments containing clioquinol, which are still available today, can occasionally cause severe irritation. Its use in topical skin preparations can stain the skin and discolour fair hair as well as staining clothing yellow, because of the iodine content.[30]

Another discontinued medicine, prescribed for digestive problems of a different kind, Prepulsid (cisapride) was a medicine that stimulated gastro-intestinal motility and was mainly used in cases of dyspepsia and reflux disease, and where gastro-intestinal motility is decreased, such as paralytic ileus. This is yet another example of a drug discontinued due to cardiac side effects, which caused prolongation of the QT interval.[31]

Dialysis dementia

In the 1970s, early kidney dialysis patients suffered from 'dialysis dementia', which led to progressive brain damage and premature death. After much research it was found that aluminium, which was used to make parts of the dialysis machine, was binding on to a substance in the blood called transferrin as the patient's blood passed through the machine, and when the blood passed back into the patient's circulation the aluminium gained access to the brain, causing the horrible side effects of the treatment. Needless to say, the machines have long since been improved so that treatment nowadays no longer carries such a risk.[32]

About ten years ago, there was a panic in the media because it was believed that high aluminium levels in the brains of elderly people might be the explanation for the onset of Alzheimer's disease. Further research showed that this was not the case. The presence of aluminium was simply due to earlier researchers using an aluminium-containing stain on the tissue samples they had examined on microscope slides. These slides used the stained brain tissue of deceased patients so that the researchers could easily identify the characteristic plaques that form in the brains of dementia sufferers, but which are not present otherwise.

Quack cures

The heyday of quack medicines was probably the eighteenth century though, even now, in the twenty-first century, quack cancer cures are still being marketed and fortunes made from them.

Krebiozen was one such preparation, a supposed cancer cure that received a great deal of publicity, particularly in the USA during the 1960s. This substance was claimed to have been obtained from the blood of horses previously injected with extract of *Actinomyces bovis*. The claims were never substantiated and in 1966 the FDA acquired some of the product and had it analysed. The preparation was found to be an amino acid in powder form, nothing special at all, and so Krebiozen was totally discredited.

In the 1970s, a substance called laetrile became popular as an alternative remedy for cancer. Laetrile was derived from apricot kernels and consisted mainly of a substance called amygdalin, which contains cyanide. It was claimed that laetrile was preferentially hydrolysed by enzymes present in the cancer cells to produce benzaldehyde and hydrogen cyanide, which would then kill the cancer cells. Fine in theory, but amygdalin does not appear to be absorbed from the gastrointestinal tract, and both normal and malignant cells contain only traces of the particular enzymes required.

The instructions provided explained that laetrile was to be taken along with a special diet. This involved taking very high doses of an assortment of vitamins and pancreatic enzymes each day. No meat, no fish, no fowl, no dairy products and no animal protein were allowed; the diet was mainly massive quantities of fruit and vegetables. Laetrile was also claimed by some quacks to be 'vitamin B17'. They claimed that a deficiency of this 'vitamin' could cause cancer but there is no evidence of this and, in fact, laetrile has no value whatsoever in human nutrition.

There were several reports of cyanide poisoning resulting from the use of laetrile, sadly including one of an 11-month-old baby who died after accidentally swallowing between one and five laetrile tablets, which contained up to 26mg of cyanide. Other reports of adverse effects were also associated with taking laetrile, especially by mouth. This quack cancer cure was banned from sale in Britain in the early 1980s.[33]

Many other spurious 'cures', such as shark cartilage, based on the incorrect notion that sharks don't get cancer, and other alternative cancer 'cures' can be found on various websites on the Internet. Unfortunately there is always someone willing to profit from the desperate.

Chapter 19

Misuse, Abuse and Historical Junkies

COCAINE, HEROIN (DIAMORPHINE) AND morphine are the most serious drugs of addiction at the start of the 21st century. There are also significant problems with the abuse of other synthetic opioids and of illicit amphetamines too. These problems are widespread, as are those of the benzodiazepines, such as the 'date-rape' drug Rohypnol, flunitrazepam, and the sleeping tablet, temazepam. Barbiturates, once a severe problem, are little used now, except in the treatment of epilepsy.

Cannabis was used medicinally in the treatment of mania, migraine and as a sedative in the early years of the twentieth century. Its non-medicinal possession and use, however, was illegal in the UK until it was decriminalised in 2005. In recent years, however, more potent strains have been grown, some of which are more than 15 times stronger than the cannabis used in the 1960s. In some cases, this increased strength has resulted in severe psychotic episodes in those who have smoked it, and where this stronger form was used in conjunction with ecstasy, it was found to cause many memory problems, because of the permanent brain damage it caused. In January 2009, cannabis was reclassified again in the UK, back to its original place as a Class B substance, with harsher penalties for growing and supplying it than when it was Class C in Schedule 1 of the Misuse of Drugs Act. The next chapter covers the legalities involved in the possession and supply of medicines, including poisons.

LSD is a much more potent hallucinogen, and its use can lead to severe psychotic states during which the user's life may be at risk. Dependence can result from continued repeated use, and this can cause extreme problems, with both physical and psychological symptoms on withdrawal.

Coke, crack, snow – abuse by any name

Cocaine, an alkaloid obtained from the Coca plant, was first isolated by the chemist, Albert Niemann in 1860 and, like opium, it quickly became a substance of abuse. Cocaine has many street names, such as coke, crack, snow, nose candy and many more. The addictive nature of cocaine is due to the rapidity with which it relieves exhaustion and lifts depression, making people feel good both mentally and physically.

Any problems the user finds with cocaine can be relieved with morphine, and this double use leads even more quickly to addiction. Even as long ago as 1921 it was estimated that 90 per cent of the opium and cocaine entering the United States was for use by addicts, and that only the remaining ten per cent was used for legitimate medicinal purposes.

Cocaine can be smoked, but is frequently taken as snuff, by inhaling the powder up the nose. Cocaine is now regarded as the most widely used drug of abuse. Initially a low dose of cocaine leads to a pleasant euphoria, but as the addiction progresses the abuse leads to rapid degeneration – morally, mentally and physically. The effects on the circulation system can also cause chest pain and even a heart attack or stroke.

Regular cocaine use can seriously damage the heart and also result in irreversible brain damage.[1] The drug works by releasing a huge amount of noradrenalin from the nerve endings, which makes the blood vessels constrict, so giving a feeling of acute tension, or a 'high'. Unfortunately, this results in damage to the blood supply of the nasal septum, which is the division in the nose between the nostrils. Eventually this leads to ulceration and perforation of the septum, flattening the nose, which will then appear to have only a single large nostril.

This problem was formerly only associated with certain types of masonry workers and the employees of chromic acid factories. But even today, some so-called 'celebrity' users have developed the large single nostril, to their great embarrassment. Plastic surgery is required to rebuild the nose and restore the nasal septum, but only once the addiction has been treated.

'Crack' cocaine is smoked and is extremely potent, with many psychological effects. Initially the effect is of euphoria, and then there is an opposite effect with a feeling of unease, depression and anxiety (dysphoria) and finally of a schizophrenic type of psychosis, with loss of control of reality. Users deal with the dysphoria by taking repeated doses to restore the euphoric sensation, and therefore need to take the drug continuously to feel even relatively well.

The effects are diminished as time passes, so users increase the dose again and again. During this time, regular users suffer from many physical symptoms, including seizures. An overdose can result in the user developing status epilepticus, with cardiac or respiratory arrest leading to death. This same pattern of use with escalating dosage leading to both physical and psychological dependence is also seen in users of amphetamines.

Those who peddle drugs seek to maximise their profits by 'cutting' their drugs with cheap inactive ingredients. In 1989, there was even a report from America of compounds containing arsenic being used to 'cut' cocaine, resulting in symptoms of arsenic poisoning occurring in those cocaine abusers who used it.[2]

The opium poppy

Opium, and the opioid analgesics in general, can lead to dependence, both physical and psychological. And abrupt withdrawal can lead to a withdrawal syndrome. The severity of withdrawal depends on many factors: the individual, the drug he or she used, the dose, the frequency and the duration of the use or abuse.

There are a number of different types of opioid receptors in the body, and each opioid acts at one or more of them. As a result it has been found that cross-tolerance and cross-dependence can occur between different opioids, because they are acting at the same receptors. If addicts receiving daily treatment miss just a couple of days of treatment, they must see the doctor for a revised (lower) dose, due to their loss of tolerance during those missed days; taking the previous dose could now be an overdose.

Tolerance diminishes rapidly after withdrawal, so babies born of drug-dependent mothers can suffer withdrawal symptoms at birth, which must be treated gently and carefully, to wean them off the drug that their mother is dependent upon.[3]

Opium was in use as a medicine for thousands of years. Certainly the ancient Egyptians used it, as did the ancient societies in India and China. In Britain, during the 18th century, demand was supplied with opium imported from Turkey and Persia, but it was very expensive because of the transport costs involved. To meet this ever growing demand, the opium poppy was grown in parts of eastern England and Scotland for over a hundred years.

This proved to be very profitable to the growers, and it pleased the customers too, as the home-grown product was so much cheaper. Apparently it grew particularly well in the area around Norwich. At this time, opium was the only drug that was used for its euphoric effects. Opium contains at least ten per cent morphine and two per cent codeine as well as a variable mixture of other alkaloids. Laudanum, the alcoholic tincture made from raw opium, was frequently the only medication to be found in many households.

Laudanum has been used by many well-known figures in the past. John Hunter (1728-93), the 'Father of Anatomy', disliked lecturing and public speaking so much that he nerved himself with 30 drops of laudanum beforehand, while a great actress faced the ordeal of the first night by means of seven drops of the same tincture.
Thomas De Quincey took his first opium, as laudanum, in 1804. He bought it from a druggist in Oxford Street, London, to relieve the pain of his facial neuralgia. By 1813, he was taking 320 grains (over 20g) of raw opium each day. This is an incredible amount of raw opium as the recommended maximum dose was two grains, although there are cases on record of those who took almost twice this amount per day and still lived to a ripe old age. In 1821, De Quincey wrote his *Confessions of an English Opium Eater* in which he praised the drug and the pleasure and serenity he derived from it.
It is said that Bramwell, brother of the Brontë sisters, read De Quincey's book, and so began to purchase opium from Bessy Hardacre, the local druggist at Haworth, whose shop was opposite the Black Bull Inn where Bramwell Brontë drank. It seems to have been somewhat fashionable for creative writers and thinkers at that time to indulge in taking opium.
The writer Lord Byron took it as Black Drop – probably the famous Kendal Black Drop, which was four times the strength of normal laudanum. This version was also taken by Coleridge, who, while being a life-long opium addict, managed somehow to keep it secret, even though he drank as much as a pint of laudanum daily.
Amongst the ladies, both Elizabeth Barrett Browning and Florence Nightingale took opium. Miss Nightingale was one of the first patients to be given opium in the form of injected morphine. Wilkie Collins used it and persuaded Charles Dickens to try it, and so Dickens also became an opium user. John Keats, the poet, and William Wilberforce, who abolished slavery, also both used it. Even the prominent politician, William Gladstone, used opium, taking laudanum to steady his nerves before addressing parliament.

In the fens where the home-grown crop had flourished so well, the fenland agricultural workers were also addicted to it. In the nineteenth century, it was possible to buy opium from any shop that cared to stock it, but as time went on there was increasing pressure from both the medical and pharmaceutical professions for some control on its supply. The doctors wanted it to be available only on prescription, which would effectively mean that only the rich, who could afford to go to the doctor, would be able to obtain a prescription for it.

This proposal caused a great fuss from the chemists in the fenland area because opium sales to the poor agricultural workers and ordinary people were such a large proportion of their total trade that such a restriction would have put most of them out of business. The Pharmaceutical Society, on their behalf, managed to win the day, persuading the government to continue to allow sales from pharmacies, and so chemists and druggists continued to sell opium to all well into the 20th century.

'Heroic' heroin

Heroin was considered to be only a variation on morphine, the opioid painkiller used for severe pain. Morphine was first isolated from opium in 1806 by an apothecary's assistant, Frederich Sertürner, who nearly killed himself in doing so, not realising how potent it was. Pierre-Jean Robiquet then isolated codeine from opium in 1832 and G F Merck isolated papaverine in 1850. In 1874, a team led by Frederick Pierce at St Mary's Hospital Medical School in London first prepared diacetylmorphine (diamorphine), which today we know as heroin. This was one of the first drugs made by modifying a natural molecule – in this case, morphine. Twenty years later, the Director of Research for the Bayer drug company, Heinrich Dreser, was seeking more powerful morphine-like drugs, and selected diamorphine. But it was not until 1898 that the Bayer drug company marketed 'Heroin', named as such because they considered its effect as nothing short of heroic. They advertised it as 'a magnificent sedative for laryngitis and coughs', and claimed it was non-addictive! Little did they know.

So 'Heroin' was once a trade name, but it is now used as the generic street name for diacetylmorphine hydrochloride, known in medicine today by the more convenient name of diamorphine. It is now recognised as being more potent than morphine and therefore more likely to cause addiction.

In 1912, the International Opium Convention was held. This convention was concerned with the abuse of opium, morphine and cocaine, as well as with smuggling and the quantities imported and exported by different countries, supposedly for medicinal use. Even ten years later, many countries, including Persia, Argentina, Colombia, Costa Rica and Paraguay, had still not unreservedly ratified this Convention. It seems that things have not changed much in the last 90 years or so either.

However, the addictive qualities of opium, morphine and

diamorphine have long been recognised. In a report printed in the *British Medical Journal* in 1921[4] regarding addiction, in the UK and the USA, diamorphine was actually vaunted as a cure for morphinism. The addicts were found to prefer it to morphine, which is hardly surprising as we now know that it is more potent and more addictive.

Between June 1922 and June 1923, more than a million dollars worth of heroin was seized from around the country by the authorities in the USA. And no doubt it was much purer than what is available on the drug scene today, 'cut to ribbons' – some of which is now found to contain as little as five per cent of the active ingredient.

Ergot abuse

Ergot is a fungal infection found on rye grains. The ergot alkaloids, derived from the fungal infection ergot, have to be carefully used in medicine. Their side effects are so severe that doses have to be kept very low in order to allow their use at all. Ergotamine is one of the alkaloids derived from ergot, which has mainly been used in the treatment of migraine. Several unfortunate cases of overdose are described in Chapter 14.

Large doses can result in severe peripheral vasoconstriction of the veins and arteries. This can be so severe that it can shut off the blood supply, and consequently may lead to the development of gangrene, particularly in the feet and legs. Prompt treatment, however, can rescue the situation and lead to a full recovery. Unfortunately, for some patients help comes too late, and the only recourse is amputation of the affected limbs.

Despite the risk of severe consequences, cases of abuse of ergotamine and related substances have occurred. Some of the other ergot alkaloids used to treat migraine are derivatives of lysergic acid, which is related to LSD, the hallucinogenic agent.

LSD and amphetamines

LSD is the shorthand name for the hallucinogenic drug lysergic acid diethylamide, also called lysergide. Drugs of this type are used for recreational purposes by drug abusers, as they induce a state of altered perception, thought or mood. Changes in visual perception are the commonest and most significant and there may be personality changes too. Although these drugs were tried out in psychiatric medicine in the middle of the twentieth century, they were not generally found to be of any use in the treatment of mental illness.[5]

Babies born to mothers who used LSD during pregnancy were

found to have a number of congenital abnormalities, including eye malformations. Psychological dependence and tolerance have both been found with hallucinogenic drugs. Some people had recurring panic attacks after using them, while others suffered from ‘flashbacks’, which either occurred spontaneously or could be induced by alcohol, stress or other drugs.

Various amphetamines have also been used as recreational drugs, for their stimulant action on the central nervous system, alleviating fatigue and producing a feeling of mental alertness and well-being. The main problem is that they are frequently taken alongside alcohol, which unfortunately creates additional problems. The acute effects can be severe, and there have been many fatalities, frequently associated with the development of an uncontrollable extremely high body temperature.

Other causes of death have been heart problems, convulsions, circulatory problems with widespread blood clotting within the veins, with muscle pain and tenderness, and acute kidney failure. Repeated use can cause liver damage and psychiatric effects such as psychosis and depression. Brain damage has also been reported. Abuse during pregnancy has led to a number of congenital abnormalities in the babies, including two cases of congenital heart disease.[6]

Ecstacy, also called Adam, E, M7M, MDMA and XTC on the street, has the chemical name methylenedioxymethamphetamine. It is not new, having been first synthesised and patented by Merck as long ago as 1912. In the 1950s, the US army even briefly investigated using it as a possible brainwashing tool during the Cold War. In the 1970s, it was used by Californian psychotherapists.

At this time, ‘uppers and downers’ were readily prescribed by doctors and many people became addicted to them. Dexamphetamine was widely prescribed by doctors in the late 1950s and early 1960s, as an ‘upper’ together with a barbiturate to be taken at night as a ‘downer’, until the introduction of the benzodiazepines, such as diazepam. Dexamphetamine acquired the popular name of ‘purple hearts’ at that time, due to the colour and shape of the tablets, which doctors prescribed ‘as easily as smarties’, until they realised how addictive the drug was.

Speed, Crystal, Crystal Meth, and Ice Meth are just a few of the street names given to a smokable form of methylamphetamine hydrochloride, which is yet another amphetamine widely used for recreational purposes. In addition to its use on the street here, this drug is used in some areas of Asia by both children and adult farm labourers to enable them to work harder and longer. When the government in Thailand enforced a crackdown, drug dealers moved on to neighbouring countries such as Cambodia. The drug traffickers often tout the drug as a 'super vitamin' in rural areas, where its users are unaware of its addictive potential. Within three to six months of regular use, users begin to run the risk of permanent brain damage. And yet, because of its guise as a 'super vitamin', even schoolchildren in Cambodia have been taking it on a daily basis.

GHB and ketamine

Gamma-hydroxybutyrate, also called GHB, GBH and liquid ecstasy, among other names, was in the past promoted on the recreational drug market for weight loss, for bodybuilding and as a sleep aid. In medicine, it is used for its hypnotic qualities, being intravenously injected as a general anaesthetic. Recreationally however, it is much more harmful. Reports of acute poisoning following illicit use led the drug authorities in America to issue warnings about its potential for abuse and the development of physical dependence.

After misuse of GHB, patients in both the UK and USA have required hospitalisation and respiratory support for periods from two hours to as long as four days, after which the symptoms may resolve spontaneously. However, deaths have occurred. The severity of the symptoms is dependent not only on the dose taken but also on the use of any other drugs simultaneously, such as alcohol, amphetamines, benzodiazepines or cannabis. Regular users have experienced a withdrawal syndrome upon discontinuing use of GHB.

Ketamine is another anaesthetic that has become a substance of abuse in the last 20 years or so. In America, the dangers associated with abuse of ketamine were first reported in 1979[7], and in the United Kingdom more recently than that. When taken by mouth or via the nose at parties and raves, this drug is known as 'Special K', 'Super K' or 'vitamin K'.

Its main effect is the production of hallucinations such as out-of-body or near-death experiences. It also causes severe loss of co-ordination, helplessness, loss of awareness of surroundings and profound analgesia, and so puts its users at severe risk of harm. Some users have experienced a state in which they no longer care whether they live or die – they think they can fly or that they are superhuman.

Hypnotic habits

Some people seem to be susceptible to the development of dependence when given certain sedatives or hypnotics. This type of dependence can also result from solvent abuse, such as 'glue sniffing', about which there will be more details later. Dependence means that there is a strong desire to continue taking the drug. Users might increase the dose and then develop both psychological and physical dependence on its effects.

Withdrawal produces a characteristic syndrome. Initial symptoms are of apprehension and weakness, followed by anxiety, headache, dizziness, tremors and vomiting. Then follow nausea, abdominal cramps, insomnia and increased heart rate. Low blood pressure and convulsions may even progress to status epilepticus after a day or two. Sometimes hallucinations and delirium tremens may develop later, after several days of withdrawal. These symptoms can be dramatically reduced in severity by a very gradual withdrawal: withdrawing the substance very slowly, over a period of days or weeks.

The barbiturates were first synthesised by Bayer in 1863, but it was not until 1903 that they began to be used as sedatives to calm nervous people and as hypnotics to help them sleep, for barbiturates depress the central nervous system. Prolonged use of barbiturates can lead to dependence of the barbiturate-alcohol type described in the previous paragraph, with both physical and psychological symptoms. The barbiturates were very useful in their day but have long since been superseded by less toxic alternatives such as the benzodiazepines, although they too have caused problems.

There are many disadvantages with the barbiturates: mainly, patients can easily develop tolerance, and so need a higher dose to produce the same effect, which can then lead to abuse. Barbiturates also cause a hangover effect in the morning when used as a sleeping pill. They can cause severe withdrawal effects, interacting with alcohol and other drugs. There are also severe toxic effects in overdose.

If given to pregnant mothers, the barbiturates will depress even the nervous systems of their babies too. These babies, like those born to heroin addicts, have been found to be very 'jittery' at birth and to have feeding difficulties. Such babies need to be given barbiturates from birth with the dose being slowly tapered off.

Today, the only medicinal use of barbiturates in the UK is that of phenobarbitone, used as an anticonvulsant in the treatment of some types of epilepsy. However, even today, there are a few old ladies still taking their barbiturate sleeping pills every night, as they have done for the last 40 years or more. It is considered to be kinder for such elderly patients to let them continue with what they know, rather than switching them to newer, shorter-acting sleeping aids.

In the past, barbiturate overdosage was a frequent cause of acute poisoning in which coma, respiratory and cardiac depression, low blood pressure and shock led to kidney failure and death. Marilyn Monroe died from an accidental overdose of her barbiturate sleeping pills. Barbiturates are used today by the Dignitas Clinic in Switzerland, given to patients who wish to end their lives at the time of their own choosing.

John Armstrong was a 26-year-old nurse at Gosport Naval Hospital, near Portsmouth. He was married to a young lady called Janet, who was only 19, and they had three children. One lunchtime in July 1955, the Armstrongs called the family doctor because their baby son Terence was ill. The doctor knew about the family since, the previous year, their oldest boy, Stephen, had died suddenly.

Baby Terence was dead when the doctor arrived at their home, so the doctor immediately notified the coroner, who sent two of his staff to collect the body. They also took away the baby's feeding bottle and a pillow on which he had vomited the previous day, for forensic examination. The pathologist carried out a post-mortem on the tiny body. He found a shrivelled red shell, like a Daphne berry skin, in the larynx and more in the stomach. He carefully put them into a bottle of formaldehyde to preserve them, as he suspected that the death was due to some sort of food poisoning.

The coroner's men went back to the house and found that there was a Daphne bush in the garden, with many berries on it. Mr Armstrong told them that the baby's pram was placed underneath it and that the highly poisonous fruit could have fallen into it. When this was reported to the pathologist he thought he had solved the problem, but, when he looked at the sample bottle, he found that the formaldehyde was now coloured red and the 'berry skins' had dissolved. Further investigations showed no signs of Daphne berries or any sign of any known poison, but careful analysis found that the red dye eosin and corn starch were present. It turned out that the red dye was used to colour the gelatine capsule shell, which had contained a barbiturate called Seconal. Further analysis found traces of this barbiturate in the vomit on the baby's pillow. Investigations at the naval hospital where the father worked revealed that 50 Seconal capsules had gone missing a year or so earlier, and had never been found.

The police now looked more carefully at the death of the other child, Stephen, the previous year, and found that his symptoms were

similar. Their daughter, Pamela, had also been taken ill, a couple of months after Stephen, with similar symptoms, but had fortunately recovered while in hospital. Baby Terence's body was then exhumed and his organs tested for the presence of barbiturates. The results were positive.
The police could not decide which parent had given the barbiturate to the baby, but some time later, Janet Armstrong admitted that her husband had got them from the hospital and they were both tried on the charge of jointly planning to commit murder by poisoning their baby. Janet was eventually acquitted, but John was found guilty, following charge and countercharge between the two of them. This was the first known murder involving the use of barbiturates.

The coming of Valium and Prozac

The benzodiazepines first appeared on the scene in the early 1960s – first Librium, then Valium, then over a dozen more. They calmed people down and helped them sleep, but were very much safer in overdose than the barbiturates had been. Doctors happily prescribed them, knowing how safe and free of side effects they were. It was not until the 1980s that the extent of the addiction they caused became known, with Ativan (lorazepam) becoming a particular problem. At that time, it was believed that up to a quarter of the UK population had taken them and that millions of people were addicted as a result.[8] Patients were slowly weaned off the offending drug by reducing the dose in tiny amounts each week or month, as appropriate for that individual. For some patients this process took more than a year.

In 1988, the first SSRI (selective serotonin reuptake inhibitor) was launched – the Prozac age had arrived, with Seroxat not far behind. Once more, this new type of happy pill was hailed as a wonder drug that was safe. There is no denying that the SSRIs are very effective antidepressants for some people, but they have their own problems too. Some patients experience suicidal feelings while taking them or experience a withdrawal syndrome when they try to stop taking them, so withdrawal must be gradual. And for some depressed people, SSRIs do absolutely nothing at all.

And alcohol

Alcohol is probably as old as life on Earth. Man no doubt discovered alcohol by accident and liked the effect it produced. We have been producing it ever since.

Alcohol depresses the nervous system, causing drowsiness, which then impairs the ability to operate machinery or drive a motor vehicle. This was not a problem for the Stone Age man, but today a couple of pints of beer will make any driver five times more likely to have an accident than if they had not had the drinks. Alcohol also interacts with many drugs to cause even greater depression of the nervous system and more severe problems.

Dependence is liable to occur in susceptible people, even sometimes accidentally. In 1975, an American medical journal reported that seven patients who were undergoing long-term treatment by haemodialysis became addicted to the vapours of denatured alcohol during their treatment. And two of them had withdrawal symptoms when they tried to stop inhaling the alcohol fumes.[9]

Alcohol can lead to memory defects and disorientation in time and space, referred to as Korsakoff's psychosis, as well as mental confusion, paralysis of the eye muscles and unsteady gait, a set of symptoms described as Wernicke's encephalopathy. It is likely that both these conditions are simply different facets of the same condition. Both conditions are treated with large doses of vitamin B1, thiamine.

Fatty deposits may occur in the liver of alcoholics, and these may go on to develop into cirrhosis. Babies born of alcoholic mothers are found to be suffering from foetal alcohol syndrome when they are born and need to be treated for it immediately after birth.[10]

At the end of 2008, new government figures revealed that every three days in the UK a child under the age of ten requires hospital treatment for the effects of drinking alcohol. These children have mental and behavioural disorders from being drunk as well as other toxic effects, including liver disease. And the number of under 18-year-olds admitted to hospital because they have drunk too much has increased by over 80 per cent in only five years, due to more and more binge drinking.

From Mandrax to Nytol

In the 1960s, methaqualone was used as a sleeping pill. It was supplied alone or in combination with the antihistamine diphenhydramine in a product called Mandrax. The antihistamine's main side effect was drowsiness, so it was included to make the product even more effective. Unfortunately methaqualone, particularly in combination with diphenhydramine, led to abuse and dependence of the barbiturate-alcohol type.

In 1967, a case of addiction was reported in a 47-year-old man.

He had taken increasing doses of Mandrax over a period of four years, until he was taking 60 150mg tablets each day. Not surprisingly, when the tablets were stopped, he developed delirium tremens, which had to be treated with large doses of an antipsychotic drug called thioridazine.[11] Another addicted patient even had a convulsion on the abrupt withdrawal of their methaqualone. Needless to say, once these problems became known the manufacturers decided to stop making Mandrax. The antihistamine diphenhydramine is still in use as a sleep aid in the 21st century, and the best-known brand in the UK is Nytol.

Solvent sniffing

A wide variety of organic solvents are used in industry, and workers may experience the effects of poisoning due to exposure to their fumes. Inhalation, ingestion or even absorption through the skin can cause problems. Solvents irritate the skin and mucous membranes and commonly affect the central nervous system. Continual exposure may result in permanent damage to the nervous system, the liver and the kidneys.

Any product that gives off volatile fumes seems liable to be abused, regardless of the risks involved, and solvents are no different. They have become substances of abuse, particularly by young people. Benzene, toluene, trichloroethane, xylene and kerosene are just a few of the better-known solvents abused by glue-sniffers and others in search of a 'high', but glues, plaster remover, cleaning products, and even petrol and aerosol propellants are all abused by sniffers.

High concentrations of trichloroethylene fumes have caused acute poisoning[12] and even fatalities in the workplace, as we have seen in the chapter on occupational hazards. Dependence was reported in medical personnel and also in factory workers who regularly inhaled trichloroethylene fumes. A 16-year-old girl who deliberately inhaled the vapour given off from some 400 to 500ml of this solvent developed coma, respiratory difficulty, muscle spasms, hyperflexia, severe vomiting and a rapid heart rate. She needed to be treated with a tracheotomy and ventilator to help her breathing, as well as with intravenous dextrose and potassium, and she eventually recovered with no residual damage.[13] However, three teenagers who inhaled the fumes of a commercial cleaning product, which contained trichloroethylene, developed acute liver damage, and two of them also had kidney damage.[14]

A 14-year-old boy who intermittently inhaled petrol vapour for four months was found on one occasion to be breathless, flushed, frightened and hallucinated.[15] Many other cases of such dependence have occurred over the years. Petrol vapour dependence can result in sudden unconsciousness and can end in death, due to liver and kidney damage. Glue sniffing of products containing toluene has also led to liver and kidney damage. Children who have been found glue sniffing may become comatose and suffer from reduced appetite, nightmares, visual problems and incoordination. Some may even have convulsions or develop psychosis as a result.

Even 40 years ago, glue sniffing was a problem. In America, of 110 deaths associated with sniffing solvents in the 1960s, 29 of them were linked to the use of trichlorethane.[16] In 1979, a 22-year-old woman with a prolonged history of drug abuse, which included regular sniffing of various glues and organic solvents, was treated for end-stage kidney failure. She was given a kidney transplant, which unfortunately failed, and this led her to return to abusing solvents. She inhaled 400ml of Zoff plaster remover over a 36-hour period. On admission to hospital she was given supportive treatment together with her regular kidney haemodialysis, and, because it was suspected that she had taken a paracetamol overdose as well, charcoal haemodialysis was also performed. Clinical recovery was complete within 48 hours and, two days later, her liver function had returned to normal, the trichloroethane having been completely eliminated.[17]

Dichloropropane is yet another organic solvent widely used in industry, and has also been used as a dry cleaning agent and agricultural defumigant. It is another victim of solvent abuse. Following the intentional inhalation of a stain remover containing it, one abuser suffered acute kidney failure, acute liver disease and circulatory problems, but after blood transfusions and haemodialysis this patient fortunately recovered.[18]

An old-fashioned sedative

In Victorian and Edwardian times many patients took potassium bromide as a sedative to help them sleep. This old-fashioned sedative is not very effective unless used in precise dosages.

The writer, Evelyn Waugh, apparently used it with a gin and crème-de-menthe chaser. As a result he suffered from a distressing period when he had auditory hallucinations, a known symptom of bromism – the name given to poisoning caused by excessive use of bromides. These details of his life were revealed by Selina Hastings in her 1994 biography of the writer. After its publication, a number of people who knew him observed that his florid appearance and foul temper might well have indicated that he suffered from chronic bromism for much of his life.

Laughing gas

Nitrous oxide has been used as an anaesthetic for longer than any other agent. This gas was discovered by the famous English chemist Sir Humphrey Davy over 200 years ago in 1800, and it was initially referred to as 'laughing gas'. When used alone, it tends to excite the patient; for a while it was fashionable to hold parties where the guests could experience the effects for themselves. It might have been considered the 'ecstasy' of its day. It was not until some 40 years later that it began to be used medicinally.

It is now used in conjunction with oxygen as a vehicle for more potent anaesthetic vapours. Although it is non-irritant and non-toxic, nitrous oxide must be given with oxygen – when used alone it causes asphyxia. In recent years, there has been a growing use of pure nitrous oxide as a recreational drug, primarily at middle-class dinner parties, but also at some nightclubs.

In January 2007, the practice of using nitrous oxide recreationally claimed its first fatality in the UK when Daniel Watts, a 23-year-old company director, was found dead at his home next to a large cylinder of the gas. Supplies are readily available from catering suppliers, as the gas is used for making whipped cream. However, such supplies carry no warnings about how nitrous oxide can cause severe injury, long-term disability and death when abused. The 'high' produced by a single inhalation is short-acting, so users tend to want to repeat the effect and can easily overdo it, not realising the importance of a supply of oxygen as well.

Accidental abusers

Not everyone who takes an overdose is intending to kill him or herself. In 1967, three boys, each being treated for leprosy, took large single doses of dapsone, a drug used to treat this disabling disease, in an attempt to simplify their treatment regime. They thought that one big dose of their medication would save them the inconvenience of many small doses. Two of them survived but the third died. And, in 1980, a man did attempt suicide by taking a massive 7.5g of dapsone. He survived but suffered permanent retinal damage to his eyes, due to the overdose.[19]

Rifampicin is one of the drugs used in the treatment of tuberculosis. During treatment, while the drug is being taken, patients are warned that this drug will colour their urine red and discolour soft contact lenses if they are worn. A man who took an overdose of 40 rifampicin tablets developed bright red pigmentation of his skin, orange-coloured sweat and a deep red colouration of his plasma and urine. Despite this, he made an uneventful recovery but was left with some minor liver damage. Another man, aged 26, took a much larger quantity of the same drug and died.[20]

The use and abuse of a wide variety of substances, not only medicines, can lead to dependence – a physical or psychological addiction (or both). Gradual withdrawal and therapy are effective treatments, but the addiction takes much longer to cure than it does to develop.

Chapter 20

Laying Down the Law on Poisons

OVER THE YEARS THERE have been many pieces of legislation relating to poisons. The examples below are a very brief summary of some of the more relevant ones up to the present day. A careful reading will show numerous instances of that well-known phrase 'the law is an ass'.

The Arsenic Act of 1851

The Arsenic Act of 1851 was introduced because so many people were dying from arsenic poisoning in the mid-nineteenth century. The government of the day decided to take action to try to reduce the year on year ever-increasing numbers. The Arsenic Act laid down a number of provisions for each and every sale of arsenic. These requirements had to be entered into a special book by the seller and included the date of the sale, the name of the purchaser, the purchaser's place of abode, his or her condition or occupation, the quantity of arsenic sold and the purpose for which it was required.

The signatures of both the purchaser and the seller were required, and if the purchaser was unknown to the seller then the sale required the presence of a witness too. The Act declared that arsenic could not be sold to anyone 'under full age', which at that time was 21. In addition, the arsenic, with a few exceptions such as for medicinal use, had to be coloured with indigo or soot to prevent it being mistaken for other harmless substances.

This Act did not restrict the sale of arsenic to the newly emerging chemists and druggists, as pharmacists were then called. It may seem very strange but any shopkeeper, of any trade, was allowed to keep and sell arsenic along with whatever other goods or provisions the shop also sold at that time.

The poisoned sweets that changed the law

In November 1858, a mass poisoning occurred in Bradford on such a scale that it eventually resulted in further legislation – although the Bradford poisonings only happened because sugar was expensive and a much cheaper, harmless alternative was readily available.

A market stallholder sold sweets to the public in small quantities.

These were supplied to him by a wholesaler who made them nearby. The sweets contained oil of peppermint in a base of sugar and gum and, because sugar was expensive, it was the custom at that time to adulterate it with an inert but very cheap substance, known locally as 'daft'. This 'daft' might have been plaster of Paris, powdered limestone, sulphate of lime or some other innocuous substance; anything would do as long as it was cheap and harmless.
Unfortunately the chemist and druggist, from whom the 'daft' was normally purchased, was ill on the fateful day. When his young assistant went to the stock room to weigh out 12 pounds of 'daft', he picked the wrong container – the one he chose contained arsenic trioxide, not a harmless substance at all.
Arsenic trioxide used to be called white arsenic for the simple reason that it is a white powder. This meant that it looked like the hundreds of other chemicals that are also white powders. The assistant's mistake went undetected, even though the resulting sweets looked unusual, and even though the sweet-maker himself became ill while making them. The market stallholder also ate one and promptly became ill himself, but this did not put him off. Because of their strange appearance, the market trader purchased the sweets at a discount. They were sold on the market at three halfpennies for two ounces (about 50g). In those far-off days, money went a lot further than it does today, and £1 then would have purchased 160 such bags of the sweets.
About 200 people were affected with symptoms of severe illness after eating the sweets, and 20 of them died, including some young children. Prompt action by the local constabulary traced the source of the poison to the sweets. The sweet-maker, the chemist and druggist and his assistant were all committed for trial on a charge of manslaughter, but were all subsequently discharged. Analysis of some of the poisoned sweets showed that each one contained 700-1,000mg of arsenic, definitely a lethal dose.

The Pharmacy Act 1868

Although somewhat delayed, the Pharmacy Act, 1868, was a direct result of the mass poisoning case of the sweets in Bradford. This 1868 Act recognised the chemist and druggist as the suitable custodian and seller of named poisons. The requirement for record keeping, introduced in the Arsenic Act of 1851, was retained, as it has been in various acts since, and up to the present day.

This Act also introduced the requirement that containers of medicines for external use should be distinguishable by touch from those used for medicines to be swallowed. At this time, the only lighting in people's homes was provided by candles or oil lamps; electric lights had yet to be invented. Consequently, there were many cases of poisoning caused by people fumbling about in the dark during the night, mistakenly taking a dose of something that was intended for external use only, and dying as a consequence. The ribbed bottle to be used for all external preparations was the answer, because it could be easily distinguished in the pitch black of night without having to light a candle.

The Workmen's Compensation Act of 1906

By the beginning of the 1900s, illness and disability caused by certain occupations had begun to be recognised. The Workmen's Compensation Act of 1906 placed a heavy liability upon certain employers who were engaged in various dangerous trades. The Act encompassed any disablement of employees by a disease incidental to and arising out of their occupation. It included anthrax (a danger to textile workers, especially those handling wool) and poisoning by lead, mercury, phosphorus or arsenic, as well as ankylostomiasis (a hookworm infection that caused anaemia).

One industrial disease, first recognised in Glasgow as long ago as 1827, was known as chrome ulcers. It was common in those working with chromium compounds, in the chrome-plating, dyeing, tanning, French polishing and calico-printing industries. The symptoms were the sudden appearance of holes in the skin, which were unbearably itchy and exposed raw flesh. Inhalation of chromic dusts could also lead to rhinitis and painless nasal ulcers, which could, in time, cause the perforation of the nasal septum.

Amendments over the years and new legislation, such as the Health and Safety at Work Act of 1974, which included the Control of Substances Hazardous to Health (COSHH) Regulations (SI1989 No1657), have greatly reduced the number of people suffering from work-related medical conditions. Today, the main hazards are chemicals used in agriculture and horticulture.

Unfortunately, many older workers were harmed long before this legislation was enacted. Many of these work-related conditions, such as asbestosis, do not cause problems until many years, even decades, later. Workers in these industries are now finding that they cannot get the compensation they are due, as their former employer or the employer's insurers have chosen to go bankrupt in order to avoid an increasing number of claims from retired workers who have become ill.

The Poisons and Pharmacy Act 1908

The Poisons and Pharmacy Act went a step further than the Pharmacy Act of 1868 and finally required that the seller of poisons be a chemist and druggist or a registered medical practitioner, and that, before delivery was made, all sales and supplies should be entered into the poison register. The following details were to be recorded: the date, the name and address of the purchaser, the name and quantity of the poison and the purpose for which it was required, together with the signature of the purchaser. And if the purchaser was unknown to the seller, the signature of a witness, who was known to the seller, was required as well. This Act also specified how the poison should be packaged and labelled.

The Venereal Diseases Act 1917

A century ago, many advertisements were for quack medicines sold by mail order that claimed to prevent, relieve or cure venereal diseases. The Venereal Diseases Act 1917 prevented any treatment of venereal disease unless it was given by a duly qualified medical practitioner. Anyone not duly qualified, but advertising such treatment to the public, would suffer a severe penalty designed to put a stop to such quackery: imprisonment, with or without hard labour, for a period of up to two years, or on summary, conviction, for a period of six months or until the payment of a fine of up to £100, a very great sum of money at that time.

The Dangerous Drugs Act 1920

The Dangerous Drugs Act 1920 related to both raw and prepared opium as well as other drugs such as cocaine, diamorphine and morphine. It laid down the regulations for the import, export, sale and supply of these products and required records of each and every transaction, similar to those required for poisons.

The Pharmacy and Poisons Act 1933

All the various pieces of legislation relating to pharmacy or to poisons which had been enacted up to this time were brought together in the Pharmacy and Poisons Act 1933, which was to remain in force until the enactment of the Medicines Act 1968, much of which did not come into force until the early 1980s, and which is still in use today. But it was decided to keep the Dangerous Drugs Act as separate legislation.

The Cancer Act 1939

As with venereal diseases and the 1917 Act, this Cancer Act was to deal with advertisements for quack medicines aimed at the general public offering to treat or prescribe remedies for cancer. It was enacted to prevent unscrupulous people from taking advantage of desperate cancer patients or their relatives by selling and supplying quack medicines purporting to cure cancer.

The Pharmacy and Medicines Act 1941

The Pharmacy and Medicines Act was the result of a long campaign begun by the British Medical Association (BMA) in 1909. A royal commission was set up by the government but World War I intervened and further attempts at legislation in the 1920s and 1930s failed to gain parliamentary time. So, it was not until the 1941 Act that advertisements for quack remedies aimed at the general public for such conditions as Bright's disease (a kidney disease), Cataract, Epilepsy, Diabetes, Glaucoma, Paralysis, Tuberculosis and Locomotor Ataxy, were finally banned.

Quack medicine was rife in the eighteenth, nineteenth and early twentieth centuries, with hundreds of remedies on sale, some of which could be purchased over the counter, but most could only be obtained by mail order. At that time, the manufacturers of medicines and remedies were not obliged to state the ingredients on the label.

The BMA wished to expose the fraudsters and so they employed a chemist to analyse a number of remedies. They asked him to ascertain the ingredients and then estimate the possible cost of them. The results were published in a book called *Secret Remedies*, published in 1909, and a further volume *More Secret Remedies* was published in 1912.

In 1783, a tax had been introduced on patent medicines, but was only payable if the ingredients were *not* disclosed and, unsurprisingly, the vast majority of manufacturers preferred to pay the tax (and pass it on to their customers) rather than expose the nature of the frequently useless ingredients they used to make their product. The system was wide open to fraud. The analyst discovered that any remedies for which massive claims were made were found to contain innocuous ingredients that cost next to nothing, and some 'medicines' contained only water.

Legislation today

Although enacted several decades ago, the current legislation in force today comprises the **Medicines Act 1968** and the **Poisons Act 1972,** together with the **Misuse of Drugs Act 1971,** which jointly regulate all retail and wholesale dealings in medicines and poisons.[1]

The Medicines Act 1968: This Act declares that all medicinal substances must be classified into one of three categories: prescription-only medicines, pharmacy medicines and those on the general sales list. Prescription-only medicines are only available on production of a prescription written by a suitably qualified person, usually a doctor or dentist, or nowadays even a nurse or pharmacist. Pharmacy medicines are those that appear on neither the prescription-only list nor the general sales list. While they do not require a prescription, they can only be purchased from a pharmacy. The sale of such pharmacy medicines must be made under the supervision of a pharmacist. The general sales list encompasses everything that is not listed as prescription only or sold as a pharmacy medicine.

The sale and supply of medicines in the United Kingdom is somewhat different from the rest of Europe, where all medicines must be obtained from a pharmacy. In the UK, medicines on the general sales list can be bought anywhere – the corner shop, the supermarket, even the petrol station!

Somewhat confusingly, a substance may be designated prescription-only when used in a medicine, but may also have non-medicinal uses. No prescription is required for a non-medicinal supply to be obtained, even though it is exactly the same substance.

Even more confusingly, some medicines may be licensed in all three categories, with only the smallest pack of the lowest strength available on the general sales list for purchase at the corner shop, larger packs being sold as pharmacy medicines and the higher strengths being only available on prescription.

The Misuse of Drugs Act 1971: Dangerous or otherwise harmful drugs, now called controlled drugs but previously called dangerous drugs, such as morphine and cocaine, even cannabis too, are all covered by this Act. Under this Act, there is a total prohibition on the possession, supply, manufacture, import or export of these drugs except as laid down in the regulations or by licence from the Secretary of State. Controlled drugs are classified into five schedules, depending on the level of control needed. There are different classes within each schedule, relating to the penalties for offences for production, possession and supply of such substances.

The use of these drugs in medicine is permitted by the Misuse of Drugs Regulations 2001, as amended. The drugs, such as morphine, must be kept in a special locked cupboard in the pharmacy or hospital ward and all transactions involving them must be recorded in the Controlled Drug Register, whenever new supplies are received as well as at the time of dispensing or administering doses of them. Despite the control in the pharmacy and hospital, once a supply is dispensed and given to the patient to take home, the medicine becomes the patient's property, and he or she can keep it anywhere and any way.

The Poisons Act 1972: This Act includes a list of specific poisons and substances containing them. Any other substances, no matter how toxic they may be, are not considered to be poisons under this Act if they are not included on the list. Some of the listed poisons are used in medicines and then become subject to the provisions of the Medicines Act 1968.

The poisons list is divided into two parts. Both Part I and Part II poisons may be sold in a pharmacy under the supervision of a pharmacist. However, Part II poisons may also be sold by registered sellers of poisons from registered premises, such as agricultural merchants. The local authority maintains a list of all registered sellers of poisons, who are only allowed to sell pre-packed containers of the Part II poisons, and some of these are only allowed in a specified form.

Some poisons may only be sold to purchasers engaged in the horticultural or agricultural business and are not available to the general public through a pharmacy. Every transaction involving poisons must still be recorded in the poison register of each registered premises, as has been the case since the Arsenic Act of 1851 was enacted.

Dr Harold Shipman – a law unto himself

Harold 'Fred' Shipman was in a totally different league than other murdering doctors. The final total of his victims will never be known, but it is believed that he murdered about 250 people, mainly women aged 75 years or older, over a period of about 25 years.

Shipman was born in 1946 into an ordinary working-class family. His mother died of lung cancer, having been given injections of morphine during the final stages of her illness. No doubt this is when the young Shipman saw its magical effect in stopping pain. Two years after she died, he went to Leeds University to study medicine. Here he met Primrose, who was only 17, and by the time they married she was already five months pregnant.

By 1974, Shipman had qualified and joined a medical practice in Todmorden, Yorkshire. The Shipmans now had two children. It was while he worked in Todmorden that his addiction to narcotics, such as pethidine, came to light. He was made to undergo a period in a drug rehabilitation centre in 1975 and, two years later, he was working as a GP once more, but now in Hyde, near Manchester.

In Hyde, he worked as a single practitioner rather than in a group practice as he had done previously. Nobody noticed over the years that a number of Dr Shipman's patients were dying, many of them while sitting in an armchair, fully clothed. He killed them by injecting opioids such as morphine and diamorphine (heroin) when he visited them at home. The symptoms of a morphine, or diamorphine, overdose have been listed in Chapter 3, but the only one that mattered here was respiratory depression – if you can't breathe, you die. A few he even killed when they visited him at his surgery. At the time of his arrest, the police found that he had stocks of diamorphine sufficient to kill 1,500 people.

It was only when Kathleen Grundy, an active 81-year-old, suddenly died and her daughter was contacted by a local firm of solicitors that suspicions were aroused. The daughter, Angela, was a solicitor herself, and always dealt with her mother's affairs. The local solicitors said that they had a copy of her mother's will, but Angela knew that she had the original. As soon as she saw the badly typed and poorly worded document the solicitors had, she knew that it was a forgery and called the police. The fake will included a bequest to Shipman of £386,000: he had murdered her for her money. Mrs Grundy's body was exhumed and a post-mortem found enough morphine in her body to kill her. The bodies of other patients were then exhumed and examined.

The police searched his home and the surgery premises, finding the typewriter that he had used to type the fake will. The Shipman house was littered with filthy clothes and old newspapers; it was a scene of total squalor, and not what you would expect to find in a doctor's home. The police investigation found that Shipman had doctored computer records, back-dating entries, making things up, even suggesting in patients' notes that they were addicted to opioids. Many of the bodies had been cremated, and the evidence they held had been lost forever.

In the trial that began late in 1999 and lasted for 57 days, Shipman was charged with only 15 of the murders. On the last day of January 2000, the jury found him guilty of all 15 as well as guilty of a charge

of forgery. The judge announced in court, after sentencing him to 15 life sentences and four years for the forgery, that he recommended that the Home Secretary ensure that Shipman spend the rest of his life in prison, never to be released.
The inquiry that followed his trial decided that he had murdered at least 215 people while practising as a GP in Hyde and Todmorden, and that he may have killed at least another 35. Shipman never explained why he did it. Was it for money? Or because he hated older women? Nobody knows, but it was most likely simply his obsession. It was only in August 2003 that the hundreds of murders attributed to Shipman appeared in the crime statistics, skewing the murder figures for all time. Shipman committed suicide in his prison cell in January 2004. He had no poison, so he hanged himself.

Though legislation has been honed throughout the centuries, it has never been able to stop the abusers – both those accidental and deliberate poisoners of the past and present.

The End

Further Information

Appendix I – Some Poison producing plants and fungi

Common Name	Genus and Species	Type of Poison Produced
Affecting the heart		
Foxglove	*Digitalis purpurea & D. lanata*	digitalis, digitoxin, digoxin
Hellebore	*Helleborus niger*	helleborin, helleborcin
Lily of the Valley	*Convallaria magalis*	convallamarin
Oleander	*Strophanthus or Acocanthera*	ouabain
Upas Tree	*Antiaris toxicaria*	heart poisons
Yellow Oleander	*Thevetia peruviana*	cardiac glycosides
Alkaloids		
Belladonna, or Deadly Night-shade	*Atropa belladonna*	atropine, hyoscyamine, and hyoscine
Bittersweet, or Woody Night-shade	*Solanum dulcamara*	solanine
Calabar or Ordeal bean	*Physostigma venenosum*	physostigmine
Castor bean	*Ricinus communis*	ricin – neurotoxin
Coca plant	*Erxylon cocythroa*	cocaine, CNS stimulant
Hemlock	*Conium maculatum*	coniine,sedative,antispasm

Henbane, Hyoscyamus	*Hyoscyamus niger*	hyoscyamine, hyoscine
Mandrake or Satan's Apple	*Mandragora officinarum*	atropine,hyoscyamine and hyoscine
Meadow saffron	*Colchicum autumnale*	colchicine
Monkshood, Wolfs-bane or Aconite	*Aconitum napellus*	aconitine, deadly poison
Opium poppy	*Papaver somniferum*	opium, morphine, codeine etc.
Pareira	*Chondrodendron tomentosum*	curare, muscle relaxant
Poison Nut, or Nux Vomica	*Strychnos nux vomica*	strychnine, CNS stimulant
Red squill	*Drimea maritime*	scilliroside, neuro-& cardiotoxic
Thornapple, or Jimson Weed	*Datura stramonium*	hyoscine, and other alkaloids
Wormwood	*Artemisia absinthium*	thujone or absinthol
Yew	*Taxa baccata*	alkaloids, v. poisonous
Purgatives and Emetics		
Akee, or Jamaican fruit tree	*Blighia sapida*	hypoglycin A, emetic
Bitter apple, or Colocynth	*Citrullus colcynthis*	colocynthin, drastic purgative

Delphinium	*Deliphinium consolida*	delphinine-emetic, purgative
Laburnum	*Cytisus laburnam*	cytisine – highly poisonous
Lupin	*Lupinus perennis*	lupinine-emetic, purgative
Cyanide producers		
Bitter Almond	*Amygdalus communis*	cyanide – in leaves & seeds
Cassava, Manioc	*Manihot utilissima*	cyanide – in the root
Cherry, Laurel	*Prunus lauro-cerasus*	cyanide – in leaves & seeds
Skin Irritants causing contact dermatitis		
Giant Hogweed	*Heracleum mantegazzianum*	irritant, worse in sun
Poison Ivy	*Rhus radicans*	urushiol, in sap, highly irritant
Poison Oak	*Rhus toxicodendron*	urushiol, in sap, highly irritant
Rue	*Ruta graveolens*	irritant in sunlight
Spurges	*Euphorbia* species	irritant dermatitis in sunlight
Stinging Nettle	*Urtica urens*	formic acid, histamine,irritant
Insecticidal, etc		
Chrysanthemum	*C.cinerariaefolium*	pyrethrum, insecticidal

Japanese Star Anise	*Illicium religiosum*	poisonous principle, sikimin
Peruvian Bark	*Cinchona succirubra*	alkaloid, quinine, anti-malarial
Ragwort, Common	*Senecio jacobaea*	hepatotoxic
Ragwort, Golden	*Senecio aureus*	hepatotoxic
Santonica	*Artemisia maritima, A.cina*	santonin, anthelmintic
Fungal Toxins		
Deadly Webcap	*Cortinarius speciosissimus*	kidney toxins – lethal
Death-Cap	*Amanita phalliodes*	ama- & phallo-toxins – lethal
Ergot	*Claviceps purpurea*	ergot, ergotamine, egometrine
Fly Agaric	*Amanita muscaria*	amatoxins, poisonous, hallucinogenic
Fungal infection	*Aspergillus flavus*	aflatoxins, e.g. mouldy peanuts
Fungal infection	*Aspergillus fumigatus*	cause of aspergillosis, a lung disease

Appendix II – Some Poison producing animals and micro-organisms

Common Name	Genus and Species	Type of Poison Produced
Snakes: - Elapids		
Black Mamba	*Dendroaspis polylepis*	neurotoxins
Inland Taipan or Fierce snake	*Oxyuranus sculuIatus*	neurotoxins
Snakes: - Vipers		
Pit viper	*Agkistrodon rhodostomo*	vasculotoxins
Snakes: - Marine		
Sea snakes	*Hydrophiidae sp*	myotoxins
Insects:		
Bees	*Apis mellifera*	melittin, apamin
Spiders:		
American black widow spider	*Latrodectus mactans*	latrotoxin
Australian funnel-web spider	*Atrax robustus*	atraxatoxin
Brown recluse spider	*Lactrodectus reclusa*	atraxatoxin
Amphibia:		
Frogs	*Phyllobates terribilis*	batrachotoxin

Frogs	*Phyllobates aurotaenia*	batrachotoxin
Frogs	*Epipedobates tricolor*	epibatidine
Californian newt	*Taricha torosa*	tetrodotoxin
Marine creatures:		
Box jellyfish (sea-wasp)	*Chironex fleckeri*	cardiotoxins
Marine snails	*Conus geographus*	conotoxins
Puffer fish (globe fish)	*Fugu rubripes*	tetrodotoxin
Porcupine fish	*Diodon*	tetrodotoxin
Marine micro-organism	*Palythoa* seaweeds	palytoxin
Marine micro-organism	*Dinoflagellate* plankton	brevetoxin, ciguatoxin, and maitotoxin
Bacteria:		
Anthrax	*Bacillus anthracis*	spores
Botulism	*Clostridium botulinum*	exotoxin
Cholera	*Vibrio cholerae*	exotoxin
Food poisoning	*Bacillus cereus*	exotoxin
Food poisoning	*Salmonella enteriditis*	exotoxin
Food poisoning	*Escherischia coli (E. coli)*	exotoxin
Food poisoning	*Staphylococcus aureus*	exotoxin

Gangrene	*Clostridium perfringens*	exotoxin
Typhoid	*Salmonella typhi*	exotoxin

Appendix III – Some Plants and Animals used in Dyeing

Common Name	Genus and Species	Colour of Pigment or dye
Plants:		
Dyer's Greenwood	*Genista tinctoria*	yellow
Wild Mignonette, Weld	*Reseda luteola*	yellow
Gamboge	*Garcinia hanburyi*	yellow, also purgative
Saffron	*Crocus sativa*	brilliant orange
Madder, Dyer's Madder	*Rubia tinctorum*	alizarin, orange-red
Tomato	*Lycopersicon esculentum*	lycopene, red caro-tenoid
Rose	*Rosa canina*	lycopene, red carotenoid (in hips)
Loosestrife family	*Lawsonia* sp.	henna, reddish-orange
Safflower, Bastard Saffron	*Carthamus tinctorius*	pinky-red
Dyer's Mulberry	*Morinda tinctoria*	pink and red
Dyer's Bugloss	*Alkanna tinctoria*	pink and red
Litmus, Dyer's Weed	*Rocella tinctoria*	red in acid, blue in alkali
Annatto	*Bixa orellana*	bixin, red/violet
Delphiniums, larkspur	*Delphinium consolida*	delphinin, ancanthocyanin

Woad	*Isatis tinctoria*	indigotin, blue
Oak	*Quercus robur*	gallic acid from the galls – black
Insects:		
Kermes (female)	*Kermes ilicis*	crimson, a rich red
Scale insect (female)	*Dactylopius coccus*	red dye – cochineal
Marine Creatures:		
Mediterranean shellfish	*Murex branderis*	vat dye – Tyrian purple
Cuttlefish	*Sepiidae sp.*	dark brown – sepia

Appendix IV – The nitty-gritty of how poisons do their worst

All living things are made up of cells, which contain and are surrounded by walls made of selectively permeable membranes. The smallest organisms such as bacteria are single cells, while more complex organisms, such as plants and animals, contain many millions or billions of cells. The cell wall membranes, which are made of phospholipids, contain channels that can be opened or closed during the cell's normal operation, so allowing or preventing passage of chemical substances necessary for functioning, such as nutrition, respiration, reproduction and excretion. Poisons can block up or force open these channels, and so disrupt the normal operation of cells. Some chemicals can simply diffuse through cell membranes or even damage them to gain access into the cells.

Membranes and cell energy production

Within each cell are more membranes, including those of the mitochondria. The mitochondrion is a structure to be found in varying numbers in every cell, and is the site of production of the neurotransmitter acetylcholine, and of energy production within the cell. Each mitochondrion is contained within a double membrane, the inner one being folded inwards to form many projections. This inner mitochondrial membrane is the site of the electron transport system, the adenosine triphosphate (ATP)-producing complex and the enzymes involved in fatty acid synthesis, known as Kreb's cycle. This complex cycle of reactions, involving enzymes, is the final step in the oxidation of proteins, fats and carbohydrates.

The electron transport chain is the name given to a series of enzymes and other proteins in living cells through which electrons are transferred and which ultimately results in the conversion of chemical energy into a readily usable and storable form. Cytochromes act as electron transfer agents in biological oxidation-reduction reactions, particularly in the mitochondria. Cytochrome c oxidase reduces oxygen to water, while conserving the free energy available. Oxidative phosphorylation is the process by which the energy released during electron transfer is coupled to the production of ATP. This process is the reason that a single molecule of glucose can provide the energy to release 38 ATP molecules. But it only happens once the reduced coenzymes have passed their hydrogen atoms on to the respiratory chain and the electron transport system. Without this energy nothing works.

Electrolytes, pumps and channels

Sodium and potassium are both essential to all life, although sodium is far less important to plants than to animals. Sodium in the body is mainly found in the fluid outside the cells, along with calcium, whereas about 95 per cent of potassium is found inside the cells. The blood contains a lot of sodium, used to maintain both the osmotic pressure and the blood pressure. The red blood cells contain the most potassium, followed by the cells of muscles and the brain. Both sodium and potassium move through specific channels in the cell membranes: sodium moves into the cell and is expelled again by way of the so-called sodium pump, while potassium leaves the cell and is then carried back in again.

The sodium pump enzyme (Na^+/K^+-ATPase) is essential to the working of our many millions of cells, particularly those of the nervous system. As much as 40 per cent of the body's total energy is spent on operating this sodium pump in our cells. In nerve cells, this flow of sodium and potassium ions is how nerve impulses are transmitted. Black mamba toxin, which is known to specifically block the potassium channels, was radioactively labelled and used as a means of mapping the layout of potassium channels in the brain. This showed that most of these channels were to be found in the hippocampus, the area of the brain responsible for learning.

Nucleic acids, proteins and phosphate

Proteins are manufactured on ribosomes within our cells. The sequence of amino acids in any particular protein is encoded in the sequence of nucleotides of the genetic material, which is usually DNA. Many toxins affect protein synthesis by damaging nucleic acid synthesis, inhibiting enzymes, damaging cell membranes, and so preventing cellular respiration. Nucleic acid synthesis damage can cause cancer.

While the poison, phosphorus, appears many times in this book, as phosphate it is essential to life, both inside and outside every cell in our bodies, as well as being a component of DNA. Most of the phosphorus in our body is present as calcium phosphate in our bones, and the rest as organophosphates, which are many and varied. Some organophosphates are the phospholipids of cell membranes, some are present as ATP, which acts as a source of energy, or as guanosine monophosphate, a messenger molecule. Phosphate also acts as a buffer, which maintains the acid/base balance in the fluid between the cells, and is involved in transporting calcium around the body.

How nerves make muscles work (or not)

The nervous system is a network of specialised cells, which carry information as nerve impulses to and from everywhere in the body. To muscles, to glands, to the walls of blood vessels, nerves of different types, whether sympathetic or parasympathetic, sensory or motor, are everywhere. Nerve impulses travel along nerve fibres. The electrical activity in the membrane of a nerve fibre passes down to the nerve ending, where it causes release of the neurotransmitter which crosses the synapse, before connecting with the receptor, and this triggers the continuation of the nerve impulse down the nerve until it reaches the motor end plate of a muscle, where it causes the muscle to contract. Neurotransmitters can be excitatory, such as acetylcholine, or inhibitory, such as glycine.

The organophosphate insecticides that are used in agriculture and horticulture are very toxic. They work by inhibiting the vital enzyme acetyl. This is the enzyme which breaks down the neurotransmitter acetylcholine into choline and acetic acid. It is found in all cholinergic nerve junctions, where it rapidly breaks down the acetylcholine released, during the transmission of a nerve impulse, so that subsequent impulses may pass later. If the neurotransmitter cannot be broken down, then the nervous system can no longer function and death results from muscle spasms and convulsions. Organophosphates inhibit other cholinesterases too, which are found in the blood and other tissues.

How toxins cause paralysis

Batrachotoxin, saxitoxin and tetrodotoxin all stop nervous transmission by affecting the passage of sodium through the sodium channels in cell membranes. Batrachotoxin is about five times more potent thantetrodotoxin, but works in the opposite way to it and saxitoxin. While saxitoxin and tetrodotoxin block the uptake of sodium ions through the sodium channels in nerve membranes, batrachotoxin enhances the transport of sodium and other ions into the nerve cells by preventing the closure of the sodium channels. They both upset the delicate balance of sodium and potassium ions and their movements across cell membranes, a balance which is essential for nervous transmission. Normally a nerve impulse results from the opening of the sodium channels, as this causes a change in the electrical voltage across the nerve cell wall. As the sodium channels open one by one along the length of the nerve, so the impulse passes along it. If batrachotoxin prevents closure, while tetrodotoxin or saxitoxin blocks the sodium channels, then the nerve, and the muscle it is attached to, cannot work. Death normally results from respiratory failure in this type of poisoning. However, as both of these toxins antagonise each other's action, in theory at least, this means that each is the antidote for the other.

Palytoxin acts by paralysing all tissues: heart muscle, skeletal muscle, smooth muscle and nerves, in addition to causing the rupture of red blood cells. These effects are all caused by the induction of pores in the cell membranes. This disrupts the normal controls by enhancing the cell membrane permeability so that various ions, such as sodium, potassium and calcium, can cross the cell membranes at will, with disastrous consequences.

Brevetoxins cause depolarisation of both nerve and muscle cell membranes, by enhancing the influx of calcium ions into the cells through the calcium channels. These complex chemicals exert their biological effects in the uncontrolled release of neurotransmitters in people who have been poisoned by eating shellfish or coral reef fish, both of which consume deep-sea plankton like the dinoflagellates. This poisoning results in the symptoms of tetany.

Conotoxins are a mixture of toxins, each working in a slightly different way to block nervous transmission. While one blocks the release of the neurotransmitter acetylcholine from the nerve ending, another blocks the receptor site that the acetylcholine should attach to and a third prevents the functioning of the sodium channels of the skeletal muscles to which the nerve is attached. This combined action of the conotoxins results in total paralysis.

Insect stings and spider bites

The stings of insects, such as bees and wasps, contain mixtures of chemicals that include melittin and apamin. Melittin causes the cells in the area of the sting to secrete histamine, which causes intense itching and pain in the area, together with local inflammation. It also induces the formation of more ion channels in the membranes of nerve cells near the site of the sting, which increases the nerve activity and the pain. Apamin is a small peptide that blocks the potassium channels, thus preventing stabilisation of the affected cells, and resulting in increased nervous excitation and pain caused by the sting. Other chemicals in the venom include enzymes, which digest the cell walls, thus allowing the pain-producing chemicals access to a larger area around the site of the sting.

A major component of Black Widow Spider venom is latrotoxin, which acts on the walls of the nerve ending. It opens channels so that sodium and calcium ions can flood into the nerve cell, triggering the release of acetylcholine. The calcium ions are the main problem as the blood vessels relax and the skin flushes due to the release of the acetylcholine. This also causes sweating, painful muscle cramps and difficulty breathing, for which the best treatment is large quantities of calcium ions, to block up the ion channels opened by the toxin.

Plants that get on your nerves

The alkaloids, atropine from Deadly Nightshade, and solanine from Woody Nightshade, can both cause temporary muscular paralysis by blocking the action of the neurotransmitter acetylcholine at the junction where the nerve joins the muscle membranes. Fortunately this action is dose related and reversible, so treatment for atropine poisoning is fast and effective. Physostigmine, from the Calabar bean, contains the antidote to atropine, as it is a reversible inhibitor of cholinesterase activity. Physostigmine has itself been used as a poison.

Aconite, also known as Monkshood, contains the alkaloid aconitine, the action of which reduces the ion selectivity of the sodium channels in cell walls. This results in an increased uptake of sodium and other ions through these channels. All parts of this perennial plant are poisonous; even a small piece of leaf, if placed in the mouth, will produce a tingling and burning sensation. The leaves and unripe fruits of Hemlock contain the alkaloid coniine and act in a similar manner to aconite.

Cyanide interferes with the oxygen uptake of cells. It does this by inhibiting an enzyme called cytochrome c oxidase, which is vital for the transport of oxygen within our cells – that is, cellular respiration. Without oxygen, we die. Treatment of this type of poisoning uses sodium nitrite to produce a state of methaemoglobinaemia, where the iron atoms in the blood pigment haemoglobin have been oxidised from the divalent ferrous form to the trivalent ferric form. Unlike haemoglobin, methaemoglobin cannot bind molecular oxygen and so cannot transport oxygen around the body. But it can combine with the cyanide ions to produce cyanmethaemoglobin. This protects the cytochrome oxidase system which is so vital for cellular respiration. As this substance, cyanmethaemoglobin, slowly breaks down, the cyanide released can then be converted, with the help of the enzyme rhodanese, to the relatively non-toxic thiocyanate, which can then be safely excreted in the urine. Sodium thiosulphate helps to speed up this process by providing an additional source of sulphur atoms for the rhodanese enzyme.

Curare acts as a competitive neuromuscular blocker by competing with the neurotransmitter acetylcholine for the receptor sites of the motor end plate on the muscle membrane, so causing paralysis. Physostigmine yet again is the antidote.

Strychnine affects the nervous system by competing with the inhibitory neurotransmitter glycine, so preventing the switching off of the nerve impulse. Strychnine causes a continual stimulatory effect, which can eventually result in death due to spasm of the respiratory muscles.

Digitalis and Ouabain are both cardiac glycosides (sugar derivatives), which act on the heart muscle and increase the force of contraction in cases of heart failure.

Protein inhibitors and neurotoxins

Ricin is a cytotoxic glycoprotein consisting of two peptide chains linked by a disulphide bond (this bond also occurs in the botulinum toxin). Ricin works in an interesting way, with one peptide chain binding to surface receptors on the outside of the cell while the other peptide chain passes into the cell, and then inhibits the production of proteins.

Highly poisonous toxins are produced by various microbes, which cause food poisoning and other problems. *Clostridium perfringens* produces at least a dozen toxins, which are each a type of protein. These toxins perform various functions: some can break down cells, some make holes in cell walls through which potassium can leak out and some necrotise and so kill the invaded cells.

Botulinum toxin, produced by the related *Clostridium botulinum,* is a neurotoxin, a protein made up of two peptide chains linked together by a disulphide (S-S) bond, which attacks the nervous system. One part of the peptide chain attaches firmly to nerve cell walls, following which the nerve cell transports the whole toxin molecule into its interior. Once inside, the second part of the chain destroys a protein called synaptobrevin, which is needed for release of a transmitter from the vesicles in which it is stored by the nerve cell. Without release of the neurotransmitter, the muscle cells attached to the nerve ending are paralysed, and if the muscles affected are vital to life, such as the lungs, death quickly follows. There are seven distinct types of botulinum toxin, identified simply by the letters A to G. Type A is considered to be one of the most poisonous of chemicals.

Botulism is an extremely serious form of food poisoning, requiring intensive life support and treatment in hospital, as it selectively affects the central nervous system by inhibiting acetylcholine-mediated neurotransmission. As well as the symptoms of vomiting and diarrhoea, the pupils become dilated, causing double or blurred vision. There is also ataxia, which causes difficulty in speech and swallowing, with shakiness and unsteady gait, and bulbar palsy can develop, resulting in a progressive flaccid muscle weakness and paralysis, which descends from the head and can result in respiratory paralysis and death within two or three days. In such fatal cases, death is often due to heart or lung failure, resulting from a malfunction, caused by the toxin, of the cardiac or respiratory centres in the brain.

Anthrax is caused by spores from *Bacillus anthracis,* which germinate in the lungs or on the skin of the host into bacilli that release three different toxins into the infected body. The first toxin is a 'protective antigen' that enables the other two toxins to enter the host's cells. The 'oedema toxin' inhibits neutrophils, the white blood cells which would normally kill invading bacteria, and which are one of our major defences against infection, and this leads to the accumulation of fluid, oedema. Finally, the 'lethal toxin' stimulates the release of tumour necrosis factor and interleukin, which is associated with anthrax sudden death.

Mycotoxins act by inhibiting DNA and protein synthesis, damaging cell membranes and inhibiting cellular respiration. These are complex cyclopeptides (the amatoxins and phallotoxins), which damage the production of nucleic acids in our cells by inhibiting some of the enzymes that are involved in their production. The ultimate cause of death is due to severe liver damage, but not before days of suffering from abdominal pain with violent vomiting and continuous diarrhoea.

Interfering with enzymes

Arsenic, cadmium, lead and mercury are all poisonous because they interfere with enzymes in our bodies. They react irreversibly with thiol groups (of a sulphur atom joined to a hydrogen atom) at certain non-active sites of enzymes, particularly affecting enzymes involved in the nervous system. Lead and mercury also bind to phosphate groups on cell membranes, causing a reduction in lipid solubility and so affecting the transport of solutes and their concentration gradients across the membranes. The presence of these metals triggers a defence mechanism, in which production of a special enzyme metallothionein is stimulated. This enzyme contains a lot of sulphur atoms to which the metal atoms can attach: seven to each enzyme molecule, which can then be transported to the kidneys for excretion.

Arsenic is a metalloid, but like the metals cadmium, lead and mercury, it attaches to sulphur-containing enzymes and, by blocking their action, causes the toxic effects seen in poisoning. The body can excrete arsenic quite easily, so low doses obtained in the normal diet can be readily disposed of.

However, other metals are more problematic. Unfortunately, the cadmium atoms are so strongly bound to the enzyme that they tend to accumulate there, causing damage to the kidneys by preventing the reabsorption of proteins, glucose and amino acids. This damage can eventually result in kidney failure. The natural rate of excretion of cadmium is so slow that it stays in the body for about 30 years. Consequently, even the smallest possible intake can cause problems. The replacement of zinc by cadmium in the testes, for example, damages viable sperm production by affecting the enzyme DNA polymerase.

Nickel is thought to cause cancer by substituting itself for zinc and magnesium atoms in this same enzyme, DNA polymerase. It is thought that this variant, because of its slightly different size, affects the behaviour of the enzyme, perhaps by causing it to bind to the wrong nucleotide, and so resulting in a rogue sequence of DNA and hence a cancerous cell.

Selenium: It is estimated that there are about 100 proteins containing selenium in the body, although the functions of only about 30 of these are currently known. It is only recently that the 21st amino acid, selenocysteine, coded by our genes, was discovered. The enzyme glutathione peroxidase contains selenium. This enzyme, which eliminates peroxides before they can form harmful free radicals, protects our intracellular structures from oxidative damage. Other selenium-containing enzymes, found in the liver and the brain, promote hormone production in the thyroid gland and profoundly affect the amount of active thyroid hormone in the circulation.

Mercury poisoning used to be quite common, but it is now rare in the Western world due to improved health and safety legislation. Mercury affects the nervous system, and all mercury compounds are extremely toxic. Mercury has a specific affinity for sulphur and so attaches itself to the sulphur atoms of certain amino acids which, if they happen to be part of an enzyme, will then de-activate this enzyme. In mercury poisoning, de-activation of the enzyme mentioned earlier, called the sodium pump (Na^+/K^+-ATPase), results in the classical symptoms of mercury poisoning: the shakes and mental disturbance.

Uncoupling oxidative phosphorylation

Dintro-o-cresol and dinitrophenol were both formerly used in agriculture, and dintro-o-cresol was claimed to be five times as potent as dinitrophenol. Their mode of action was to increase metabolism by uncoupling oxidative phosphorylation in the cells. This meant that metabolism within the cells ran out of control. These substances came to be used medicinally, as treatment for obesity, in the 1930s. Because of their mode of action they were very dangerous to use and fatalities occurred due to the induced heat-stroke and cerebral oedema (swelling of the brain) caused, which then led to respiratory and cardiac failure and thus death.

This uncoupling, and subsequent loss of control in the cells, can also happen in an aspirin overdose. The hyperventilation that can occur results in anaerobic metabolism, together with the production of lactate and heat. Lactic acid is produced when cells break down glucose anaerobically – that is, without the use of oxygen – to produce energy. These symptoms are all stimulated by the uncoupling of oxidative phosphorylation, which can then lead on to the development of metabolic acidosis. The effects are dose related and potentially fatal.

Lactic acid can also be produced in the tissues when they receive too little oxygen due to an impairment of the blood supply, as occurs in a heart attack or shock or, less dramatically, when you are running and get a 'stitch'. Normally lactic acid is removed from the blood by the liver; however, if this is not possible, lactic acidosis results.

How 'statins' lower cholesterol

The main group of drugs used to treat high cholesterol are called the 'statins'. These drugs act by inhibiting an enzyme in the liver, called 3-hydroxy-3-methylglutaryl coenzyme A reductase (or HMG-CoA reductase for short). This inhibition reduces the total cholesterol, as well as the low density lipoprotein (LDL) cholesterol concentration and the very low density lipoprotein (VLDL) concentrations in the plasma and also tends to reduce the triglycerides, while increasing the high density lipoprotein (HDL) cholesterol concentrations (the 'good' cholesterol).

The statins actually work by stimulating an increase in the low density lipoprotein receptors on hepatocyte membranes in the liver. This increased number of receptors means that more low density lipoproteins (LDLs) can be removed from the circulation, so reducing the cholesterol level.

Cholesterol is synthesised by the liver, mainly during the night, so the lipid-lowering drugs called 'statins' are taken during the evening, so that they have maximum effect.

How drugs prolong the QT interval

A number of drugs, mentioned in Chapter 18, have been removed from use as they were shown to prolong the QT interval of the heart beat trace taken with an electrocardiogram. These drugs were found to inhibit the potassium channels in cardiac tissues, and it was this effect that caused the prolongation of the QT interval, which was a risk factor in the development of cardiac arrhythmias.

Appendix V – Deficiency and excess of vitamins and minerals

Vitamins

Vitamins are needed in very small amounts for healthy growth and development. They cannot be made by the body, and so must be obtained from diet. A healthy, varied diet will supply all the vitamins (and minerals) you need to stay healthy. However, you can have too much of a good thing and **hypervitaminosis** is the name given to excessive consumption of vitamins, which has become fashionable with some groups of people, particularly in the USA.

While this is not serious with water-soluble vitamins where any excess is simply excreted in the urine, it can be very serious if fat-soluble vitamins are taken to excess, as the body will store them in the body fat, resulting in a toxic build-up. Vitamins A, D, and E are all fat soluble and so liable to cause a toxic build-up in excess. All the other vitamins are water soluble. Vitamins A, C and E act as antioxidants, helping to protect us by reducing the likelihood of serious diseases, such as heart disease and possibly even some types of cancer. Vitamins generally act as co-enzymes and, without them, certain enzymes would not function properly.

Vitamin A is fat soluble and is very toxic in overdose. Liver is a rich food source of vitamin A, so rich that expectant mothers are advised not to eat it during pregnancy because high doses of vitamin A have been associated with **birth defects** in the past. Oily fish such as cod, halibut and salmon are all rich sources of both vitamins A and D.

A precursor of vitamin A, beta-carotene, is the orange pigment found in carrots, which is converted in the body to the active form, known as retinol. Beta-carotene is also present in red, orange and yellow fruit and vegetables. Vitamin A is essential for growth, for the maintenance of soft mucous tissue and particularly for vision in dim light, so eating carrots really can help you see in the dark!

Yet again, it should be noted that it is important not to cross the line between benefit and excess: in the 1960s a woman in America died from a vitamin A overdose. She had decided to lose weight by restricting her diet to simply drinking carrot juice. When she died, her flesh was found to be tinted a strange orange colour, due to the build-up of beta-carotene in her tissues.

As well as excess, deficiency of vitamin A can also cause trouble, as it leads to stunted growth and visual problems, initially affecting only **night blindness,** but, if not treated, progressing to more serious optical disorders and eventually complete blindness.

Vitamin B is not a single entity but is composed of a number of vitamins in what is called the B group complex. They are all water soluble and all act as co-enzymes – this means that, without them, certain enzymes would not function properly. Although not chemically related to each other, the B group vitamins are often found together in the same types of food, such as milk and cereals.
Thiamine, vitamin B_1, is found in liver, pork, wholegrain cereals, pulses, nuts, milk and bread. Thiamine is involved in carbohydrate metabolism, and a deficiency of this vitamin leads to the condition known as beriberi. This nutritional disorder is widespread, even today, in the rice-eating communities of the world, where polished rice is eaten. Polished rice has had the seed coat removed and discarded, but unfortunately it is the seed coat that is rich in thiamine.

Two forms of this disorder can be distinguished: **wet beriberi** is so called because there is an accumulation of fluid and swelling (oedema) within the tissues, while with **dry beriberi** there is extreme emaciation. Both forms cause nervous degeneration and often result in death from heart failure. Thiamine deficiency is often seen in chronic alcoholics or as a result of prolonged deliberate fasting, as in anorexia.
Riboflavin, vitamin B_2, is important in tissue respiration and essential for the utilisation of energy from food. Excessive doses of riboflavin result in the urine becoming a bright yellow colour, which has been known to interfere with some laboratory tests. A deficiency results in a condition with symptoms including inflammation of the tongue and lips, with sores at the corners of the mouth. Such a condition is soon remedied by eating green vegetables, fish, liver or eggs.
Pyridoxine, vitamin B_6, is crucial for the production of a vital co-enzyme called pyridoxal phosphate, essential for carbohydrate and fat metabolism and for the formation of haemoglobin. It is also involved in the functioning of the nervous system. An excess can cause problems leading to the development of **severe peripheral neuropathies**, with tingling, pins and needles and loss of sensation in the fingers and toes. A deficiency of pyridoxine can lead to the development of a type of anaemia, dermatitis and even convulsions. Deficiency is rare, as pyridoxine is found in most foods; however, a deficiency state can be drug induced or can result from an inborn error of metabolism.

Genetics has an important effect on metabolism. Certain racial groups of people tend to metabolise some drugs at a much slower rate than others, which can result in a toxic build-up of the drug or its metabolites. A normal dose for one person who metabolises the drug rapidly would be harmfully toxic to someone else who metabolises it slowly. Such a patient, if given a drug called isoniazid, used in the treatment of tuberculosis, would need not only a lower dose of isoniazid but, additionally, pyridoxine to ensure that they did not suffer such toxic effects as convulsions. Such **slow metabolisers** have been found to respond better to treatment with isoniazid, but they are also more likely to develop peripheral neuropathy, with tingling fingers and toes. And they are found to be at greater risk of an adverse effect on their liver function in some cases too.

Cyanocobalamin, vitamin B_{12}, as the name suggests, contains both cyanide and cobalt. Vital for the synthesis of nucleic acids, the maintenance of myelin that sheaths the nerves and the proper functioning of folic acid, another of the B group vitamins, vitamin B_{12}, can only be absorbed from the gut in the presence of a substance called **intrinsic factor.**

This substance is normally secreted by the stomach, but if the body fails to secrete it, then a deficiency of vitamin B_{12} results, quite simply because the gut cannot absorb it. This leads to the development of **pernicious anaemia**, and results in the degeneration of the nervous system if not treated with injections of vitamin B_{12}. This vitamin is only produced by certain micro-organisms and is only contained in foods of animal origin including meat, liver, fish and eggs. Therefore, strict vegetarians, and vegans particularly, need to supplement their diet with this vitamin to avoid the development of pernicious anaemia, due to a lack of this vitamin.

Biotin, sometimes referred to as vitamin H or co-enzyme R, is another member of the B group complex essential for the metabolism of fat and protein. Egg yolk and liver are rich sources of biotin, so a deficiency of this vitamin is rare. However, patients receiving long-term total parenteral nutrition can become deficient in biotin, as can those with certain inherited metabolic disorders.

Folic acid is another member of the B group complex. It is important in the synthesis of nucleic acids and is interdependent with cyanocobalamin. A deficiency of one can lead to a deficiency of the other. A deficiency of folic acid can lead to **megaloblastic anaemia** in which red blood cells do not form or mature properly. Foods rich in folic acid are green vegetables, pulses, liver, wheatgerm and yeast extract.

It was only about 20 years ago that the importance of folic acid in the diet of pregnant mothers was fully realised. Today, women intending to become pregnant are recommended to take a daily tablet of 400 micrograms of folic acid for the three months before conception and the first three months of the pregnancy to prevent **neural tube defects**, such as spina bifida and other congenital malformations like **cleft lip and cleft palate** in their baby.

Nicotinic acid, vitamin B3 is found in oily fish, meat, whole cereals and nuts. It is also called niacin and is interchangeable with nicotinamide, as both forms are equally active in cellular respiration. Nicotinic acid can be made in the body from the amino acid tryptophan (see the section on protein later in this chapter), and is converted to the active form – nicotinamide – in the body. A deficiency of either nicotinic acid or tryptophan leads to **pellagra**, a condition commonly found in maize-eating communities. This is because the nicotinic acid in maize is only present in a chemically bound form that prevents its absorption from the gut. This binding makes the vitamin unavailable to those who eat the maize. The symptoms of pellagra are a scaly type of dermatitis, together with diarrhoea and depression.

Pantothenic acid, or **vitamin B5,** another member of the B complex group, is a component part of coenzyme A, which is important in the body's use of carbohydrates, fats and proteins and also in the metabolism of fatty acids. No deficiency state has ever been seen thanks to the presence of pantothenic acid in so many foodstuffs.

Ascorbic acid, vitamin C, is a water-soluble vitamin that has antioxidant properties. It is essential for the synthesis of collagen, the maintenance of healthy connective tissues and the integrity of cell walls. Citrus fruit and vegetables are all rich sources of this vitamin. Deficiency of vitamin C leads to **scurvy**, a condition common to sailors centuries ago caused by the lack of fresh fruit and vegetables in their diet during long sea voyages.

Vitamin D is a fat-soluble vitamin that enhances the absorption of calcium and phosphorus from the intestine and promotes their deposition in bone. The daily dose measured in micrograms is essential to growth and development. A good dietary source of both vitamins A and D is to be found in fish oils.

Vitamin D occurs in two forms: ergocalciferol, from plants, and cholecalciferol, produced by the action of sunlight on our skin. So, a deficiency can be due to either poor diet or to lack of sufficient sunlight, or both, resulting in decalcified bones and the development of **rickets** and **osteomalacia**. Based on findings that the number of cases of multiple sclerosis increases the further north sufferers live in the northern hemisphere, recent and ongoing research suggests that the condition may be caused by the interaction of a genetic variant together with lack of sunlight, and hence a deficiency of vitamin D production in the skin. However, vitamin D is toxic in large doses.

Vitamin E is a generic term that refers to any one of a group of chemically related fat-soluble compounds. These compounds are antioxidants, thought to stabilise the cell membranes by preventing the oxidation of their unsaturated fatty acid components, thus preventing deterioration in healthy cells. Otherwise this might result in the deposition of abnormally large amounts of fat within the cells. Such a build-up of fat predisposes to thrombosis. It is for this reason that it was suggested that taking a daily supplement of vitamin E might reduce the risk of coronary thrombosis in those with heart disease. Coronary thrombosis is the formation of a blood clot in one of the coronary artieries, which obstructs the flow of blood to part of the heart and results in the death of a segment of heart muscle; that is, the patient has a heart attack.

Large doses of vitamin E can cause diarrhoea, abdominal pain and other gastro-intestinal disturbances. The absorption of vitamin E from the gut is dependent on the presence of bile and also on normal pancreatic function. Vitamin E deficiency is uncommon, but can occur in rare cases where there is a problem with fat absorption due to an inherited disorder.

Vitamin K is another fat-soluble vitamin that exists in two forms: phytomenadione, of plant origin, and menaquinone, of animal origin. Vitamin K is vital for the formation of the blood-clotting factor prothrombin, and of several other clotting factors. This vitamin is synthesised by bacteria (friendly bacteria) in the large intestine and is also present in green, leafy vegetables, so a dietary deficiency is rare. All newborn babies are given an injection of this vitamin shortly after birth to ensure they have a sufficient supply to deal with any bleeding that may have occurred internally during delivery.

Minerals:

Certain minerals are essential to the human diet, frequently acting as cofactors in enzyme systems or as part of other complex molecules. The minerals required are calcium, phosphorus, potassium, sodium, iron, chlorine, sulphur and magnesium, in addition to trace elements, including manganese, zinc, copper, iodine, cobalt, selenium, molybdenum, chromium and silicon. Problems may arise when the body is unable to use them as nature intended.

Calcium is essential for the normal development and functioning of the body. It is an important constituent of bones and teeth. About 99 per cent of the body's calcium is to be found in the bones, mainly as calcium phosphate. Some of the remaining one per cent is to be found in the bloodstream, where its level is maintained by hormones. The presence of calcium is essential for many metabolic processes, including blood clotting, nerve function and the contraction and relaxation of muscles.

The parathyroid glands, which are situated in the neck, are responsible for the secretion of the parathyroid hormone, which regulates both the calcium and phosphate levels in the body and also controls their distribution in the blood and the bones. **Hypercalcaemia**, a high blood level of calcium, may, rarely, be due to vitamin D toxicity, but may also be caused by hyperparathyroidism, acute renal failure, malignancy or even by chewing too many calcium carbonate antacid tablets; this can result in the so-called milk-alkali syndrome. Conversely, in **hypocalcaemia**, a decrease in plasma levels is seen, which may be due to vitamin D deficiency, chronic renal failure or low parathyroid levels. A low calcium blood level can lead to tetany, while an excess can result in the formation of calculi, that is, stones, in the gall bladder or kidneys.

Phosphorus is mainly found as calcium phosphate in the bones and teeth, but phosphorus is also vital in compounds involved in energy conversion and storage in the body. Osteomalacia and rickets are both caused by vitamin D deficiency, to which phosphate deficiency may also contribute. **Phosphate deficiency** can occur in premature babies who have been exclusively fed on human breast milk; they need to be given supplements to help them grow well. Phosphate is usually given medicinally as sodium phosphate or potassium phosphate, both of which are readily absorbed by the body.

Phosphorus poisoning can result in destruction of tissue, due to excessive amounts of phosphorus in the system, and can affect the liver, kidneys, muscles, bones and cardiovascular system. A raised plasma-phosphorus concentration in the blood is usually associated with kidney failure, but it can also be due to decreased levels of parathyroid hormone. Low phosphate levels can be caused by an inherited X-linked genetic disorder, or increased parathyroid levels, insufficient absorption of phosphate or even the previously mentioned vitamin D deficiency.

Sodium is continually being lost from the bloodstream as it passes through the kidneys and, although much is recycled, some is still lost via the urine, faeces and sweat, so a regular supply from the diet is required. Sodium chloride – salt – comes from mineral deposits of

rock salt or as sea salt which was extracted from sea water. Table salt is sodium chloride with a little magnesium carbonate added to make it free flowing. Iodised salt has a little potassium iodide added to it.
A **deficiency of sodium**, which can occur as a result of severe diarrhoea or vomiting, through profuse sweating or the prolonged use of diuretic drugs, causes muscle spasms or cramp, weakness, tiredness, dizziness, confusion, fainting and palpitations. Treatment is very simple – by replacement with more salt. Normal saline is a 0.9 per cent solution of sodium chloride in water. This particular strength, being isotonic with body fluids, is used intravenously as it does not upset the body's fluid balance.

In tropical countries, salt saves millions of lives when used as a treatment for diarrhoea. **Rehydration** sachets of sodium and potassium salts, together with glucose, are dissolved in boiled water and given to those suffering from diarrhoea and cholera to rehydrate them. If left untreated, dehydration can be rapidly fatal. Glucose is included to provide the body with an energy source, needed by the gut, to absorb both the salts and the water.

People who have **high blood pressure**, **heart trouble** or **kidney disease** are usually told to reduce their salt intake as a means of reducing the sodium content of their diet. This is because if the heart or kidneys are diseased, they are less able to remove excess salt. They compensate for this by retaining extra water, which in turn results in an increase of the blood pressure, to the detriment of the patient.

The emetic effect of a sudden increase in sodium has traditionally been used in the form of salty water to make people vomit following ingestion of poisons, although today this is considered to be a dangerous practice.
Iron is essential to life. We each have about four grams of iron in our body, about enough to make a small nail, over half of which is contained in the **haemoglobin** within the red blood cells. The rest is distributed around the body as myohaemoglobin, in the muscles, in cytochromes and as iron storage substances in the body tissues. Meat, particularly liver, is a rich dietary source of iron.

Iron deficiency is given the medical name anaemia, but there are a number of different types, each with a different cause, some of which are genetic, as also are some of the diseases caused by too much iron in the body. Treatment is dependent on the cause and, in many cases, is symptomatic.
Iron-deficiency anaemia is caused by blood loss, malabsorption or inadequate iron intake.

Megaloblastic anaemia is caused by a deficiency of folic acid, or of vitamin B_{12}, or malabsorption because of a lack of intrinsic factor, as in **pernicious anaemia,** already explained earlier in this appendix under cyanocobalamin – vitamin B_{12}.

Aplastic anaemia can be inherited, but may be acquired from a variety of causes, such as from the effects of cytotoxic drugs, radiotherapy during treatment for cancer, auto-immune reactions, idiosyncratic reactions to other drugs or even viral infections, such as hepatitis C. If left untreated, this type of anaemia can be fatal; however, a bone-marrow transplant, from a suitable donor, will restore bone-marrow function.

Haemolytic anaemia can be congenital or acquired, either as a disorder of the immune system or as a result of toxicity caused by such toxins as copper and arsenic, snake venoms or even some drugs. Sickle-cell disease and thalassaemia are hereditary blood diseases due to genetic defects in the haemoglobin, each caused by a different abnormality of the protein part of the haemoglobin molecule. Treatment in each case is by the administration of repeated blood transfusions. But while this replaces the lost sickle cells, so treating the anaemia produced in sickle cell disease, there is an iron build-up in thalassaemia which is treated with iron chelators, mentioned in the chapter on treatment.

Enzyme defects, such as various forms of glucose-6-phosphate dehydrogenase deficiency, are X-linked hereditary conditions, in which the absence of the enzyme glucose-6-phosphate dehydrogenase, normally part of carbohydrate metabolism, results in the breakdown of red blood cells (haemolysis). This can happen after exposure to infection or to oxidants in some foodstuffs and certain drugs such as methylene blue, which have been mentioned in the treatment chapter.

Treatment involves identifying and avoiding those agents that trigger the haemolytic anaemia, as well as treating attacks symptomatically. There are several varieties of this type of anaemia, linked to racial origin: an African version which is sensitive to some antimalarial drugs, an Oriental one and a European version, including **Favism**, which occurs in parts of the Mediterranean and Iran. Favism causes red blood cells to become sensitive to a chemical in broad beans (fava beans). It destroys the red blood cells leading to severe haemolytic anaemia, which is treated with blood transfusions.

Too much iron: There are disease states caused by both an excess of iron as well as by a deficiency. A genetic disorder, haemochromatosis, also known as **bronze diabetes**, is an iron storage disease that can lead to a build-up of iron in the body and results in the impairment of, and damage to, many organs, including the liver, pancreas and other endocrine organs. The build-up is manifest as a bronze colour to the skin, together with diabetes and liver failure.

Patients who suffer from **thalassaemia,** an inherited anaemic disease, also have problems due to a build-up of iron because of their treatment. They have an abnormality of the protein part of the haemoglobin molecule, which can be treated by the administration of multiple blood transfusions. Unfortunately this also gives the patients extra iron that is not needed, so they also take drugs to help them get rid of this build-up of excess iron.

Magnesium is another mineral essential to life. We each have about 25g of it in our body, concentrated mainly in the bones. Magnesium is a cofactor in at least 90 enzyme systems needed for the proper functioning of our muscles and nerves. Our main dietary source of magnesium is green, leafy vegetables.

Too little magnesium can cause problems, such as a deficiency state that can occur in alcoholism. Lack of magnesium can lead to lethargy, irritation, depression and personality changes – symptoms similar to **delirium tremens.**

Potassium is found mainly inside but also to a lesser extent outside every cell in our body. This mineral is essential to life and, together with sodium, is vital for the normal functioning of the nervous system and muscle. A deficiency state can occur when the kidneys malfunction, during starvation and in patients taking long-term loop diuretics like frusemide; indeed, the latter is such a problem that some diuretic preparations are formulated with potassium included to provide a built-in replacement for this loss. Severe dehydration, such as from a bad bout of diarrhoea, can cause a temporary deficiency of potassium, accompanied by a feeling of weakness. This resultant muscle weakness can affect the heart if severe, causing irregular beating and even cardiac arrest and death.

Too much potassium can also cause problems. The excess potassium tends to depress the central nervous system, and large doses can paralyse it, causing convulsions, diarrhoea, kidney failure and cardiac arrest. Patients have died from **cardiac arrest** following a direct injection of potassium chloride straight into a vein – it needs to be added to 500ml of normal saline and then slowly infused over a period of several hours.

Indeed, injection of potassium chloride directly into a vein is the lethal injection used as a form of capital punishment in the USA, with death resulting from cardiac arrest.

The injection of potassium chloride has also been used in cases of multiple pregnancy, where medical staff considered that selective reduction was necessary. This is achieved by the injection of potassium chloride into the chosen foetal thorax *in utero* to abolish its cardiac activity without affecting the other babies, which are then allowed to continue to term.

Copper is another essential element that we acquire from our diet. This mineral is good at what it does but is dangerous in excess or deficiency. It is also a vital component of a number of enzymes, without which we cannot live.

A build-up of copper can occur in **Wilson's disease,** an inborn metabolic disorder caused by the body's inability to utilise copper properly. Free copper is deposited in the liver, causing jaundice and cirrhosis. It also accumulates in the brain, resulting in mental retardation and other symptoms similar to those of Parkinson's disease. The corneas in the eyes become stained with a characteristic brown ring.

The cause of this condition is a congenital deficiency of a copper-containing protein, caeruloplasmin, normally present in the blood plasma. This protein normally forms a non-toxic complex with copper, which can then be excreted.

Menke's disease is an X-linked genetic disorder, caused by chronic lack of copper. This is because the body is unable to make the copper-transporting protein required, which leads to retarded growth, cerebral degeneration and death in infancy.

Zinc is yet another mineral essential to life, and high concentrations are to be found in the eye, prostate, muscle, kidney and liver. Semen is rich in zinc, and a deficiency of this mineral results in low sperm counts. As zinc is involved in many enzyme systems and transcription factors in our cells, a deficiency of zinc in the long term will result in stunted growth, retarded sexual development and taste disturbance.

However, control of zinc can be used for the good. The rare genetic disorder **acrodermatitis enteropathica**, which used to be fatal to all babies born with it, can now be successfully treated with zinc salts. Alcoholics frequently need to take zinc supplements as the enzyme alcohol dehydrogenase, involved in breaking down alcohol, contains a zinc atom. Due to alcoholics' excessive consumption of alcohol, they need more of this enzyme and therefore more zinc. As always though, it's a tricky balance: very large doses of zinc supplements may inhibit the absorption of copper and so cause other problems.

In Bangladesh, mothers given zinc supplements during pregnancy gave birth to healthier babies, who were less likely to die from diarrhoea or pneumonia, both major killers in the developing world. A weekly dose of zinc reduced the death rate by 85 per cent. So simple, yet so effective.

Iodine is essential not only for normal human growth and development, but also for a normal life span. Most of the 30mg of iodine in our bodies is found in the thyroid gland, where it is used to make two hormones, thyroxine and tri-iodothyronine, which are both involved in regulating our metabolism.

Hypothyroidism, caused by a deficiency of iodine, makes a person feel listless and cold, and tends to cause weight gain. This condition can eventually lead to the development of a goitre, where the thyroid gland enlarges, causing a swelling of the neck, in an attempt to produce more thyroid hormones.

At the opposite extreme, too much thyroid hormone causes **hyperthyroidism**, resulting in restlessness and hyperactivity. Hyperthyroidism, also called Graves disease, after the Irish physician Dr R. J. Graves (1850-1932), is also referred to as thyrotoxicosis and can result in the development of exophthalmic goitre, with a swollen neck and bulging eyes resulting from the enlarged thyroid gland. This condition tends to result in weight loss. The symptoms of thyrotoxicosis can be caused by simple overgrowth and thus over-activity of the thyroid gland, by a hormone-secreting benign tumour or by carcinoma of the thyroid gland.

Treatment may be by surgical removal of the thyroid gland, by administration of radio-active iodine to destroy part of the gland or by use of drugs such as carbimazole, which interfere with the production of the thyroid hormones. Whichever treatment is used, the goal is to end up with the patient slightly deficient in thyroxine, as this can then be easily treated with daily thyroxine tablets.

Iodine is particularly important for the development of the nervous system: babies born to mothers deficient in iodine during pregnancy have been found to suffer from **cretinism.** Cretinism is a syndrome of dwarfism and mental retardation, with a coarseness of skin and facial features due to a congenital lack of the thyroid hormone from birth. However, too much iodine can cause problems too, as it can lead to iodine poisoning, the main features of which, if mild, are an odour on the breath and staining of the mouth. Seaweed, also called kelp and bladderwrack, is a rich source of iodine. It is included in a number of herbal slimming products, on the premise that an increased iodine intake might speed up the body's metabolism sufficiently to help with the desired weight loss.

Selenium was only proved to be essential to life in 1975. It is stored by the body in red blood cells, the liver, spleen, heart and nails and then converted in the tissues to metabolically active forms. It is excreted mainly in the urine, with a little also in the faeces.

Only required in very small amounts of less than half a milligram per day, it is possible to overdose without even realising. As little as five mg taken as a single dose will produce toxic symptoms such as foul breath and body odour. An overdose of selenium causes loss of hair, nail changes, diarrhoea, dermatitis, a garlic odour to the breath, fatigue and peripheral tingling fingers and toes.

Selenium is present in many foods, mainly in the form of amino acid compounds, selenomethionine and selenocysteine. However, selenium intake in the UK has dropped markedly since we stopped importing wheat from North America and began using European-grown wheat instead. The selenium in European soils was stripped out during the last ice age, but this was not the case in North America.

A dietary deficiency of selenium is associated with a form of cardiomyopathy called **Kershan disease,** endemic in one part of China where little selenium is present in the soil and thus little is present in the foodstuffs grown there.

Sodium selenite is used as a source of selenium for those patients who develop deficiency states. This can be the case following prolonged periods of total parenteral nutrition. Low selenium levels have been shown to lead to low sperm counts, and men who took a selenium supplement could double their sperm count and so increase their fertility. Studies have also shown a correlation between low selenium intake and risk of prostate cancer – another good reason for men to take a selenium supplement.

Trace elements:

As already mentioned, the body is a complex machine that requires so many different ingredients in the proper quantities to function and maintain balance. Trace elements are just that, only a minute quantity being required, but, nonetheless, absolutely essential.

Manganese is involved in a number of our enzyme systems and is stored in the brain, kidneys, pancreas and liver. Accumulation of manganese can occur in patients, particularly children, receiving long-term total parenteral nutrition. If manganese dust is inhaled by miners in badly ventilated mine workings, it results in brain damage and other symptoms that appear to be very similar to those of patients diagnosed with Parkinson's disease.

Cobalt is a vital trace element that forms part of vitamin B_{12}. A deficiency of this vitamin can lead to pernicious anaemia.

Molydenum is needed for healthy teeth, for the functioning of some enzymes and also for the male sexual libido.

Chromium is involved in enzyme systems for carbohydrate and fat metabolism as well as for the production and action of insulin.

Silicon is needed for healthy skin, hair and nails.

Fluorine is an important trace element which, when incorporated in tooth enamel as a fluoride – a compound of fluorine – makes teeth more resistant to dental decay. In addition to creating healthy teeth, fluoride is also needed for bones, muscles and blood vessels. Our water supplies contain about one part per million of fluoride, due either to the natural mineral content or the addition of the element by the water supplier. In acute fluoride poisoning, when sodium fluoride is taken by mouth, it produces the very corrosive hydrofluoric acid in the stomach. Other effects of fluoride poisoning include tremors, tetany and convulsions, leading to cardiac and respiratory arrest. Death results within two to four hours following a dose of as little as 5-10mg of sodium fluoride.

Appendix VI – A Lethal Lexicon: where the words come from

Many of the terms used in this book to describe harmful substances and their actions have interesting origins, some of which are explained in this list of lethal lexicon.

Fatal
Today, 'fatal' tends to mean causing death, although it has a wider definition as something belonging to, appointed by, or announcing fate; that is, the inevitable destiny, whether good or bad. It is usually regarded as the result of ill-fortune or accident, leading to death.
The word is derived from the Latin, *fatalis* from *fatum,* meaning a prediction, or from *fatus* meaning spoken.

Lethal
Means deadly, mortal or death-dealing.
The word is from the Latin *letalis,* from *letum*, meaning death.

Mortal
Means liable to death, certain to die at some future time.
It also means causing death, deadly, fatal.
The word is from the Latin *mortalis,* from *mori*, to die.

Morbidity
Relates to the relative frequency of occurrence of a disease.

Mortality
Refers to the number of deaths that occur at a given time, in a given group, or from a given cause.

Noxious
Is yet another term meaning poisonous, harmful or unwholesome.
From the Latin *nocere,* to hurt.

Poison
A poison is a substance that either destroys life or impairs health.
In large enough doses almost any substance acts as a poison, but this term is generally used only to refer to substances that are harmful in quite small quantities, such as arsenic and cyanide. In chemistry, the term poison is also used to describe a substance that inhibits the activity of a catalyst as well as to describe a substance that absorbs

neutrons and so affects the workings of a nuclear reactor.
The word is derived from the Latin *potio,* meaning a draught,or from the verb *potare*, meaning to drink.

Potion
An associated word meaning a draught of medicine or poison; comes from the same Latin source as poison.

Toxic
Refers to anything that is poisonous; it is derived surprisingly from the Greek *toxon,* a bow, *toxikos,* for the bow and *toxikon,* arrow poison.

Toxin
Nowadays, refers to a specific poison of organic origin, in particular one which is produced by a living organism, especially a bacterium. In the body, toxins act as antigens and special antibodies.
Toxins may be endotoxins or exotoxins.
Endotoxins are poisons, generally harmful to all body tissues, and they are to be found contained within certain Gram-negative bacteria and are only released when the bacterial cell dies and disintegrates.
Exotoxins are more specific, highly potent poisons, which are harmful only to a limited range of tissues. They are produced by bacteria and secreted into the surrounding medium. Exotoxins tend to be unstable and can be made inactive by heat, light and chemicals; well-known examples are the exotoxins which cause tetanus, botulism and diphtheria.

Antitoxin
An antiserum, containing an antibody which will specifically bind to and neutralize a toxin, such as the diphtheria and tetanus antitoxins.

Toxoid
A toxin treated in such a way that its toxic properties have been removed while retaining its ability to stimulate the production of antibodies.

Venom
A poisonous fluid secreted by certain snakes, spiders, scorpions and other creatures, introduced into the victim by means of a bite or sting. Some venoms produce only a little local pain and swelling, while others can produce far more severe problems, which may

ultimately be fatal. The word is derived from the Latin *venunum,* meaning poison.

Victim
Someone who is subjected to death, suffering or ill-treatment.
The word is derived from the Latin *victima,* a beast for sacrifice.

Mithridate
Is an antidote to poison. It takes its name from Mithridates VI, also known as Eupator, who was the king of Pontus and Bithynia (120-63 BC).
He was famed for his herbal skills and was said to have made himself immune to poison by the constant use of antidotes. Some plants seem to have become associated with this type of antidote, including Mithridate mustard (field pennycress), and Mithridate pepperwort (field cress).
Antidotes were also called theriacs or treacles.

Theriac
The Ancient Greeks produced a variety of theriacs, initially to snakebites. One theriac called Galene, the prototype of this medieval 'antidote', contained 64 different constituents, most notably the flesh of skinned, roasted vipers. This was supposed to combat the effects of poison. Since it took 40 days to make Galene, and up to 12 years for it to mature before use, those making it were often tempted to cut corners or dilute it.
The great medical authority of the time, Galen, used it for other disorders, in addition to combatting the effects of poison, although he did stress the need for extreme caution during both the production of theriac and the use of it. The apothecaries of Venice were ordered in 1268 to prepare all supplies of theriac in the presence of three leading physicians and to store them for at least six months before putting them on sale.

Treacle
Is another antidote.
The best known was Venice Treacle (Theriaca Andromachi), a compound which was claimed to be composed of some 64 drugs in honey.
Culpeper gives a list of the ingredients for this treacle and several other antidotes, including Mithridate, London Treacle and Matthiolus's Great.

Derived from the Greek *theriake,* wild beast, especially a poisonous snake.

Antidote
Remedy against Poison and Pestilence. These preparations took weeks to prepare and then needed time to mature, sometimes several years, before they could be used. They were not an instant cure by any means. Today an antidote is a drug which counteracts the effects of a poison.
Derived from the Greek *antidotes,* from *anti*, against, and *didonai,* to give.

Poison crops up in everyday speech and colloquialisms with astonishing frequency, as the following examples show:

A poison pill
Containing a fast-acting poison, such as cyanide, was traditionally used as a means of suicide by captives in some countries, especially spies during times of conflict.

Also a poison pill
In the business world today, this is a ploy used by a company to try to deter a bidder when faced with an unwelcome takeover bid.

A poison pen letter
Is a malicious letter, usually anonymous, written with the intention of frightening, disturbing or warning someone about something.

One man's meat is another man's poison
Means that what is palatable or beneficial to one person may be distasteful or harmful to another.

What's your poison,
Meaning what would you like to drink, is not quite as light as it appears to be as alcohol is indeed a poison itself in sufficient quantity.

A poisoned chalice
refers to an honour or award which is likely to prove a disadvantage to its recipient. It also refers to a cup or chalice literally containing a lethal drink, as in Shakespeare's *Hamlet*, where Hamlet prepares such a drink to kill Gertrude.

Abbreviations

AIDS	– acquired immune deficiency syndrome
ATP	– adenosine triphosphate
BAL	– British Anti-Lewisite
BMA	– British Medical Association
BMJ	– *British Medical Journal*
CJD	– Creutzfeldt-Jakob Disease
COSHH	– Control of Substances Hazardous to Health
CSM	– Committee on Safety of Medicines
DDT	– dicophane, a powerful insecticide
DNA	– deoxyribonucleic acid
DNP	– dinitrophenol
DOH	– Department of Health
DVT	– deep vein thrombosis
FAO	– Food and Agriculture Organization (United Nations)
FDA	– Food and Drug Agency
GBH /GHB	– gamma-hydroxybutyrate
HDL	– high density lipoprotein
HIV	– human immunodeficiency virus
IUD	– intrauterine device
IVF	– in vitro fertilisation
LDL	– low density lipoprotein
LSD	– lysergic acid diethylamide
MCA	– Medicines Control Agency
MDMA	– methylenedioxymethamphetamine
ME	– myalgic encephalomyelitis
MHRA	– Medicines and Healthcare Products Regulatory Agency
MIC	– methyl isocyanate
NHS	– National Health Service
NSAID	– non-steroidal anti-inflammatory drug
PABA	– para-aminobenzoic acid
PCB	– polychlorinated biphenyl
PKU	– phenylketonuria
RNA	– ribonucleic acid
SMON	– sub-acute myelo-optic neuropathy
SSRI	– selective serotonin re-uptake inhibitor
TCDD	– 2,3,7,8-tetrachlorodibenzo-p-dioxin
TNT	– trinitrotoluene
TPN	– total parenteral nutrition

VLDL	– very low density lipoprotein
WHO	– World Health Organization

Glossary

Abdomen – the portion of the trunk below the diaphragm.
Abortion – the expulsion or removal of an embryo or foetus from the uterus at a time when it is incapable of independent survival.
Abortifacient – a substance that will induce an abortion or miscarriage.
Absorption – the uptake of fluids and other substances by the tissues.
Acid – sharp or sour taste; also chemically, an acid combines with a base to form a salt.
Acid-base balance – the levels of carbonic acid and bicarbonate in the blood must be maintained at a constant ratio of 1:20. Any disturbance results in a state of acidosis or alkalosis in the blood and tissues. This may be caused by lung or kidney problems, as these organs both play an important role in the regulation of the body's acid-base balance.
Acidosis – a condition in which the acidity of the body fluids is abnormally high, caused by failure of the mechanisms, in the lungs or kidneys, which normally maintain the acid-base balance.
Active – effective or potent, ready.
Activate – to make active, to stimulate or increase biological activity.
Acute – an illness of rapid onset, severe symptoms and brief duration.
-aemia – relating to a biochemical condition of the blood.
Adeno – relating to a gland.
Adipose tissue – fibrous connective tissue packed full of fat cells.
Adsorption – the adhesion of a thin layer of molecules of a substance to the surface of a solid or liquid.
Aerobe – an organism that requires oxygen for life and growth.
Agent – something that causes an effect.
Agranulocytosis – a severe deficiency of certain blood cells, the neutrophils, due to damage to the bone marrow, where blood cells are made; may be caused by toxic drugs or chemicals.
Albumin – a water-soluble protein, made by the liver and present in blood plasma. Can indicate kidney or heart disease if found in the urine.
Alcohol – a type of organic chemical in which a hydroxyl group (-OH) is substituted for a hydrogen atom.
The alcohol we drink is ethyl alcohol, C_2H_5OH.
Other alcohols, such as methyl alcohol, CH_3OH, are poisonous.

-algia – a suffix meaning pain.
Alkali – a basic substance that combines with an acid to form a salt.
Alkaloid – a nitrogen-containing substance produced by a plant. They are many and varied, and have potent effects on the body. Many alkaloids are used as drugs, such as morphine, quinine and atropine.
Alkalosis – an illness in which the blood and tissues become too alkaline, due to disturbance of the acid-base balance.
-amine – an organic compound containing an amino ($-NH_2$) group, e.g. histamine.
Amino acids – the fundamental components of proteins.
Amphetamine – a very addictive type of stimulant drug, which lessens fatigue and produces a feeling of mental alertness and well-being.
Anabolic – building up from simple to complex substances.
Anaemia – a reduction in the quantity of haemoglobin in the blood. There are many causes but the symptoms are tiredness and breathlessness, pallor and poor resistance to infection.
Anaerobe – an organism which can live and grow without oxygen.
Anaphylactic shock – an extreme and generalised allergic reaction in which there is widespread release of histamine, which causes massive swelling, resulting in breathing difficulties, heart failure, circulatory collapse and sometimes death.
Ankylostomiasis – a disease caused by a widely distributed parasitic hookworm that lives in the small intestine of humans in Europe, America, Africa and Asia.
Anorectic – a drug that reduces the appetite.
Anorexia – loss of appetite.
Antagonism – having the opposite effect.
Anti- – against.
Antibody – a special type of protein in the blood; a globulin which circulates and attacks antigens to render them harmless. An essential part of our immune system.
Anticoagulant – a drug that prevents blood from clotting.
Anticonvulsant – a drug that prevents convulsions, used in treating epilepsy.
Antidote – a drug that counteracts the effects of poison.
Antigen – any substance the body regards as foreign or potentially dangerous, against which it produces an antibody.
Anti-inflammatory – reduces or prevents inflammation.
Antipyretic – reduces fever by lowering body temperature.
Antiserum – a fluid that contains antibodies.

Antitoxin – an antibody produced by the body to counteract a toxin.
Aplastic anaemia – anaemia due to bone marrow failure.
Apoplexy – a stroke.
Argyria – deposition of silver in the skin and other tissues, resulting from chronic exposure.
Arthritis – inflammation of one or more joints.
Aspirate – to withdraw fluid from the body by suction.
Asphyxia – suffocation.
ATP – adenosine triphosphate, a compound that acts as an energy store within body cells, which is needed for muscular contraction.
Atrophy – wasting away.
Auxins – plant hormones or growth substances.

Base – reacts with an acid to produce a salt.
Bacteria – a type of micro-organism, widely distributed and the cause of many illnesses and infections.
Barbiturate – a highly addictive sedative drug.
Bends – compressed air illness, suffered by divers. If a diver surfaces too quickly, nitrogen that has dissolved in the bloodstream forms bubbles, causing pain in the joints; this causes damage as it leads to blocking the small blood vessels of the body, including those in the brain. This condition can be lethal without prompt treatment in a decompression chamber.
Benign – harmless.
Bilharziasis – schistosomiasis, an illness caused by a parasitic blood fluke. The larval stage is transmitted by water-snails in sewage-contaminated water in the Americas, Africa and Asia.
Blood pressure – the pressure of blood against the main arteries. Low blood pressure caused by shock or circulatory failure is called hypotension. In contrast, hypertension is raised blood pressure and if untreated can result in a stroke and other vascular damage.
Bound – some substances, such as metals, become bound to chelators and so are effectively removed, as they are then no longer available in the free state to damage organs; similarly some drugs are largely circulating in the body in the bound form and the drug's action is due only to the small amount that is circulating in the free state.
Brimstone – the old name for sulphur.
Buffer – a substance or solution that maintains the acidity or alkalinity (pH) even when diluted or when more acid or alkali is added. The bicarbonate system is the main buffer in the body.

Calcification – build-up of calcium in the tissues.
Calculus/calculi – a stone, or stones, formed in the body, usually composed largely of calcium phosphate or oxalate; they form in the gall bladder, the kidneys and bladder and even sometimes in the salivary glands. They are very painful and need to be removed surgically, although today they are usually broken up into tiny pieces using ultrasonic devices.
Cancer – a malignant tumour or growth, arising from the abnormal and uncontrolled division of cells, which then destroys the surrounding tissues.
Cantharides – Spanish Fly, the dried bodies of the blister beetle, *Lytta vesicatoria*; contains cantharadin, a toxic and highly irritant chemical, which on contact causes blistering of the skin.
Carbohydrate – a source of energy, such as sugars and starch, manufactured by plants and an essential part of the diet in humans.
Carbolic acid – an old-fashioned disinfectant, no longer in use, now called phenol.
Carboxyhaemoglobin – this substance is formed when carbon monoxide combines with haemoglobin in the blood. The blood is then incapable of transporting oxygen to the tissues; this is the cause of death of carbon monoxide poisoning.
Carcinogen – any substance, which when exposed to living tissue, causes cancer.
Carcinoma – cancer.
Cardiac – relating to, or affecting, the heart.
Cardiomyopathy – chronic disorder affecting the muscles of the heart.
Cardiovascular – pertaining to the heart and blood supply.
Carminative – relieves flatulence; used for gastric discomfort and colic.
Catabolic – breaking down complex substances into simple ones.
Cataract – an opacity in the lens of the eye which causes blurred vision and ultimately blindness.
Cathartic – laxative.
Cellular respiration – see Electron Transport.
Chelator – a chemical that binds on to metals; chelators are used in the treatment of poisoning. Some form the active centres of enzymes.
Cholesterol – a fat-like material present in the blood and most other tissues of the body, especially thc ncrvous system.
Chronic – long-standing; frequently a disease state of long duration with gradual onset and very slow changes.
-cidal – causes death.

Circulation – the heart and blood supply, arteries, veins and capillaries.
CJD – Creutzfeldt-Jakob Disease. A rare, rapidly progressive, neurological disease for which there is, as yet, no effective treatment.
Clotting /clotting factors – solidifying of blood; chemicals essential to stop bleeding. See Coagulate and Coagulation factors.
Coagulate – the process of changing blood from liquid to solid by clotting.
Coagulation factors – a number of chemicals which are all essential for the coagulation of blood. Some people, such as haemophiliacs, lack one or more of these factors and so their blood is unable to clot without an injection of the missing clotting factors.
Co-enzyme – an organic compound that plays an essential role in a reaction that is catalysed by an enzyme.
Co-factor – a substance which must be present in a sufficient quantity before certain enzymes can act.
Colic – severe abdominal pain, usually with fluctuating waves of pain.
Coma – a state of unconsciousness, which is unrousable.
Congestion – accumulation of fluid, such as blood or lymph.
Consumption – the old name for tuberculosis.
Contamination – pollution or infection.
Convulsion – a seizure due to an involuntary contraction of the muscles, which produces contortions of the body and limbs.
Copperas – any one of copper, iron or zinc sulphate, used as an adulterant in beer.
Coumarin – a substance produced by some plants or synthetically, used as an anti-coagulant.
Counter-irritant – a substance that causes irritation when applied to the skin and is used to relieve more deep-seated pain.
Cramp – a painful contraction of a muscle, which may be caused by a chemical imbalance of the salts in the tissues.
Crustacean – marine creature having a shell, e.g. shrimps, sea snails.
Cumulative – additional, increasing in quantity.
Cyanosis – a bluish discoloration of the tissues due to lack of oxygen.
Cystitis – inflammation of the bladder and urethra.
Cytochrome – a compound of a protein linked to haem; these are enzymes involved in electron transfer. See also Electron transfer chain.
Cytokines – proteins released by cells when activated by an antigen,

acting as enhancing mediators for the immune response through interaction with specific cell surface receptors on leucocytes. There are several types: interkeukins, lymphokines, interferons and tumour necrosis factors.

Deficiency – lack of.
Defoliant – a chemical which, when applied to plants, causes their leaves to drop off.
Degeneration – deterioration and loss of specialised function of cells.
Delirium – an acute disorder of brain function due to toxicity, infection or a metabolic disorder or deficiency.
Dependence – physical and/or psychological effects produced by the habitual taking of certain drugs, with a compulsion to continue taking such drugs.
Depression – a reduction of normal activity; a psychiatric illness.
Diabetes – a metabolic disorder causing great thirst and much urine.
Dialysis – the separation of substances from a solution, as in the process of filtering waste from the blood of patients with kidney failure.
Diastolic – in measuring blood pressure, the lower number is the diastolic pressure, that is, the pressure between heartbeats when the heart muscle relaxes.
Dislocation – displacement from the normal position of a joint.
Distension – stretching and swelling.
Diuresis – increased secretion of urine by the kidneys.
Diuretic – a drug that promotes the excretion of salts and water by the kidneys.
DNA – deoxyribonucleic acid – the genetic material found in the nucleus of the cell, which controls heredity.
Dracontiasis – a tropical disease caused by the parasitic nematode *Dracunculus medinensis,* the guinea worm, which is transmitted to humans via contaminated drinking water.
Dyspepsia – indigestion.

Eczema – a common itchy skin disease treated with judicious use of corticosteroid creams and ointments, but mainly by the use of emollients.
Electron transfer chain – a method of converting chemical energy into a readily usable and storable form; the respiratory chain is the most important, as without oxygen we die.
Elevated – increased.

Emaciation – wasting of the body, due to malnutrition, cancer, tuberculosis or parasitic worms.
Emesis – vomiting.
Emollient – an agent that soothes and softens the skin.
Encephalitis – inflammation of the brain.
Encephalopathy – a disease affecting the functioning of the brain.
Endemic – frequently occurring in a particular region or population, usually of a disease.
Endotoxin – a poison contained by certain bacteria, released only when the bacterial cell is broken down or dies.
Enterotoxin – a poisonous substance acting on the gastrointestinal tract, causing vomiting, diarrhoea and abdominal pain.
Enzyme – a biological catalyst, that is, a protein, which in very small amounts speeds up the rate of a biological reaction, without being used up itself.
Eosinophilia – an increase in the number of white blood cells, called eosinophils, in the blood. Caused by certain drugs, allergies, parasitic infestations and some forms of leukaemia.
Epidemic – a sudden outbreak of infectious disease that spreads rapidly in a population.
Erythema multiforme – eruptions following a herpes simple infection; involves widespread flushing of the skin due to dilatation of the blood capillaries.
Erythropoietic – causing production of red blood cells in the bone marrow.
Eschar – a scab.
Excretion – removal of the waste products of metabolism from the body: carbon dioxide and water from the lungs, urine from the kidneys, faeces from the gastrointestinal tract and water and salts from sweat glands in the skin.
Exophthalmic – causing protrusion of the eyeball.
Exotoxin – a poison produced by a bacterial cell and secreted into the surrounding medium, as in botulism and tetanus.
Exudate – slow escape of liquid.

Factor – a substance that is essential to a physiological process.
Fats – the main form of stored energy in the body. It also acts as an insulation material and is one of the main constituents of our diet, which must contain an adequate supply of essential fatty acids.
Fatty acids – the basic constituents of many important lipids, including the triglycerides. Some fatty acids are essential, as we cannot synthesise them, and so they must be obtained from our diet.

Favism – an inherited enzyme deficiency causing red blood cells to become sensitive to a chemical in broad beans (fava beans). It destroys the red blood cells leading to severe haemolytic anaemia and requiring blood transfusion.
Ferritin – an iron-protein complex that is one of the body's storage forms of iron.
Fibre – anatomically, a fibre is a threadlike structure such as a nerve, muscle or collagen fibre; dietetically, fibre is an important constituent of our diet, also called roughage, a non-digestable component needed for the proper functioning of the gastrointestinal tract.
Fibrosis – thickening and scarring of connective tissue, usually due to inflammation or injury.
Fluid ounce – see Ounce.
Fracture – break.
Free – not bound.
Fulminant – a condition or symptom, rapid, severe and of short duration.
Fungi – simple organisms, including yeasts, moulds and mushrooms.
Fungicide – an agent that kills fungi.

Gammaglobulin – consists mainly of immunoglobulins, a deficiency that results in an increased susceptibility to infections.
Gastric – pertaining to the stomach.
Gastrostomy – a surgical procedure in which an opening from the outside to the stomach is created, to facilitate feeding when swallowing food is impossible due to disease or to obstruction of the oesphagus.
Glycosuria – presence of glucose in the urine in abnormal amounts.
Goitre – a swelling of the neck resulting from enlargement of the thyroid gland.
Gout – an extremely painful condition, caused by uric acid crystals forming in the joints, particularly in the feet.
Grain – an old unit of mass, formerly used in medicine, equivalent to approximately 65mg. It was 1/7,000th part of a pound weight.
Granuloma – a localised collection of cells, usually produced in response to an infection or foreign body.
Griping – severe abdominal pains
Gynaecomastia – enlargement of the breasts in men, caused by some drugs as a side effect, but mainly caused by a hormonal imbalance or hormone therapy.

Haem – an iron-containing pigment (a porphyrin) that combines with the protein globin to form haemoglobin.
Haem- – relating to blood.
Haematemesis – vomiting blood.
Haemochromatosis – a hereditary disorder called iron storage disease or bronze diabetes, which causes damage and functional impairment to many organs.
Haemodialysis – blood flows on one side of a semi-permeable membrane, while dialysis fluid flows in the opposite direction on the other side of the membrane, removing accumulated waste products. In kidney failure, the patient's blood supply must be connected to the machine for a number of hours, several times a week.
Haemoglobin – the oxygen-carrying substance in red blood cells.
Haemolysis/haemolytic – destruction of red blood cells or having that effect.
Haemoperfusion – a method of removing unwanted substances from the blood by passing it over a highly permeable synthetic membrane. In the treatment of poisoning, a material such as activated charcoal lies on the other side of the membrane to adsorb the unwanted substances, as the blood passes over the membrane.
Haemopoiesis – the process of producing blood cells and platelets.
Haemorrhage – bleeding, tends to be used to refer to severe blood loss.
Haemosiderin – a storage form of iron with a protein shell.
Hallucinogen – a substance which causes hallucinations.
Halogen – any of a specific group of elements, namely fluorine, chlorine, bromine, iodine or astatine.
Hayfever – allergic rhinitis – inflammation of nasal mucous membranes caused by allergy.
Hepatic – pertaining to the liver.
Hormesis – a controversial theory that states that low-dose stimulation by poisons can lead to high-dose inhibition of the effects of poisoning.
Host – a plant or animal upon which a parasite lives.
Hyper- – increased or excessive.
Hyperglycaemia – increased or high blood sugar level, as in diabetes.
Hyperlipidaemia – a high level of fats in the blood, e.g. cholesterol.
Hypertension – high blood pressure.
Hyperthermia – high body temperature.
Hypo- – lack, deficiency or decreased.
Hypogammaglobulinaemia – a deficiency of gammaglobulin in the

blood. It can be inherited or acquired, as in lymphomas.
Hypoglycaemia – low blood sugar level.
Hypotension – low blood pressure.
Hypothermia – low body temperature.
Hypoxia – low oxygen level in the tissues.

Iatrogenic – an unforeseen or inevitable side effect, resulting from the treatment given for a condition.
Immunomodulatory – effect of modifying the immune response in order to treat or prevent a disease.
Incidence – frequency or range of occurrence.
Indicator – a substance that shows the presence of a chemical.
Inflammation – in response to injury, involves pain, heat, redness, swelling and loss of function.
Infusion – the slow injection of a substance usually into a vein, as with a saline drip.
Ingest – take in food.
Inhale – breathe in.
Inhibition – reduction or prevention of function.
Injection – administration by a syringe.
Insecticide – a poison specific to insects.
Insomnia – inability to sleep.
Interaction – effect produced by the mutual action of two or more substances.
Interleukin – one of a family of proteins that control some aspects of the immune response and haemopoiesis.
Intoxication – symptoms of poisoning due to ingestion of any toxic material, such as alcohol or heavy metals.
Intra- – within.
Intramuscular – into a muscle, a type of injection.
Intrathecal – a type of injection, directly into the cerebrospinal fluid contained in the meninges surrounding the spinal cord.
Intravascular – within a blood vessel.
Intravenous – a type of injection, directly into the bloodstream via a vein.
Involuntary muscle – not under conscious control, e.g. in gut, stomach, blood vessels.
Irreversible – cannot be reversed.
Irritability – typical response of certain tissues to a stimulus.
Irritant – causing irritation, e.g. nettles are an irritant and cause pain and swelling.
Isotonic – solutions which have the same osmotic pressure.

-itis – inflammation of an organ or tissue, e.g. arthritis.
In vitro fertilization – fertilization of an ovum outside the body, commonly referred to as a technique to produce a 'test-tube' baby.

Kidney – one of the pair of organs which excrete nitrogenous wastes, principally urea, from the blood. Only one working kidney is needed to survive.

Lachrymatory – causing the eyes to water.
Lactose – milk sugar.
Laryngeal – affecting the larynx.
Larvae – the pre-adult stage of life in some insects and nematodes. Caterpillars are the larvae of butterflies.
Larvicide – a poison specific to the larval stage of insect life.
Latency – period of delay in development.
Lavage – washing out part of the body with water or a medicated solution.
Lesion – area of damaged tissue, caused by disease or wounding.
Leukaemia – one of various malignant diseases in which there is increased production of some white blood cells, the leucocytes, as well as suppression of the other white cells, the red cells and platelets in the blood.
Lipoprotein – a type of protein, found in the blood and lymph, which is combined with fats and other lipids.
Luminescent – glows in the dark.
Lymphoma – a malignant tumour of the lymph glands.
-lysis – breakdown or remission.

Malignant – life-threatening if untreated, e.g. cancer.
Malleability – capable of being beaten or rolled into shape.
Mammary – pertaining to the breast.
ME – myalgic encephalopathy – chronic fatigue syndrome.
Medulla – the inner region of any tissue or organ, also the name of the part at the base of the brain.
Medullary paralysis – stopping of major pathways for nerve impulses, and centres for circulation, salivation, swallowing and respiration.
Metabolism – the chemical and physical processes in the body which maintain growth and functioning.
Methaemoglobinaemia – a substance formed when the iron present in haemoglobin is oxidised from the ferrous to the ferric form. It cannot bind with or transport oxygen to the tissues. It can be

produced by some drugs or may be an inherited condition.
Microgram – a millionth part of a gram.
Milligram – a thousandth part of a gram.
Minerals – naturally occurring, inorganic substances essential for life.
Miscarriage – the loss of the foetus during pregnancy; an abortion.
Mitochondria – structures in the cytoplasm of every cell; they are the site of energy production within the cell.
Molecule – the smallest particle of a substance; it is composed of a group of atoms.
Morels and false morels – mushrooms and toadstools.
Motility – the ability to move.
MRSA – methicillin (or multiple) resistant staphylococcus aureus, a very severe infection.
Mucosa – moist membrane linings of many body cavities and tubular structures, including the nasal passages, respiratory tract, the mouth and the gastro-intestinal tract.
Muscle – a tissue that can contract and relax, so producing movement.
Myalgia – painful muscles.
Myo- – pertaining to the muscles.
Myoglobin – an iron-containing protein, similar to haemoglobin, which is found in muscle cells. It can bind to oxygen and so acts as an oxygen reservoir within the muscle cells.

Narcosis – a state of reduced consciousness or total unconsciousness caused by the use of narcotic drugs, such as opium.
Necro – death of.
Necrosis – the death of some or all the cells of an organ or tissue.
Nematode – a roundworm; there are many types, such as hookworm, guinea worm and threadworm, which are parasitic in plants, animals and humans.
Nephro- – relating to the kidneys.
Nervous system – a network of specialised cells throughout the body, which pass information around the body. The brain and spinal cord form the central nervous system, and everything else is the peripheral nervous system, which includes the autonomic nervous system and comprises the sympathetic and the parasympathetic nervous systems.
Neuritis – inflammation of one or more nerves.
Neuro- – pertaining to the nervous system.
Neuropathy – a disease of the peripheral nerves.

Nostrum – a traditional remedy of the pharmacist's own compounding.
NSAIDs – non-steroidal anti-inflammatory drugs, e.g. ibuprofen.

Oedema – swelling caused by the accumulation of fluid in the tissues.
Onchocerciasis – a tropical disease of the skin and underlying tissues, caused by a parasitic worm. It leads to the growth of fibrous nodular tumours around the worms; later bacterial infections can lead to abscesses. Infestation can also lead to blindness.
Opiate / opioid – acting like opium, can be natural or synthetic.
Optical – relating to sight and the eye.
Osmosis – the passage of a fluid from a less concentrated to a more concentrated solution through a semi-permeable membrane.
Osmotic pressure – the pressure exerted during osmosis; the more concentrated the solution the higher the osmotic pressure.
Osteoporosis – a condition in which there is loss of bony tissue causing brittle bones which are liable to fracture.
Ounce – equal to 480 grains and a twelfth part of a pound troy, or one sixteenth part of a pound avoirdupois. A fluid ounce was the volume of distilled water which weighed an ounce. There are 20 fluid ounces to a pint in the UK, but only 16 to the US pint.
Oxidation – the process of oxidising, reacting with oxygen.**Oxyhaemoglobin** – haemoglobin with oxygen bound to it.

Paralyticileus – functional obstruction of the ileus (small intestine) due to loss of peristatic movement.
Paraesthesia – abnormal sensation, such as pins and needles.
Parasympathetic nervous system – enervates the blood vessels and most internal organs. This system works in balance with the sympathetic nervous system, frequently opposing its actions.
Parenteral – administered by a route other than by mouth, e.g. by injection.
Path- – a prefix denoting disease.
-pathy – a suffix denoting disease.
Peptide – a molecule formed by the bonding together of two or more amino acids.
Peri- – about, near or enclosing.
Pericarditis – inflammation of the membranous sac (pericardium) around the heart.
Peripheral – not central. See Nervous system.
Peritoneal dialysis – a method of dialysis that uses the peritoneum

(membrane of the abdominal cavity) as a semi-permeable membrane. Used when haemodialysis is not appropriate.
Pesticide – a poison specific to pests.
Petechiae – small, round, flat dark-red spots caused by bleeding into the skin or beneath the mucous membrane.
Phenol – also known as carbolic acid, a strong disinfectant, no longer in use.
Phenylketonuria – an inherited defect of protein metabolism.
Photophobia – an abnormal reaction to light, where such exposure causes intense discomfort of the eyes. It may be associated with dilatation of the pupils as a result of eye drops, or with migraine, or infections e.g. measles, German measles, meningitis.
Photosensitivity – an abnormal reaction of the skin to sunlight.
Placenta – an organ that develops within the uterus to supply the foetus with food and oxygen and eliminate waste during pregnancy.
Placebo – a medicine that is ineffective but may help to relieve a condition because the patient has faith in its powers. New drugs are always tested against placebos in clinical trials.
Pleura – covering of the lungs and inner surface of the chest wall.
Pleuro – a prefix denoting the pleura or the side of the body.
Plumbism – lead poisoning.
Poly – many.
Polypharmacy – many medicines all being prescribed and taken by a single patient – a prime source of problems.
Porphyria – a group of rare inherited disorders resulting from errors of metabolism caused by enzyme deficiencies in the production of haem.
Porphyrins – pigments derived from porphin, which form chelates with metals, such as iron, copper, cobalt and zinc. These chelates are constituents of important body chemicals such as haemoglobin, myoglobin and the cytochromes.
Potentiation – to make more effective or more potent.
Precursor – existing before, and being converted to the active form.
Prophylaxis – a means of preventing disease, e.g. immunisation.
Proteins – nitrogen containing organic compounds made up of one or more chains of amino acids, which are synthesised in the body. They are essential constituents of the body, forming structural tissues, like muscles, as well as enzymes and hormones.
Psychosis – a mental disorder in which there is loss of contact with reality, e.g. schizophrenia.
Pulmonary – pertaining to the lungs.
Pungent – strong smelling or tasting.

Purgative – a drug that stimulates or increases the frequency of bowel evacuation.
Pyrexia – high temperature, fever.

Quack medicine – claims to cure everything but contains nothing of medicinal value.
QT interval – the interval shown on an electrocardiogram (ECG) that contains the deflections produced by contraction of the heart's ventricles.
Quickening – the first movement of the foetus felt by the mother, usually at about 16 weeks of pregnancy.

Raynaud's disease – a condition of unknown cause in which the arteries of the fingers are unduly reactive and enter spasm involuntarily when the hands are cold. A similar condition results from use of vibratory tools and the ingestion of ergot derivatives and may result in gangrene.
Receptor – a specialised group of cells, which can detect changes in the environment and trigger nervous impulses in the sensory nervous system.
Renal – pertaining to the kidney.
Reversible – able to be reversed or changed.
Rhabdomyolysis – a condition resulting in breakdown of the muscle fibres and progressive weakness.
Rhinitis – inflammation of the nasal mucous membranes, e.g. caused by a virus or allergy.
Ringworm – a fungal infection of the skin, scalp and nails.
RNA – ribonucleic acid – found in the nucleus and cytoplasm of cells; it is involved in the synthesis of proteins.

Scabies – a skin infection caused by a mite that tunnels into the skin to lay eggs; causes severe itching, particularly at night.
Schistosomiasis – an illness caused by a parasitic blood fluke. The larval stage is transmitted by water-snails, in water contaminated with sewage, in the Americas, Africa and Asia. Also called bilharziasis.
Septicaemia – blood poisoning that can result in widespread destruction of the tissues following the absorption of infective organisms or toxins from the bloodstream.
Shock – condition associated with circulatory collapse, when the arterial blood pressure is too low to maintain an adequate blood supply to the tissues.

Sign – an indication of illness observed by a doctor, but that is not apparent to the patient. See also Symptom.
Smooth muscle – also called involuntary muscle; produces slow, long-term contractions of which we are not consciously aware, as in the stomach and intestines, blood vessels and the bladder.
Spasm – a sustained muscular contraction which is involuntary.
-static – stopped.
Status epilepticus – a medical emergency in which there are repeated epileptic convulsions without return of consciousness between them.
Sterilisation – destruction of micro-organisms by heat, radiation, chemicals or filtration.
Stevens-Johnson syndrome – a rare skin condition characterised by severe blisters and bleeding in the mucous membranes of the eyes, lips, mouth, nasal passages and genitals. This syndrome is believed to be a life-threatening form of erythema multiforme and in many cases is thought to be caused by a drug reaction.
Stomatitis – inflammation of the mucous lining of the mouth.
Stricture – narrowing of any tubular part of the body, e.g. oesophagus.
Stridor – the noise of breathing, which is louder and harsher than a wheeze, heard when the trachea or larynx is obstructed.
Surfactant – a wetting agent, e.g. detergent.
Sympathetic nervous system – enervates the blood vessels, heart, lungs, intestines and other abdominal organs, also the sweat glands, salivary glands and genitals. This system works in balance with the parasympathetic nervous system, frequently opposing its actions.
Symptoms – an indicator of illness noticed by the patient. See Sign.
Syndrome – a collection of signs and symptoms indicative of a specific disorder.
Synergist – a substance (e.g. a drug) that interacts with another substance to produce an increased effect, which is greater than would be achieved by the sum of the effects of each substance if given separately.
Synergy – increased effect, greater than the sum of each part separately.
Synthetic – man-made.
Syphilis – a sexually transmitted disease caused by a bacterium.
Systolic – the highest pressure when the ventricles of the heart are contracting during a heart beat. See Blood pressure and Diastolic.

Tachycardia – an increased heart rate, above normal.

Teratogen – an agent that damages the embryo or foetus by causing developmental abnormalities, in utero.
Tetany – muscular spasm and twitching, due to low blood calcium level, particularly of the face, hands and feet.
Thiol group – a sulphur atom combined with a hydrogen atom (-SH).
Thrombosis – a blood clot with a vein.
Thyrotoxicosis – a syndrome caused by excess thyroid hormone.
Total parenteral nutrition – a form of artificial feeding in which all nutrients, carbohydrates, proteins, fats, vitamins and minerals are administered by way of a tube, either intravenously or by using a naso-gastric tube.
-toxic – poisonous.
Toxin – a poison produced by a living organism, e.g. a bacterium.
Toxicity – the state or degree of poisoning.
Trace element (or micronutrient) – a substance needed in a very small amount for normal healthy growth; often components or activators of enzymes.
Tracheostomy – an operation in which a hole is made into the trachea through the neck to relieve an obstruction to breathing.
Trimer – a substance formed from three molecules of a monomer.
Tumour – an abnormal swelling or growth, which may be benign or malignant.
Turpentine – an essential oil, extracted from long-leaf pine oil, by steam distillation. Turpentine susbtitute is a petroleum fraction of similar boiling point.

-uria – pertaining to urine or to urination.
Urine – fluid excreted by the kidneys, contains many of the body's waste products.

Vasculo- – pertaining to blood vessels.
Vaso- – pertaining to blood vessels.
Vasoconstriction – decrease in the diameter of blood vessels, especially arteries.
Vasodilatation – increase in the diameter of blood vessels, especially arteries.
Vasospasm – sustained involuntary muscular contraction of the blood vessels, such as in the hands in Raynaud's disease.
Vector – anything that transmits disease.
Vertigo – a disabling sensation of spinning or tilting that is a symptom of disease in the inner ear or of the vestibular nerve.

Vitamins – essential food factors, required in small amounts for healthy growth and development. They cannot be synthesised by the body and are thus essential constituents of our diet.
Voluntary muscle – under conscious control, e.g. skeletal muscles involved in movement.

References

Chapter 1
1.Luke E. Lancet 1962; 1: 10
2.Calabrese E. Crit Rev Toxicol 2001; 31: 637-48

Chapter 2
1. WHO. Environmental Health Criteria 18. Geneva: 1981
2. HSE. Toxicity Review 21, London HMSO 1989
3. Baer R L. J Allergy & Clin Immunol 1973;51: 35
4. WHO Drug Inf 1995; 9: 159-60
5. Ibels L S, Pollock C A. Med Toxicol 1986; 1: 387-410
6. Campbell W A. Chem Brit 1990; 26: 558
7. Dathan J G, Harvey C C. Br Med J 1965; 1: 1181
8. Florentine M J, Sanfilippo D J. Clin Pharm 1991; 10: 213-21
9. Moncada S, Palmer RM and Higgs EA. Pharmacol Rev June 1991; 43(2):109-42
10.Martindale 32nd Ed. 1999; 1618: ii
11.Clark R F, et al. JAMA 1996; 275: 1087-8
12.Ransome W, et al. N Eng J Med 1961; 264: 384
13.Bryce-Smith D. Chem Br 1989; 25: 783-6

Chapter 3
1.Potters Herbal Cyclopaedia 2003
2.A Modern Herbal Grieve Mrs M 1973
3.Mathieson H S. Abstr World Med 1947; 1: 481
4.Thurstone D, Taylor K. Pharm J 1984; 223: 63
5.Henry J, Volans G. Br Med J 1984; 289: 990-3
6.Martindale, 32nd Ed. 1999; 1407:i
7.Knight B. Br Med J 1979; 1: 350
8.Lampe K F. Ann Rev Pharmacol Toxicol 1980; 20: 15
9.Köppel C. Toxicon 1993; 31: 1513-40
10.King B. Lancet 1979; ii: 1411
11.Nyathi CB et al. Cent Afr J Med 1989;35:542-5

Chapter 4
1.Berressem P. Chem. Brit. 1999;35,40.
2.Warrell D A, Fenner P J. Brit Med Bull 1993; 49: 423-39.
3.Catterall W A. Ann Rev Pharmacol Toxicol 1980; 20 :15
4.Buchwald H D et al. Science 1984;144:1100.
5.Prince R C. Trends Pharmacol Sci 1988; 13: 46.
6.Whittaker V P. Trends Pharmacol Sci 1990; 11: 8

7.Narahashi T, and Wu C H. Ann Rev Pharmacol Toxicol 1988;28:141

Chapter 5
1.CSM/MCA Current Problems 1994; 20: 2
2.CSM. Current Problems 1994; 20: 9-11
3.Scott D K, Vale J A. Pharm J 1990; 245: 95-9
4.Meredith T, Vale A. Br Med J 1988; 296: 77-9
5.White R F, Proctor S P. 1997; 349: 1239-43
6.Steenland K. Br Med J 1996; 312: 1312-13
7.Bismuth C et al. Drug Safety 1990; 5:243-51
8.Henry J A. Br Med J 1992; 305: 5-6
9.Dept. Health; Dietary reference values. HMSO 1991
10.Iwamoto N, et al. Lancet 1994;343:357
11.Lonn E et al, JAMA 2005; 293:1297
12.WHO.Trace elements in human nutrition and health. Geneva, 1996
13.Bushinsky D A, Monk R D. Lancet 1998; 352:306-11
14.Harju e. Clin Pharmacokinetics 1989; 17: 69-89
15.McLean R M. Am J Med 1994; 96: 63-76
16.Burr W. Clin Obstet Gynecol 1986; 13:227-90
17.O'Donohue J, et al. Eur J Gastroenterol Hepatol 1993; 5: 561-2
18.Sarkar B, et al. J Paediatr 1993; 123: 828-30
19.MRC. Arch Dis Child 1993; 68: 426-7
20.Elder G H, et al. Lancet 1997; 349: 1613-17
21.Ferguson A. Prescribers J 1997; 37: 206-12

Chapter 6

1.Bayly G R, Ferner R E. Prescribers J 1995; 35: 12-17
2.Webb D. Br J Hosp Med 1993; 49: 493-6
3.Ferner R. Prescribers J 1993; 33: 45050
4.Guthrie S K. Ann Pharmaco Ther 1990; 24: 721-34
5.Steven W, et al. Int J Clin Pharmac Biopharm 1974; 10:1
6.Tush G M, Anstead M I. Ann Pharmaco Ther 1997; 31: 441-4
7.Coleman M D, Coleman N A. Drug Safety 1996; 14:394-405
8.Tefferi A, et al. Am J Med 1990; 88:184-8
9.Nauman J, Wolff J. Am J Med 1993; 94: 524-32
10.Lewis C. Drug Safety 1994; 11: 153-62
11.Proudfoot A T, et al. Med Toxicol 1986; 1: 83-100
12.McCarthy J T, et al. Quart J Med 1990; 74: 257-76
13.Stenmer K L. Pharmacol Ther 1976; 1: 157-60

14.Brewer G J. Drugs 1995; 50: 240-9
15.Dodds C, McKnight C. Br Med J 1985; 291: 785-6
16.Pak C Y C. J Clin Pharmacol 1979; 19: 451-7
17.Nagler J, et al. J Occup Med 1978; 20: 414
18.Stolshek B S, et al. Med Toxicol 1988; 3: 167-71
19.Eddleston M, Warrell D A. Q J Med 1999;92:483-485
20.Elder G H, et al. Lancet 1997; 349: 1613-17
21.Gibbins R L. Br Med J 1993; 306: 600-1
22.Anon. Drug Ther Bull 1989; 27: 39-40
23.Buckley B M, Vale J A. Prescribers J 1986; 26: 110-15
24.Home T W. Med J Aust 1988; 148: 540
25.Smith T A, Figge H L. Am J Hosp Pharm 1991; 48: 2190-6
26Warrell D A, Fenner P J. Br Med J 1993; 49:423-39
27.Ewan P W. Prescribers J 1997; 37: 125-32

Chapter 7

1.Rosencranz HS, Carr HS. Br Med J 1971; 3: 702-3
2. FAO/WHO. WHO Tech Rep Ser 309, 1965
3. Mason HJ. Hum Exp Toxicol 1990; 9: 91-4
4. HSE. Toxicity Review 21, London HMSO 1989

Chapter 8

1.Homan P. Pharm J 1999; 263: 1007
2. Anon. Br Med J 1913; i: 36 & ii:506
3. Anon. Lancet 1921; ii: 1117
4. Dale HH. Ibid 1921; ii: 112
5. Sheridan M. Sunday Times 2006; 26 March
6. Hamilton A. New Eng J Med 1936; 215: 426
7. Gilbert D. Br Med J 1937; i: 1157
8. Schwartz L. Amer J Publ Hlth 1936; 26: 586
9. Anon. Br Med J 1916; ii: 842 & Lancet 1916; ii: 1026
10.Beach FXM, et al Br J Ind Med 1969; 26: 231
11.Sterekhova NP, et al. Abstr Wrld Med 1971; 45:542
12.Oliver Sir T. Br Med J 1921; ii: 111
13.Scott R, et al. Lancet 1976; 2: 396
14.Owen D. Br Med J 1934; ii: 833
15.Macauley MB, Mant AK. J R Army Med Cps 1964;110: 27
16.Villar TG, et al. Practitioner 1974; 213: 281
17.Petering HG, Tepper LB. Pharmac Ther 1976; 1: 131
18.Lockhart LP. Brit Med J 1933; i:1694
19.Faulkner JM J Amer Med Ass 1933; i: 1694
20.Israels MCG, Susman W. Lancet 1934; i:509

21. Yearb Pharm 1924;524
22. Cooper WC et al. Am Ind Hyg Ass J 1964; 25:431
23. Ainslie Walker JT. Lancet 1925; i: 1163

Chapter 9

1. McCarthy M. Lancet 1993; 342: 362
2. WHO. Environmental Health Criteria No.83, Geneva: WHO 1989
3. WHO. ibid No.2, Geneva: WHO 1976
4. WHO. Tech Rep Ser Wld Hlth Org No.443; 1970
5. Van Laethen Y, Lopes C. Drugs 1996; 52: 861-9
6. WHO. Tech Rep Ser Wld Hlth Org No.830; 1993
7. ibid No.513; 1973
8. ibid No.720; 1985
9. Martindale & Westacott 18thEd. Vol II 1925, 33
10. WHO. Tech Rep Ser Wld Hlth Org No.525; 1973
11. IPCS Health & Safety Guide No.55, Geneva WHO 1991
12. Villar TG, et al. Practitioner 1974; 213: 281
13. WHO. Environmental Health Criteria No.39, Geneva Who 1984
14. Sawada Y, et al. Lancet 1988; 1: 299
15. Bidstrup PL, Payne DJH. Br Med J 1951; ii: 16

Chapter 10

1. Accum F. Treatise on Adulteration of Food & Culinary Poisons. London 1820, pii.
2. Hassall AH. Food & its Adulterations. London 1855
3. Farré M, et al. Lancet 1989; ii: 1524
4. Firth D, Bentley JR. ibid 1921; ii: 901
5. Huxtable R. J Am Med Ass 1977; 238: 1233
6. McLean EK. Pharmac Rev 1970; 22: 429
7. Hogan GR, et al. Lancet 1978; 1: 561
8. King B.ibid 1979; ii:1411
9. Partington CN. Br Med J 1950; ii: 1097
10. Anon. ibid 1933; ii: 579
11. Stewart TH, et al. S Afr Med J 1969; 43: 200
12. Gordon RS. Lancet 1981; 2: 1171-2
13. Kennedy N, et al. Arch Dis Child 1997; 76: 367-8
14. Anon. Pharm J 1936; ii: 214 & Lancet 1936; ii: 1153
15. Henderson WR, Raskin NH. Ibid 1972; 2: 1162
16. Williams B. Med J Aust 1972; 2: 390
17. Wilson LG. Med J Aust 1976; 1: 505
18. Moder K G, Hurley D L. Mayo Clin Proc 1990; 65: 1587-94
19. Anon. J Am Med Ass 1949; 139: 588-688

20.Cochrane WJ, Smith RP. Canad Med Ass J 1940; 42: 23
21.Anon. Med Lett 1974; 16: 75
22.Anon. Lancet 1990; 336: 366
23.Dive A, et al. Hum Exp Toxicol 1994; 13: 271-4
24.Keeney AH, Melinkoff SM. J Am Med Ass 1951; 146: 401
25.Anon. Br Med J 1922; ii: 1273
26.Haris P et al.J Env. MonitoringDOI:10.1039/b500932d; 2005
27.Ibels LS, Pollock CA. Med Toxicol 1986; 1: 387-410
28.WHO. Tech Rep Ser Wld Hlth Org No462, 1971
29.Fielder RJ, Dale EA. Toxicity Review No 7, London HMSO 1983
30.de Groot, et al. Lancet 1987; i: 1084
31.Jones RE, et al. Mon Bull Minst Hlth 1957; 16: 241
32.Anon. Rep Med Offr Minist Hlth, London 1935: 157
33.Anon. Br Med J 1918; i: 511

Chapter 11
1.Anon. Br Med J 1912; ii: 1570
2.Anon. Lancet 1956; 2: 182
3.Wilcox Sir WH, Br Med J 1922; ii: 371
4. Feinglass EJ. New Eng J Med 1973; 288: 828
5. WHO. Environmental Health Criteria No.85, Geneva WHO 1989
6. Memon NA, Davidson AR. Br Med J 1981; 282:1033
7. Lee B. ibid 1981; 282: 1321
8. Richardson AP. J Pharmacol 1937; 60:101
9. Church LE. Br Dent J 1976; 141: 234
10.Naik RB, et al. Postgrad Med J 1980; 56: 451-6
11.Newns G. Lancet 1949; ii: 964
12.Athanasion M, et al. J Pediatr 1997; 130: 680-1
13.Coskey RJ. Archs Dem 1974; 109: 96
14.Loeb FX, King TL. Am J Dis Child 1974; 128: 256
15.Anon. Lancet 1904; ii: 1439
16.Anon. Br Med J 1912; i: 183
17.Turtle WRM, Dolan T. Lancet 1922; ii: 1273
18.Anon. Analyst 1939; 679
19.Balassa J. Br Med J 1970; 2: 589
20.Maitland FP. ibid 1931; ii: 77
21.Wahlberg P, Nyman D. Lancet 1969; 2: 215
22.Love EB, Miller AA. ibid 1951; i: 1306
23.Foxell AWH. Br Med J 1951; i: 397
24.Kemohan RJ. ibid 1949; i: 888
25.Foxell AWH. Br Med J 1951; i: 397
26.Barnes RL, Wilkinson DS. ibid 1973; 4: 466

27.Clayton TM. ibid 1915; i: 208

Chapter 12
1.Woltman HW. Clin Jl Jan 1923; 24: 48
2.Bartleman EL, Dukes C. Br Med J 1936; i: 528
3.Anon. Ibid 1922; i:373
4. Israels MCG, Susman W. Lancet 1934; i: 509
5. Martindale 22nd Ed. 1941; Vol I: 288
6. Jones F Emrys. Br Med J 1913; ii: 849
7. Goldman L, et al. J Am Med Ass 1948; 137: 354
8. Srivastava PC, Varadi S. Br Med J 1968; 1: 578
9. Warley MA, et al. Ibid 1968; 1:117
10.Hardwick N, et al. Br J Dermatol 1989; 120: 229-38
11.Barr RD, et al. Br Med J 1972; 2: 131-4
12.Todd DJ. Br J Dermatol 1994; 131:751-66
13.Bluhm R, et al. J Am Med Ass 1990; 264: 1141-2
14.Tanew A et al. J Am Acad Dermatol 1988; 18:333-8
15.Hughes CG. J Am Acad Dermatol 1983; 9: 770
16.Bloem JJ, van der Waal L. Oral Surgery 1974; 38:675
17.thiels C, Dumke K. Fortschr Röntgenstre 1977; 126: 173
18.Gabriel SE, et al. N Eng J Med 1994; 330: 1697-1702
19.Anon. Med Lett Drugs Ther 1985; 27: 54-5
20.Clarke E W , et al. Contact Dermatitis 1977; 3 : 69-74

Chapter 13
1.Anon. Pharm J 1927; i: 361
2.Watson EH. J Am Med Ass 1945; 129: 333
3.Young EG. Canad Med Ass J 1949; 61: 447
4.Birch J. Br Med J 1928; i: 177 & Lancet 1928; i: 287
5.Anon. Prescriber, July 1913
6. Laurie NM. Canad Med Ass J 1950; 63: 298
7. Sibert JR. Br Med J 1973; 1: 803
8. Lundell E, Nordman R. Ann Clin Res 1973; 5: 404
9. Rogers SCF, et al. Br J Derm 1978; 98: 559
10.Anon. J Am Med Ass 1937; ii: 1160
11.Paisseau G. Br Med J Epith 1934; ii: 61
12.South African Med Record July 1912 per Lancet 1912; ii:471
13.Datham JG, Harvey CC. Br Med J 1965; 1: 1181
14.Ross AT. J Am Med Ass 1964; 188:830
15.Husband P, McKellar WJD. Archs Dis Chld Hlth 1970; 45: 264
16.Anon. Pharm J 1949; ii: 95
17.Anon. Lancet 1910; ii: 1693

18. Sinniah D, Baskaran G. Lancet 1981; 1: 487-9
19. Anon. Pharm J 1911; ii: 407
20. Gunn JA. Br Med J 1917; i: 579
21. Anon. Lancet 1949; i: 314
22. Homblass A. J Am Med Ass 1975; 231: 245
23. Scally CM. Br Med J 1936; i; 311
24. Duggan PJ. Ibid 1937; i: 918
25. Anon. Lancet 1917; ii: 162
26. Willis HW. J Pediat 1937; 65
27. Anon. J Am Pharm Ass 1939; 389
28. Anon. Br Med J 1912; i: 724
29. Lester BM. J Am Acad Child Adoles Psychiatry 1993; 32: 1253-5
30. Knowles JA. J Pediatr 1965; 66: 1068
31. Gordon I, Whitehead TP. Lancet 1949; ii:647
32. Hart CW, Naunton RF. J Am Med Ass 1964; 190:392
33. Matz GJ, Naunton RF. Ibid; 1968; 206: 910
34. Paufique L, Magnard P. Bull Soc Ophthal Fr 1969; 69: 466
35. DiMaio JM Henry LD. Sth Med J 1974; 67: 1031
36. Lugo G, et al. Am J Dis Child 1969; 117: 328
37. Heinonen OP, et al. Birth Defects & Drugs in Pregnancy, Littleton MA. Publishing Sciences Group. 1977; p296
38. Axton JHM. Postgrad Med J 1972; 48: 417
39. Willimott SG. Lancet 1931; ii: 1133
40. Evans ANW, et al. Practitioner 1980; 224: 315
41. Anon. Br Med J 1969; 3: 408
42. Thompson J. ibid 1950; i: 645
43. Prain JH. ibid 1949; ii: 1019
44. Smith RP, et al. New Eng J Med 1950; 342: 641
45. Anon. Lancet 1961; 1: 869
46. deCastro FJ, et al. Clin Toxicol 1977; 10: 287
47. Jue SG. Drug Intell & Clin Pharm 1976; 10: 52
48. Cronin AJ, et al. Br Med J 1979; 1: 722
49. Walter DC, et al. AmjDis Child 1980; 134: 202
50. Besson-Leaud M, et al. Annls Pédiat Paris 1977; 24: 363
51. Hutchinson A, Kilham H. Med J Aust 1978; 2: 335
52. Anon. Lancet 1910; ii: 884
53. Vale JA, et al. Postgrad Med J 1976; 52: 598
54. Harif M, et al. Br Med J 1995; 311: 88-91
55. Lewis DR, et al. Ibid 1939; i:1283
56. Anon. Practitioner 1968; 200: 319
57. Newns G. Lancet 1949; ii: 964
58. Zueler WW, Apt L. J Am Med Ass 1949; 141:185

59.Simon FA, Pickering LK. J Am Med Ass 1976; 235: 1343
60.Anon. Practitioner 1968; 200: 179

Chapter 14
1.Thomas JG. Lancet 1962; 1: 222
2.Spurlock BW, Dailey TM. New Eng J Med 1990; 323: 1845-6
3. Williams B. Med J aust 1972; 2: 390
4. Smith ILF. Br Dent J 1968; 125: 304
5. Papa CM, Shelley WB. J Am Med Ass 1964; 189: 546
6. O'Mullane NM, et al. Lancet 1982; i: 1121
7. Boruchow IB. Cancer 1966; 19; 541
8. Boucher BJ, Wright JT. Postgrad Med J 1973; 49: 106
9. Mathieson HS. Abstr World Med 1947; 1: 481
10.Alexander E, et al. New Eng J Med 1946; 234: 258
11.Hall AJ. Lancet 1934; i: 595
12.Anon. ibid 1911; i: 162
13.Anon. Ibid 1910; ii:1270
14.Talukder M. Indian Med Gaz 1935; 628
15.Ginn HE, et al. J Am Med Ass 1968; 203: 230
16.Morales FH, Rivers RD. Trop Dis Bull 1946; 43: 230
17.Kemohan RJ. Lancet 1949; i: 888
18.Anon. Br Med J 1912; ii: 350, 1470
19.Dvoracova I. Pharm J 1968; 1: 212
20.Andrews CH. Lancet 1921; ii: 654
21.Rosin RD. Br Med J 1967: 4: 33
22.Ewart WB, et al. Can Med Ass J 1978; 118: 1199
23.Anon. Lancet 1934; i: 652
24.Simpson IA. Trop Dis Bull 1936; 634
25.Yater WM, Cobill JA. J Am Med Ass 1936; i: 1635
26.Anon. J Am Med Ass 1925; ii: 555
27.Eccles Smith R. Lancet 1922; ii: 1359
28.Slee TJ, et al. Acta Med Scand 1979; 205: 463
29.Anon. Br Med J 1909; ii: 1803
30.Cowan GAB. ibid 1947; i:452
31.Simon N, Harley J. J Am Med Ass 1967; 200: 254
32.Petros H, MacMillan AL. Br J Derm 1973; 88: 505
33.Roenigk HR, Handel D. J Am Med Ass 1974; 227: 959
34.Anon. Analyst 1926; 313Bell RD. Br Med J 1936; i: 886
35.Kennedy CC, Lynas HA. Lancet 1949; ii: 650
36.Bowen DAL, et al. Br Med J 1961; 1: 1262

Chapter 15
1.Jones EJ, et al. Can Med Ass J 198 9; 99: 10
2.Dunn MA, Siddell FR. J Am Med Ass 1989; 262: 649-52
3. World MJ. Lancet 1995; 346: 260-1

Chapter 16
1.Berkowitz R L, et al. Am J Obstet Gynecol 1996; 174: 1265-72
2.Roe RB. Lancet 1913; i: 1527
3.Watt JD. Br Med J 1951; i: 759
4.Anon. Br Med J Epit 1924; i:57
5.Tompsett JNM. Lancet 1938; i: 994
6.Lowenberg K. J Am Med Ass 1938; i:573
7.Dalgaard JB, Jakobsen J. Pharm Dig 1963; 27: 559
8.Vartan CK, Discombe J. Br Med J 1940; i: 525
9.Howetson WM. ibid 1933; ii: 170
10.Pille G, et al. Trop Dis Bull 1959; 56: 412
11.DiMaio JM Henry LD. Sth Med J 1974; 67: 1031
12.Rajan N, et al. Post Grad Med J 1985; 61: 35-6
13.Anon. Br Med J 1936; i: 363
14.Ducassou JL. Abstr World Med 1947; 1: 588
15.Edwards AC, Thomas ID. Lancet 1978; 1: 92
16.Graham DL. Archs intern Med 1977; 137: 1051
17.Gillespie RD. Lancet 1934; i: 337
18.Chang DK, Tainter ML. J Am Med Ass 1936; i: 1386
19.Haider I, Oswald I. Br Med J 1970; 2: 318
20.Gitelson S, et al. Diabetes 1966; 15: 810
21.Roberts CJC, Marshall FPF. Br Med J 1976; 1: 20
22.Vasey RH, Karayannopoulos SJ.Br Med J 1972; 1: 112

Chapter 17
1.Soignet SL, et al. N Eng J Med 1998; 339: 1341-8
2. FDI/WHO. FDI World 1995; 4 July/Aug: 9-10
3. Seal D, et al. Lancet 1991; 338: 316-18
4. Kowalsky SF, Dixon DM. Clin Pharm 1991; 10: 179-94
5. Hooper DC, Wolfson JS. N Eng J Med 1991; 324: 384-94
6. Aronson JK, Reynolds DJM. Br Med J 1992; 305: 1273-6
7. Pharm J 1921; ii: 315
8. ibid : 344
9. Stoll A. Pharm J 1965; 1: 605
10.Mouren P, et al. Presse Méd 1969; 77: 505
11.Anon, Br Med J E 1915; i: 19
12.Karalliedde L. Br J Anaesth 1995; 75: 500

13.Hanif M, et al. Br Med J 1995; 311:88-91
14.Seah CS, et al. Med J Aust 1995; 2: 424
15.Szuler IM, et al. Can Med Ass J 1979; 120: 168
16.Robinson TJ. Br Med J 1975; 3: 139
17.Datta DV. Lancet 1977; 1: 484 & 903
18.Datta DV, et al. Ibid 1979; 2: 641
19.Anon. Lancet 1976; 2: 1148
20.Chan H, et al. Clin Toxicol 1977: 10: 273
21.Lightfoot J, et al. Clin Toxicol 1977; 10: 273
22.ibid 1978; 239:1037
23.Kew J, et al. Br Med J 1993; 306: 506-7
24.Saper R B, et al. JAMA 2008; 300(8): 915-923

Chapter 18
1.Report on Pub Hlth & Med SubjectsNo.112, London HMSO 1964
2. Schuler U, Ehninger G. Drug Safety 1995; 12: 364-9
3. Gordon JN et al. Gut 2005;54:540-545
4. Merrell Pharmaceuticals. Lancet 1983; i: 1395
5. Keefe M. Prescribers J 1995; 35:71-6
6. Wingfield M. Br Med J 1991; 302: 1414-5
7. Bigby M, Stem R. J Am Acad Dermatol 1985; 12 : 866-76
8. CSM/MCA. Current Problems 1993; 19: 10-11
9. Stewart JT, et al. Br Med J 1985; 290: 787-8
10.Hall SM. Ibid 1994; 309: 411
11.Abrahams C. Lancet 1976; 2: 346
12.Lacey Rw, et al. Br Med J 1985; 291: 481
13.Yunis AA. Ann Rev Pharmacol Toxicol 1988; 28:83-100
14.Anon. Br Med J 1958; 1: 515
15.Anon. J Am Med Ass 1978; 240: 513
16.DOH. Drug Misuse & dependence. London HMSO 1991
17.CSM/MCA Current Problems 1997; 23: 12-4
18.ibid 1994; 20: 2
19.CSM. Current Problems 11, 1983
20.CSM Br Med J 1986; 293: 41
21.CSM/MCA Current Problems 1993; 19: 10-11
22.ibid. 1993; 19:9
23.ibid. 1997; 23: 7
24.UGDP.JAMA 1971;217:777-784
25.CSM/MCA Current Problems 1997; 23: 13
26.CSM, Adverse Reactions Series No 11, January 1975
27.Rodriguez ML, et al. Ann Intern Med 2000; 132: 598
28.CSM/MCA. Current Problems 1996: 22: 2

29.Hansson O,Herxheimer A. Lancet 1981; 1:450
30.Nakae K, et al. Lancet 1973; 1: 171
31.CSM/MCA Current Problems 1998;24:11
32.Ackrill P, et al. ibid 1980; ii:692-3
33.Chandler RF, et al. Pharm J 1984; 232: 330-2

Chapter 19
1.Johanson C-E, Fischman MW. Pharmacol Review 1989; 41: 3-52
2.Lombard J, et al. N Eng J Med 1989; 320 :869
3.Gerard C,et al. Br J Hosp Med 1990; 43: 138-41
4.Dixon WE. Br Med J 1921; ii: 821
5.Leikin JB, et al. Med Toxicol Adverse Drug Exp 1989; 4: 324-50
6. Benozzi F, Mazzoli M. Lancet 1991; 338: 1520
7. Anon. FDA Drugs Bull 1979; 9: 24
8. Nutt DJ. Br J Hosp Med 1996; 55: 187-91
9. de Santo NG, et al. J Am Med Ass 1975; 234: 841
10.Hanson JW. Br J Hosp Med 1977; 18: 126
11.Ewart RBL, Priest RG. Br Med J 1967; 3: 92
12.Ashton CH. Ibid 1990; 300: 135-6
13.Warembourg H, et al. Lille Méd 1964; 9: 192
14.Baerg RD, Kimberg DV. Ann Inter Med 1970; 73: 713
15.Bethell MF. Br Med J 1965; 2:276
16.Bass M. J Am Med Ass 1970; 212: 2075
17.Nathan AW, Toseland PA. Br J Clin Pharmac 1979; 8:284
18.Locatelli F, pozzi C. Lancet 1983; ii: 220
19.Sturt J. Papua New Guinea Med J 1967; 10:97
20.Broadwell RO, et al. J Am Med Ass 1978; 240: 2283

Chapter 20
1.Appelbe G, Wingfield J; Dale and Appelbe's
Pharmacy Law and Ethics, 7th Edition, Pharmaceutical Press 2003.

Bibliography

People
Brewer's Dictionary of Phrase and Fable, Charles, 1999
Brewer's Rogues, Villains and Eccentrics, William Donaldson, Cassell, 2002
Chambers Biographical Dictionary, Chambers, 1997

Life
Mrs. Beeton's Book of Household Management, reprint Chancellor Press, 1982
The New Female Instructor, (1834), revised ed. Roster's, London, 1988
What the Victorians did for Us, Adam Hart-Davis, Headline, 2001
London, The Biography, Peter Ackroyd, Chatto & Windus, 2000
The Great Stink of London, Stephen Halliday, Sutton, 1999
1700, Scenes from London Life, Maureen Waller, Hodder & Stoughton, 2000
What the Tudors and Stuarts did for Us, Adam Hart-Davis, Boxtree, 2002
Elizabeth's London, Liza Picard, Weidenfeld & Nicolson, 2003

History
Science, A History 1543-2001, John Gribbin, Allen Lane, 2002
Hutchison Dictionary of Scientific Biography, Helicon Publishing, 2000
Blood and Guts, A Short History of Medicine, Roy Porter, Allen Lane, 2002
Cambridge Illustrated History of Medicine, ed. Roy Porter, CUP, 1996
Madness, A Brief History, Roy Porter, OUP, 2002

Chemistry
Nature's Building Blocks, An A-Z Guide to the Elements, John Emsley, OUP, 2001
H_2O, The Biography of Water, Phillip Ball, Phoenix, 1999
The Shocking History of Phosphorus, John Emsley, Macmillan, 2000
Five past Midnight in Bhopal, Dominique Lapierre & Javier Moro, Scribner, 2002
Silent Spring, Rachel Carson, Penguin, 1962
The Elements of Murder, John Emsley, OUP, 2005
Venomous Earth, Andrew Meharg, Palgrave Macmillan, 2005

Doctors and Other Scientists

Quacks, Roy Porter, Tempus, 2000
Royal Poxes and Potions, Raymond Lamont-Brown, Sutton, 2001
Landmarks in Western Science, Peter Whitfield, The British Library, 1999
Science and the Rise of Technology since 1800, Ed Russell & Goodman, OUP, 1972

Medicines

Medicine and Society in Later Medieval England, Carole Rawcliffe, Sutton, 1995
Secret Remedies, British Medical Association, 1909
More Secret Remedies, British Medical Association, 1912
The Extra Pharmacopoeia, Martindale & Westacott, VI , 18th Ed., H K Lewis & Co., 1924
The Extra Pharmacopoeia, Martindale & Westacott, VII, 18th Ed., H K Lewis & Co., 1925
The Extra Pharmacopoeia, Martindale, VolI, 22nd Ed., The Pharmoceutical Press, 1941
The Extra Pharmacopoeia, Martindale, VolI, 23rd Ed., The Pharmoceutical Press, 1952
The Extra Pharmacopoeia, Martindale, VolII, 23rd Ed., The Pharmoceutical Press, 1955
The Extra Pharmacopoeia, Martindale, 28th Ed., The Pharmoceutical Press, 1982
The Extra Pharmacopoeia, Martindale, 29th Ed., The Pharmoceutical Press, 1989
Martindale, The Complete Drug Reference, 32nd Ed., The Pharmoceutical Press, 1999
Martindale, The Complete Drug Reference, 33nd Ed., The Pharmoceutical Press, 2002

Poisonous Plants

The Complete Herbal and English Physician Enlarged, Culpeper, 1653
A Modern Herbal, Mrs M Grieve, Jonathen Cape, 1931
Potter's Herbal Cyclopaedia, Dr Elizabeth Willianson, Daniel, 2003
Discovering the Folklore of Plants, Margaret Baker, Shire, 1999
Herbs in Magic and Alchemy, C L Zalewski, Prism Press, 1990
A Druids Herbal, Ellen Evert Hopman, Destiny Books, Vermont USA, 1995

Plants of the Gods, Richard Evans Schultes & Albert Hoffman, Healing Arts, 1992
Old Wives Tales, Mary Chamberlain, Virago 1981

Pigments and Dyestuffs

Bright Earth, the Invention of Colour, Phillip Ball, Viking 2001
Colour, Travels through the Paint-box, Victoria Finlay, Hodder&Stoughton 2002
Mauve, Simon Garfield, Faber & Faber, 2000

Legislation and Forensic Investigation

Appelbe G, Wingfield J; Dale and Appelbe's
Pharmacy Law and Ethics, 7th Edition, Pharmaceutical Press 2003.
The World's Worst Medical Mistakes, Martin Fido, Parragon, 1996
Poisons of the Past, Mary Kilbourne Matossian, Yale Univ. USA, 1989
The Poison Principle, Gail Bell, Macmillan, 2002
Hidden Evidence, David Owen, New Burlington Books, London, 2000
More Secrets of the Dead, Hugh Miller, Channel 4 Books, 2001

Miscellaneous

Why Moths Hate Thomas Edison, Ed. Hampton Sides, Norton & Co., 2001
Can Reindeer Fly? The Science of Christmas, Roger Highfield, Phoenix, 2002

Websites

There are millions of websites on every subject under the sun. Below are a few sites which may be of interest for those readers requiring further information on particular topics.

mass poisoning	www.bhopal.net
cyanide poisoning	www.cnweb.com
lead poisoning	www.parentsplace.com/health/
alcohol abuse	www.alcoholconcern.org.uk
product tampering	www.facsnet.org
Tylenol murders	www.personal.psu
biochemistry behind Viagra	www.in-cites.com/papers/dr-salvador-moncada.html
quack medicines	www.quackwatch.org
Dignity in Dying	www.dignityindying.org.uk

For information about medicines generally, including side effects:

Department of Health	www.doh.gov.uk
British National Formulary for Children	www.bnfc.org
Medicine Net	www.medicinenet.com

Search engines such as 'Ask' or 'Google' will find sites for other items.

Index